Mechatronic Futures

Peter Hehenberger · David Bradley
Editors

Mechatronic Futures

Further Challenges and Solutions for Mechatronic Systems and their Designers

Second Edition

Springer

Editors
Peter Hehenberger
School of Engineering
University of Applied Sciences Upper Austria
Wels, Austria

David Bradley
Abertay University
Dundee, UK

ISBN 978-3-031-83570-4 ISBN 978-3-031-83571-1 (eBook)
https://doi.org/10.1007/978-3-031-83571-1

This Springer imprint is published by the registered company Springer Nature Switzerland AG
The registered company address is: Gewerbestrasse 11, 6330 Cham, Switzerland

Foreword

Mechatronics—The Sequel

It was an honour for me to be asked by the editors, Prof. David Bradley and Prof. Peter Hehenberger, to write the preface to this second edition of the book *Mechatronic Futures*, the reason being that I was involved very early in the genesis and evolution of the mechatronics idea. Here, I present my personal account of how I lived through the birth of mechatronics, how I saw mechatronics evolve and how I see that mechatronics will evolve in the near future. It may show why this second edition of *Mechatronic Futures* is a must read.

How It Began

The *Leitmotiv* for my interest in mechatronics was—and still is—my strong conviction that system thinking should be at the basis of all our actions, in daily life in general, but also in engineering science and practice. This is exactly what mechatronics provides: a solid basis for designing, manufacturing and exploiting high-performance complex machines and systems, by a synergetic cross-fertilization between the different engineering disciplines involved, such as mechanical engineering, control engineering, microelectronics and computer science. Concurrent—as opposed to sequential—or simultaneous engineering are frequently used terms that express the same idea.

It is not surprising to find the roots of the mechatronics idea in Japan. Harmony (nagomi, wa) between the different aspects of life is a profound concept in Japanese culture, in daily life (food, housing, social life), but also in professional life where collaboration and teamwork are held to be important to create a positive and productive work environment. The integrative, holistic mechatronics approach has enabled Japan to dominate the machine tool, automotive and consumer electronics market from the 1970s on.

In the late 1970s–early 1980s, I had good contacts with colleagues from Toyohashi University of Technology in Japan, at that time a newly established university pioneering the mechatronics idea in education and research. Inspired by these contacts, in 1986 we started at KU Leuven, as a pioneer in Europe, a postgraduate programme on mechatronics, next to our research programme on robotics that we had already had since 1974. From 1990 on, mechatronics became a very successful study programme in the mechanical, electrical and computer science departments at KU Leuven. The message spread quickly, as in the period 1985–1990 I was invited in several universities around the world (the Netherlands, Mexico, Poland, UK and Indonesia) to advice on how to set up study programmes in mechatronics.

What It Is

In the beginning years, the term '*mechatronics*' was not warmly welcomed, particularly in the USA. The opponents considered it as just another name for '*electromechanics*'. The integrative power of the concept escaped them completely. Gradually, the awareness of the value of the mechatronics design philosophy enlightened the spirits in many engineering schools all over the world and study and research programmes emerged as mushrooms. Some even realized that the term '*mechatronics*' became too narrow. In the Philips Research labs in Eindhoven, the Netherlands, strong advocates of the mechatronics idea from the outset, the term '*mechaphysics*' was launched, to indicate that more disciplines were involved. The present terminology used around Eindhoven, where the mechatronics-inspired industry is flourishing, is '*High Tech Systems*'. Another term used nowadays is cyber-physical systems (CPS) to express the same idea, putting it in a wider context by emphasizing the interconnection of different mechatronic systems (the Internet of Things (IoT)). The Industrie 4.0 term, launched in Germany, refers to the IoT applied to manufacturing.

To me, mechatronics was—and still is—the science of motion control: generation, synchronization and suppression of motion. Motion is an essential characteristic of machines. Mechatronics is about motion in machines (CNC machine tool, robot, DVD-player, car, disc drive, wafer stepper, copying machine, camera, etc.) and in machine systems (factory, railway system, airline, hospital, etc.).

A mechatronic system (machine, machine system) consists of a hierarchy of levels: (i) module (component) level (i.e. hydraulic cylinder), (ii) machine level (i.e. robot), (iii) machine system level (i.e. car assembly plant, the Belgian railway system). Each level contains: (i) one or several controller-power amplifier-actuator-sensor loops), (ii) motion controller (sequence and trajectory), (iii) task programming module (or human/machine interface (HMI)). Hence, a mechatronic system has a fractal structure. It is the consequent realization of the '*sense/think/act*'—loop at multiple levels.

The benefits of the mechatronics design approach [1] (at the machine level) are manifold: (i) enables compensation of mechanical deficiencies (electronic gearbox) resulting in high (motion) accuracy (i.e. distributed '*super*' telescope), (ii) enables robust performance (adaptive or '*smart*' systems), by incorporating self-diagnosis, self-repair and even self-learning (self-X) loops, (iii) product flexibility and scalability can be enhanced by introducing modularity and reconfigurability. Mechatronic systems can be considered as '*living*' or information-orientated machines (i.e. adaptive optics, active car suspension, large telescope with active mirrors).

Towards '*Mechatronics Societies*'

The '*system of systems*' is formed of interconnected intelligent complex machine and/or infrastructure systems, as encountered in traffic systems, smart (power) grids, networked production, intelligent manufacturing, smart multi-modal transport and logistics, supply networks, hospital logistics and operations management, which are all mechatronic systems *pur sang*. At KU Leuven, we call them '*Mechatronics Societies*'. They are complex systems (systems of systems) of interconnected components/machines/machine systems (resources), containing actuators, sensors, controllers and task programming systems (HMIs) or human users [2].

Execution control of such systems is challenging. Hierarchical control, as still found in most factories, is not suitable anymore, because of its lack of robustness and reliability. Distributed multi-agent control is the way to go. The system agents are described (modelled) in a structural, rather than a functional way, in order to guarantee scalability. As an example, a manufacturing company can be described by a resource agent (describing the equipment, infrastructure, personnel of the company), a product agent (describing the knowledge on how a specific task has to be executed) and an order agent (describing the task to be executed). Communication between these agents is needed to execute the task optimally. Suitable control approaches can thereby be used. Bioinspired coordination and control algorithms, such as ant colony optimization, are becoming popular [2].

The main challenge when moving from traditional mechatronic systems towards mechatronics societies is how to deal with the increased complexity in controlling the overall system, by developing: (i) modular, scalable, extendable system architectures and (ii) distributed (self)-optimizing proactive coordination and control systems, with predictive power and (socially) acceptable behaviour [2].

Digital Twins

The explosion of the complexity of mechatronic products, systems and societies can only be dealt with by advanced digitalization. Central here is the concept of '*comprehensive Digital Twin*'. A digital twin (DT) is the digital version of a physical entity (product, production system, hospital and employee). A comprehensive DT describes, in digital form, the complete lifecycle of a product: design, production, quality control, use, dismantling, recycling. A DT should be '*the single source of truth*' that means it should at each moment be the perfect digital rendering of the product or process it represents. The form of a DT depends on the application it is used for. A DT of a factory that should serve to simulate process planning is completely different from a DT of a wind turbine gearbox that is to be used to predict the stresses in the gearbox shaft.

Digital twins can serve multiple purposes. The real-time emulation of the state of a machine can optimize the working conditions to minimize energy consumption or can predict the maintenance intervals of the machine, or the remaining lifetime of roller bearings. An important application of DTs is their use as virtual sensors, to '*measure*' difficult- or impossible-to-measure quantities, such as force, power, energy and wear.

The previous examples show that the DTs have to run at the same rate or faster as/than the process they emulate, to give them predictive power. This requires sufficient computing power that is not always available yet. As an example, real-time finite-element programmes are still not generally available.

AI and Mechatronics

Mechatronics leads to more '*intelligent*'—I prefer '*smart*'—products and machines. Intelligence has to be understood here as giving the mechatronic system some cognitive capabilities, such as autonomous decision-making based on advanced pattern recognition and pattern classification algorithms. Machines with general human intelligence, such as expressing emotions, consciousness, self-awareness, are not expected in the near future. Some, including me, doubt if ever machines will become conscious or self-aware, that is, being somebody. The fast evolution of explainable artificial intelligence (AI) will contribute in future to the development of machines that explain the reasons behind their decisions. AI in its future forms, like generative AI (GenAI), will enhance the use of digital twins, making use of the huge amounts of data that become available in the Cloud on virtually every imaginable subject. Databases serving as expert systems are already assisting medical specialists in their diagnoses. It is expected that similar systems will increasingly become available in industrial settings. Deep learning will, by its multi-layer neural network structure, enhance and speed up the extraction of the features from the raw input data, compared to earlier machine learning systems. Applications in mechatronics and robotics abound for diagnostic and quality control tasks, for adaptive control, for easing grasping and assembly, etc.

Reinforcement learning is an iterative learning method where an action is rewarded or punished depending on its outcome. A problem with this learning method is that it requires a (very) large training set in order to be successful. Another problem is that, when training takes place in the real working environment, failures may not cause damage to the working space. This can be avoided when training takes place in a simulated environment, using digital twins. A bright future lies ahead for reinforcement learning in mechatronic systems, for instance to train assembly operations using robots.

Precision-/Micro-/Nano-mechatronics

Moore's law for microelectronics, stating that the number of transistors on a chip doubles every two years or even faster, has a counterpart in mechatronics: Taniguchi's law. This latter states that the mechanical precision of mechatronics systems doubles every two years. It can be stated that Moore's law can only continue to be satisfied thanks to the development of machine precision as expressed by Taniguchi's law. The present 3 nm chips thank their existence to the Extreme-UV wafer steppers, developed by ASML. Advanced air bearing technologies and the availability of position measurement sensors of sub-nanometer resolution, next to advanced position control algorithms, contribute to this development. We can rightfully say that mechatronics, by its integrated vision, has contributed significantly to the development of microelectronics and many other industries (optical, aerospace).

The necessity to incorporate more functionality in smaller spaces necessitates the development of new small-size actuators and sensors. Scale laws reveal that on the microscale, alternative actuator systems can be more useful than traditional actuators. Piezoelectric and electrostatic actuators are good examples of the trend. New actuation principles, such as inchworm and resonant drives become possible, yielding sub-nanometer resolution at very low and very high speeds.

Design of Mechatronic Systems

Designing mechatronic systems implies designing an optimal mechanical structure and an optimal motion controller. Traditional sequential design: first mechanical structure optimization and subsequently the controller optimization cannot lead to an optimal structure. Simultaneous optimization of mechanical and control behaviour is imperative. This can be achieved by a '*mechatronic compiler*' whereby general functional requirements are translated in a semi-automatic way into an optimal mechatronic system. Such a compiler has two modules: (i) the conceptual design module, where first the optimal configuration of the system to be built, consisting of different building blocks, is determined, using optimization criteria, such as maximal working space, maximal static and dynamic stiffness, followed by (ii) the detailed design, where the optimization parameters of the combined digital twins of the optimal mechanical configuration and of the controller are optimized using a multiple-criteria optimization strategy. This leads to a better optimum than sequential optimization.

The use of the digital twins of the optimized mechatronic system can be used to verify and validate the behaviour of the optimized system, making physical prototyping superfluous. This is particularly important for '*testing*' aerospace structures.

It requires however accurate real-time DTs, which are presently not yet available. Continued research is required to fill this gap.

Application Areas

The mechatronics design methodology has proven, is proving and will continue to prove its added value in the most diverse areas of human activity, industrial and societal. It started by improving product and machine design, where performance and precision could be dramatically increased. It continued by invading the world of mechatronic societies. First in manufacturing, where complete plants, factories, networked manufacturing systems and enterprises could be optimized. Particularly, ICT-based manufacturing processes, such as 3D-printing and related processes, can benefit from the mechatronic approach. Non-manufacturing applications can be found in logistic systems, robots and robot fleets, open-air engineering (open-pit mining, road construction, harvesting), intelligent traffic and transportation systems, railway operations, smart grids, smart homes, e-health and integrated care.

Epilogue

Mechatronics is still alive and kicking as an integrated approach to designing and operating superior products, machines, machine systems and mechatronic societies. But there is still a lot to achieve. The performance of our mechatronic systems can still dramatically be improved, in terms of precision, yield, scalability, miniaturization, interconnectability, reconfigurability, human/machine interfaces, decentralized coordination and control. Particularly, the non-manufacturing applications deserve our special attention in future, to make technology serve humanity in a better way.

I sincerely hope that this new edition of the book *Mechatronic Futures* can inspire you, dear reader, to contribute to making our world a better place.

Prof. Hendrik Van Brussel
KU Leuven
Leuven, Belgium

References

1. Van Brussel H (1996) Mechatronics—a powerful concurrent engineering framework. IEEE/ASME Trans Mech 1(2):127–136
2. Valckenaers P, Van Brussel H (2016) Design for the unexpected. From holonic manufacturing systems towards a humane mechatronics society. Butterworth-Heinemann

Prof. Hendrik Van Brussel is Emeritus Professor in Mechatronics and Automation at the Department of Mechanical Engineering, KU Leuven, Belgium. He was a pioneer in robotics research in Europe and an active promoter of mechatronics as a new paradigm for machine design. His research interests include robotics, mechatronics, holonic manufacturing systems, precision engineering.

Prof. Van Brussel is Fellow of Society of Manufacturing Engineers (SME) and Life Fellow of Institute of Electrical and Electronics Engineering (IEEE), honorary doctor from Rheinisch-Westfälische Technische Hochschule (RWTH), Aachen, Germany, from '*Politehnica*' University in Bucarest, Romania, and from '*Transilvania*' University Braşov, Romania.

He is Member of KVAB (Royal Flemish Academy of Belgium for Science and the Arts), Foreign Member of the Royal Swedish Academy of Engineering Sciences (IVA), Foreign Member of the US National Academy of Engineering (NAE), Honorary Member of the Hungarian Academy of Sciences (MTA).

He was President of euspen (European Society for Precision Engineering and Nanotechnology) and of CIRP (International Academy for Production Engineering).

The original version of the book has been revised. A correction to this book can be found at https://doi.org/10.1007/978-3-031-83571-1_22

Contents

Editors and Contributors

About the Editors

Peter Hehenberger is Professor for Integrated Product Development, Head of the '*Smart Mechatronics Engineering*' Research Group and Scientific Partner of the '*Institute for Agile Transformation*' at the School of Engineering, University of Applied Sciences Upper Austria, Wels/Austria.

Prior to this appointment, he was Assistant Professor and Deputy Head at the Institute of Mechatronic Design and Production of the Johannes Kepler University of Linz (JKU), Austria, and Deputy Research Area Coordinator for '*Process Modelling and Mechatronic Design*' at the Linz Center of Mechatronics (LCM). In 2015, he was invited as Visiting Professor at the Universite de Technologie de Compiegne, Departement Genie des Systemes Mecaniques, France.

He completed his Diploma and Doctorate in Mechatronics from JKU in 2000 and 2004, respectively. In 2016, he received the postdoctoral lecturing qualification (venia docendi) for the subject area '*Engineering Design*'.

Core research interests cover '*Sustainability in Development and Production of Mechatronic Systems*' and '*Engineering Processes*' along with topics on '*Digital Transformation*'.

He has published over 150 peer-reviewed papers in international journals and conference proceedings and has been a guest editor for three journal special issues. He serves as an Editorial Board Member for the *IFAC Journal Mechatronics*, *Computer-Aided Design and Applications* and *International Journal of Product Lifecycle Management* (IJPLM). In addition, he has authored or co-authored 7 books related to his research field.

He has practical experience with industry and design projects and has been a key researcher and manager of several successful national and international research projects. He holds lectures on the topics product development, engineering projects, model-based systems engineering, product lifecycle management, mechanical engineering, mechatronic design, technical drawings, computer-aided design, simulation and manufacturing.

Among other scientific service activities, Dr. Hehenberger organized international Workshops on Mechatronic Design (Linz 2012, Paris 2013, Leuven 2014, Milano 2015, Paris 2016), Smart Mechatronics Engineering (Naples 2019, Naples 2020, Online 2021–2022, Wels 2023) and was General Chair of the 20th International Conference on Research and Education in Mechatronics (REM 2019) in Wels/Austria.

Dr. Hehenberger is active in the research community where he refers to his research area of '*Mechatronic Design*' as group member of IFIP WG5.1 '*Global Product development for the whole life-cycle*' and IFAC TC 4.2 '*Mechatronic Systems*'. He is Austrian representative and Communications Publications and Events Officer for IFIP TC 5 '*Information Technology Applications*'. He also co-chaired a Special Interest Group (SIG) on '*Methodologies for Design, Integration, Modelling and Simulation of Cyber Physical Systems*' between 2015 and 2019 within the Design Society.

David Bradley having taught at Lancaster University, Bangor University and Abertay University in the UK, Professor Emeritus David Bradley has been a visiting fellow at the Technical University of Denmark and a professorial research consultant at Sheffield University and Barnsley Hospital and a research engineer, while on sabbatical leave, at New Zealand Electricity. His prior research interests have included engineering design, power system harmonics, construction robotics, eHealth and mHealth technologies and the modelling of hydro-power stations.

He is currently Professor Emeritus at Abertay University in the UK and has a long-standing involvement in mechatronics teaching and research which began in the mid-1980s when he was involved in the establishment of an M.Sc. programme at Lancaster University. He has also acted as a consultant on mechatronics course development in, among others, South Africa, Mauritius, Singapore and Colombia and has been a visiting lecturer on mechatronics programmes in Colombia, Singapore, South Africa as well as at several locations in Europe. A co-founder of the UK Mechatronics Forum, later the Mechatronics Forum, he was also the originator of the Mechatronics Forum Conference series.

Though no longer directly involved in academic life, he maintains close links with the mechatronics community acting as a consultant and advisor to various research and teaching groups as well as a continuing involvement in reviewing, both articles and research grants.

His interests now centre around the ways in which mechatronics has evolved, and indeed continues to evolve, in respect of developments in technology and their impact on society, as for instance in achieving sustainability, within the wider context of mechatronics.

He has been the author, lead author or co-author of 6 engineering textbooks and this book is the fifth of a series of mechatronic books for Springer for which he has been a contributing editor. In addition, he has been involved in several further book chapters along with some 50 papers in academic journals and a further 150 at conferences.

Contributors

Ghaith Al-Refai earned his BSc in Electrical and Electronics Engineering from Jordan University of Science and Technology, Irbid, Jordan, in 2010. He subsequently received his MSc and PhD degrees in Electrical and Computer Engineering from Oakland University, Michigan, USA, in 2014 and 2018, respectively. He is currently Assistant Professor in the Mechatronics Engineering Department at the German Jordanian University, Amman, Jordan.

His research interests include autonomous driving, computer vision, machine intelligence, AI, and robotics. Additionally, he actively reviews for various journals and conferences in the field of mechatronics engineering.

Marwa Al-Faouri was born in Ein Al-Basha, Jordan. She received a BSc degree in Mechatronics Engineering from the University of Jordan, Amman, Jordan, in 2013, with the senior project specializing in system design and automation in pharmaceutical industries. Eng. AlFaouri has acquired practical in-industry experience in building management systems between 2013 and 2015. She is currently Research and Teaching Assistant in the Mechatronics Engineering Department at the German Jordanian University, Amman, Jordan. Eng. AlFaouri has several publications in well-reputed journals and conferences. Her research interests are focused on robotics, Industry 4.0. and AI, and building services.

Natheer Almtireen was born in Amman, Jordan. He received his BSc degree in Mechatronics Engineering from the University of Jordan, Amman, Jordan, in 2010 and his MSc degree in Advanced Mechanical Engineering from the University of Southampton, Southampton, UK, in 2012. He earned his PhD degree in Mechanical Engineering from Karlsruhe Institute of Technology, Karlsruhe, Germany, in 2022.

He is currently Assistant Professor in the Mechatronics Engineering Department at the German Jordanian University, Amman, Jordan. His research interests focus on mechatronics systems design and control, Engineering systems modelling and simulation, and fluidic and energy systems. He serves as Active Reviewer for several journals and conferences in the areas of mechanical and mechatronics engineering.

Georgios Andrikopoulos is Assistant Professor in mechatronics, specializing in the field of robot design, and Director of the Robot Design Lab. His research is centred around the design, development, and control of autonomous robotic systems, with a focus on interdisciplinary applications. These applications include soft robots for safe interactions, exoskeletons designed to assist in human motion, as well as hybrid robotic solutions for industrial inspection purposes.

He has been involved in the creation, technical implementation, scientific or overall management of several European and National R&D projects, including EU Horizon (FETOPEN CSA, FETOPEN RIA, ICT), Vetenskapsrådet (Swedish Research Council), and digital futures, among others. He has also initiated multiple national and international research collaborations, including an ongoing collaboration with HONDA Research Institute of Japan.

His scientific work includes more than 60 articles published in high-impact international journals and conferences in the fields of interest, while listed in IVA's (Royal Swedish Academy of Engineering Sciences) top-100 innovations of 2021, top-25 most popular publications in Elsevier's Control Engineering Practice, and the top-10 most downloaded articles in Springer's Meccanica journal.

He has served as International Program Committee Member and Technical Track Chair for multiple international conferences while being an active reviewer for several high-impact international journals and conferences (including Science Robotics, IEEE Trans. on Mechatronics, IEEE Trans. on Industrial Electronics, among others). He is currently Member of the Socially Intelligent Robotics Consortium, Member of the Swedish Society of Biomechanics, and Member of IEEE (CSS, RAS, IES).

Fredrik Asplund received his PhD in 2014 and is currently employed as Associate Professor at KTH Royal Institute of Technology on the topic of cyber-physical systems with a focus on safety practice. He is Docent in cyber-physical systems since 2022.

Research interests include innovation management, system safety, empirical software engineering, cognitive systems engineering, and engineering education in the CPS domains. In pursuit of these interests Dr. Asplund leads the FIRE and WILDCAT projects, 3-year projects with project partners from Sweden, Estonia, and Ukraine that focus on identifying and handling the risk of rural wildfires. Dr. Asplund was Post-doctoral Researcher within the mobility for growth programme and is now Marie Curie and VINNMER Fellow. His service to the academic community includes being Elected Member of the Faculty Board at the School of Industrial Engineering and Management at KTH.

Jorge Azorin-Lopez received a degree in Computer Engineering in 2001 and a PhD degree in Computer Science at the University of Alicante (Spain) in 2007. Since 2001, he has been Faculty Member of the Department of Computer Technology at the same university, where he is currently Full Professor and the Academic Secretary. He was Deputy Director of research in that department and of the University Institute of Computer Research. Also, he is Quality Coordinator of the PhD programme in Computer Science at UA. He has supervised 6 PhD students.

His current research interests include 3D computer vision, machine learning, deep learning, ambient intelligence, and human activity analysis. In these lines of research, Dr. Azorín has worked in 27 research projects (8 of them as coordinator) funded by national, regional, and local public and private entities. He has authored more than 150 contributions in several journals, conferences, and book chapters. He has served as Reviewer to numerous scientific journals and international conferences and chair sessions in top conferences.

He was awarded a Post-doctoral Research Fellowship by the Spanish Ministry of Science for research at the University of Edinburgh. Moreover, he has been Invited Professor at the CNRS Lab Hubert Curien in France since 2015 and has given invited talks in Europe and South America. He is Member of the Excellence European Society for Cognitive System (EuCognition) and associations as the International

Neural Network Society and the Spanish Society of Architecture and Computer Technology.

Liliana-Laura Badita-Voicu graduated from the University of Bucharest, the Faculty of Physics, with a specialization of biophysics, in 2004 and received her Master's degree in Biodynamics at the Faculty of Biology, in 2007. She obtained her PhD degree in Mechanical Engineering, from the Romanian Academy of Sciences, the Institute of Solid Mechanics, in 2012.

She joined the Micro/Nano-mechatronic Technologies Department of the National Institute of Research and Development in Mechatronics and Measurement Technique in Bucharest, Romania, in 2007. Currently, she is Coordinator of the Laboratory for Process Micro- and Nano-Technologies and Vice-President of the Scientific Council. Her research areas cover tribology, micro/nanostructures, thin films, atomic force microscopy, mechanical engineering, biophysics, and medical physics. She participated in a series of courses and seminars and received numerous scholarships in various research centres in Europe and Asia, in these areas.

Dr. Badita-Voicu has coordinated 5 research projects funded by national programmes and has also participated in several national and European funded research and development projects. She is Associate Editor of the International Journal of Mechatronics and Applied Mechanics—IJOMAM, Author and Co-Author of 4 published books, 4 book chapters, more than 30 papers published in WoS indexed scientific journals, 60 papers published in other scientific journals, and 40 papers presented and published in the proceedings of international and national scientific events.

David Blackwood is Emeritus Professor of Sustainable Development at the University of Abertay. He is Chartered Engineer and Member of the Institution of Civil Engineers. He graduated from Dundee Institute of Technology (now Abertay University), and following five years in the water industry, he joined the faculty at Abertay University. During this time, he gained an MSc in Construction Management from Loughborough University and progressed to several senior roles including Head of Division of Natural and Built Environment and Co-Director of the Urban Water Technology Research Centre.

David's research interests focus on sustainable development, particularly in the areas of decision-making, stakeholder engagement, and sustainability assessment. He has published over 80 peer reviewed journal and conference papers, is Co-Author of one book and two book chapters, and has produced over 20 major research project reports for a range of funders and clients, including EPSRC, the Scottish Government, UK Water Industry Research Limited, Veola Water and Fife and Dundee City Councils.

Stefan Boschert studied physics at University of Freiburg i.Br., Germany, where he finished his PhD on the numerical simulation of semiconductor melts under microgravity in 1999. After his PhD, he worked for an engineering consulting company in Basel, Switzerland, and joined Siemens AG in 2004. As Technical Expert, he led several public funded projects and supported the Siemens business units in research and development projects. His main topic of expertise is 3D multiphysics simulations.

Currently, he focuses his research on developing simulation solutions supporting the operation of physical assets, thus combining physics-based simulation methods with data-driven approaches. As Principal Key Expert, he coordinates the research activities for modelling and simulation in design and engineering of the technology field Simulation and Digital Twins.

Jordan H. Boyle is Assistant Professor in the Materializing Futures section in the Faculty of Industrial Desin Engineering at TU Delft, the Netherlands. He did his BSc and MSc in Electrical Engineering at the University of Cape Town, South Africa, and his PhD in Computational Neuroscience at the School of Computing, University of Leeds, UK, and was Research Fellow, then Lecturer, then Associate Professor in Mechanical Engineering also at Leeds. His research interests revolve around mobile robotics, with a particular focus on swarms, bioinspired solutions, human robot interaction, and the design and fabrication of novel robot platforms.

Matthieu Bricogne is Assistant Professor at the Mechanical Engineering Department/Roberval laboratory in Université de Technologie de Compiègne (UTC). He obtained his MSc in Mechanical Engineering and a Mechanical Engineering Degree at UTC in 2004. He has worked for 5 years for Dassault Systèmes, an industrially leading company specializing in 3D and Product Lifecycle Management (PLM) software. He obtained his PhD degree in Mechanical Engineering from the Ecole de Technologie Supérieure in Montréal (ETS) and UTC in 2015.

His main research topics are collaborative design, mechatronics system design, knowledge-based engineering within a PLM environment, and Building Information Modelling (BIM). Since 2013, he has been Vice-Director of the ANR LabCom DIMEXP (DIgital MockUp multi-EXPertises) joint laboratory UTC/DeltaCAD, whose main objective is to develop methods and models to improve the integration of multidisciplinary expertise and heterogeneous data in advanced digital environments. Since 2023, he has been In Charge of UTC's digital transformation.

Dejiu Chen has been Associate Professor in Embedded Control Systems since 2013, with the Mechatronics Division, Department of Engineering Design, KTH Royal Institute of Technology, Sweden. He received his MSc degree in Mechatronics from KTH in 1998 and his PhD degree in Mechatronics from KTH in 2004. His current research focuses on developing novel strategies and innovative industrial frameworks for advanced situation awareness and self-management in intelligent cyber-physical systems. He has worked for Enea Data AB, Sweden, as Senior Technical Instructor during 2007–2009. Dejiu Chen has been Principal Investigator for various international research projects. He is Senior Member of IEEE.

Natalie Coull is Head of the Department of Cybersecurity and Computing at Abertay University and Director of their NCSC-recognized Academic Centre of Excellence in Cyber Security Education (ACE-CSE gold award). Natalie was responsible for developing Abertay's bachelor's programme in Digital Forensics and the NCSC Certified Masters programme in Ethical Hacking. Natalie has research expertise in secure coding, digital watermarks, and computer memory-based security concerns. Working with colleagues at Roehampton University, and the University of the West of England, Natalie is one of the founders of CSE Connect—an organization created to bring together those interested in cybersecurity education across the UK and provide a platform for sharing best practice and resources with the wider academic community. Natalie is also Co-Director of Abertay's Cyberquarter.

Lynne Coventry is Director of Abertay University's Cyberquarter. This interdisciplinary cybersecurity research and innovation centre brings the Scottish government, businesses, and universities closer together to research, innovate, share knowledge to improve Scotland's cyber resilience. Lynne previously worked at Northumbria University, holding a number of positions including Director of Northumbria's NCSC-accredited Academic Centre of Excellence in cybersecurity research. As Interdisciplinary Researcher herself, bringing together different perspectives and methodologies to address global cybersecurity issues is not new to her. Her background in psychology and computing is an ideal starting place to explore human aspects of cybersecurity. Lynne has research expertise in authentication systems, human cybersecurity interaction, cybersecurity behaviours, and barriers to secure behaviour.

Abbas A. Dehghani Graduate of the University of Leeds, UK, is Professor of Bio-Mechatronics and Medical Robotics in the School of Mechanical Engineering at the University of Leeds. His research interests include bio-robotics, intelligent control, sensors, and actuators using an integrated system design approach where functional materials, sensors, actuators, and control are brought together to develop intelligent systems/devices. His research focus is mobility and intelligent control in robots and also development of devices to support mobility in patients and the growing ageing population as well as wearable robotic systems for enhancing human capabilities. His current research includes the design and development of intelligent robotic exoskeletons, soft robotics, and artificial limbs.

Hisham El-Moaqet received a PhD degree in Mechanical Engineering from the University of Michigan, Ann Arbor, MI, USA, in 2015. He is currently Assistant Professor with the Department of Mechatronics Engineering at the German Jordanian University, Amman, Jordan. His current research interests include dynamic systems identification and control, machine learning, physiological systems, and mechatronics systems design.

Benoît Eynard is currently Leader of Industrial Engineering team of Roberval research laboratory and Holder of PTC Industrial Chair at the Université de Technologie de Compiègne (UTC), France. In 2007, he joined UTC as Full Professor for managing the Department of Mechanical Systems Engineering until 2012. Previously, he was Assistant Professor at the Université de Technologie de Troyes where he managed the MSc programme on Information Technology for Mechanical Engineering.

From 2013 to 2020, he was General Chairman of the division on Factories of the Future of the French Society of Mechanical Engineering also known as the S.MART academic group.

He is currently Editor-in-Chief of the International Journal of Product Lifecycle Management (IJPLM). He is also acting as Editorial Board Member of 5 other international journals.

He is also Member of the design society where he serves as Advisory Board Member, and he led for three years the special interest group on '*Design Methods for Cyber-Physical Systems*.' He is Member of the IFIP working group 5.1. dealing with '*Global Product development for the whole life cycle*.'

In 1999, he obtained his PhD on Computer-Integrated Manufacturing from the University of Bordeaux. Now, he is Recognized Researcher in product lifecycle management, collaborative design, systems engineering, mechatronic design, digital factory, manufacturing process management, cyber-physical production systems, eco-design, and sustainable manufacturing. He has published 120 papers in international journals and more than 250 international conferences. He also has been Guest Editor for about 20 journal special issues and academic books.

Lei Feng is Associate Professor in the Mechatronics division, Department of Engineering Design, KTH Royal Institute of Technology, in Stockholm, Sweden. He received BS and MS degrees from the Department of Mechanical and Electronic Engineering, Xi'an Jiaotong University, Xi'an, China, in 1998 and 2001, respectively, and a PhD degree from the Systems Control Group, Department of Electrical and Computer Engineering, University of Toronto, Toronto, ON, Canada, in 2007.

His main research interests include methods and applications of model-based systems engineering (MBSE), formal methods for the verification and control synthesis of cyber-physical systems, a combination of model-predictive control and deep learning, energy management control of mechatronic systems, autonomous driving, trustworthy edge computing systems and applications, novel mechatronic systems enabled by 3D/4D printing, etc.

Ian Ferguson is Professor of computing at Abertay University in Dundee, Scotland, where he leads the Security Research Group and has a particular interest in digital forensics. He gained his PhD in Software Engineering from the University of Sunderland in 1998. His work in the fields of mobile and ubiquitous computing in the smartLab group of the University of Strathclyde led to an interest in cybersecurity and in particular the challenge of improving methods, techniques, and tools for digital forensics.

Member of the '*Policing in a Digital World*' Professional Reference Group, he has provided Expert Witness consultancy in the digital forensics field to various defence solicitors and to Police Scotland.

He has supervised twenty successful PhDs, external examined six, and has recently led the successful bid to attract the £18M cyberquarter KE project to Abertay.

Elias Flening holds a PhD and is Lecturer in Project Management at the Division of Network and Systems Engineering (NSE) at KTH in Stockholm, with a focus on the interplay between perceived complexity and expertise within engineering development work. His doctoral research scrutinized the inherent challenges within engineering development projects to investigate the relationship between perceptions of expertise and complexity in the lived experiences of engineers.

His interest in organizational research fundamentally stems from two points of care: First, a strong curiosity about the low-level and seemingly mundane details of work as a lived experience; second, the well-established positive correlation between objective measures of performance and subjective well-being. Simply put, he finds intellectual stimulation in investigating the concrete details as to how engineers are faring at work.

Andres Fuster-Guillo received the B.S. degree in Computer Science Engineering from Polytechnic University of Valencia (Spain) in 1995 and the PhD degree in Computer Science at the University of Alicante (Spain) in 2003. Since 1997, he has been Member of the Faculty of the Department of Computer Science Technology and Computation at the University of Alicante, where he is currently full professor.

He was Deputy Coordinator of the Polytechnic School at the University of Alicante for seven years and Director of the Secretariat for Information Technology at the University of Alicante for four years. During this period, he has coordinated and participated in several strategic technology projects: Open University (transparency portal and open data), UACloud, Smart University, among others. Dr. Andrés Fuster-Guilló has worked in more than 20 research projects (5 of them as coordinator) funded by national, regional, and local public and private entities.

He has authored more than 90 contributions in several journals, conferences, and book chapters. He has participated as Reviewer in several scientific journals and international conferences. He was awarded a Post-doctoral Research Fellowship by the Spanish Ministry of Science and Education at the University of Edinburgh. His current areas of research interest include 3D computer vision, machine learning, deep learning, ambient intelligence, human activity analysis, and automated visual inspection.

Nahuel E. Garcia-D'Urso received the B.S. degree in Computer Science Engineering and the M.S. degree in Data Science at the University of Alicante (Spain). Currently, he is PhD Student at the Department of Computer Technology in the same university and collaborates in research regarding the Tech4Diet project in a multidisciplinary environment along with nutrition experts of the Faculty of Health Sciences. His areas of research interests include 3D computer vision, machine learning, data science, and deep learning.

Daniel Gilmour is Professor of sustainability assessment and enhancement at Abertay University. He leads the Sustainable Development and Inclusive Living Research Challenge Space, one of four strategic research areas in the university. He is Leader of The Transformations in Society and Environmental Policy theme within the Scottish Alliance for Geoscience, Environment and Society (SAGES) a Scottish pooling of over 650 of Scotland's top researchers, to ensure that the latest scientific advancements are effectively translated to meet the needs of policymakers, industry, businesses, and other users.

Daniel's research focuses on the enhancement of sustainability in the built environment with interests in the fields of sustainability assessment, decision support, and public participation in decision-making, in particular sustainable service provision. He leads a multidisciplinary team of researchers with an interest in sustainability assessment, enhancement, and communication. Drawing on established areas of expertise within university, the work includes developing visualizations and interactive tools to support wiser and inclusive decision-making on a range of projects within local government and the water industry. He is Chartered Water and Environmental Manager and Senior Fellow of the Higher Education Academy.

Martin Edin Grimheden received his PhD in 2006. He is Associate Professor in Mechatronics at KTH Royal Institute of Technology and is currently Head of the Department of Engineering Design. His research interests include learning in mechatronics.

Maki K. Habib is Globally Recognized Expert in robotics, mechatronics, and intelligent systems, with over 30 years of experience in academia and industry. His journey began at the University of Technology—Baghdad, where he earned his B.Sc. in Control and Systems Engineering and M.Sc. in Computer Engineering. He received his Doctor of Engineering Sciences in Intelligent and Autonomous Robotics from the University of Tsukuba-Japan.

Prof. Habib has held prestigious positions globally, including research roles at RIKEN and RISO-Laboratories in Japan, visiting researcher positions at EPFL, Switzerland, and academic roles as Associate Professor and Professor at Monash and Swinburne Universities, KARIS-Korea, and Sago University-Japan, respectively. He is Professor of robotics and mechatronics at the American University in Cairo (AUC), where he led the Robotics, Control, and Smart Systems (RCSS) master programme. His research spans intelligent autonomous systems, robot learning, industrial automation, and bioinspired robotics, with applications in Industries 4.0 and 5.0, healthcare, Nano/micro-robotics, and AI-driven technologies and applications-based machine learning techniques.

Prolific Author, Prof. Habib has published over 24 books, 38 book chapters, 90 journal papers, and 245 conference papers. He is African Academy of Sciences Fellow and IEEE Senior Life Member. He also serves as Editor-in-Chief of Elsevier's AI Encyclopedia and the International Journal of Artificial Intelligence and Machine Learning. His global contributions to robotics and AI have earned him recognition among the top 2% most impactful scientists globally by Elsevier/Stanford University since 2020. His work has influenced both academia and industry worldwide.

Robert W. Kay is Professor of advanced manufacturing at the University of Leeds. RK has been involved in manufacturing processes research both in academia and industry since 2001. Robert W. Kay's career started in the electronics manufacturing sector where during his PhD, he developed a novel microfabrication process which led him to create the spin-out company, MicroStencil Limited. During Robert

W. Kay's 7-year period as Chief Technical Officer, he established a UK-based production line alongside manufacturing partnerships in Asia and built a diverse customer base. In 2012 RK started his 1st academic posting where his research involves using digital fabrication techniques such as 3D printing to produce functional devices and products. RK now has a range of funded projects in the field both looking at fundamental process developments and industry-sponsored research. All of Robert W. Kay's work is multidisciplinary and involves collaborative teams from other universities, industry, research centres and medical professionals. Robert W. Kay's current research application areas include the manufacturing of bespoke and complex devices covering topics such as engineered ceramics, medical devices, wireless communications, sensor systems and electronic components, robotics and autonomous systems.

Esfandiar Khaleghi is Chartered Engineer (CEng) with significant experience in both industry and academia. As Fellow of the Higher Education Academy (FHEA), he is dedicated to promoting academic excellence and advancing research in engineering.

Mr Khaleghi holds a Bachelor's degree in Mechanical Engineering from Amirkabir University of Technology and a Master's degree with Honours in Biomedical Engineering from the University of Birmingham. His interdisciplinary education, which spans engineering and biomedical sciences, has shaped his innovative approach to technology, particularly at the intersection of human well-being and engineering. Further studies at Stanford University's Technology Ventures Program in Design Thinking have enhanced his skills in creative problem-solving and technology-driven solutions.

Currently completing a PhD at Kingston University London, Mr. Khaleghi's research focuses on developing techniques to measure and parameterize emotional and physical states for use in Advanced Driver Assistance Systems (ADAS). His work aims to enhance automotive safety by integrating artificial intelligence (AI) with human-centric systems.

Mr. Khaleghi is Programme Leader for the BSc in Automotive and Motorsport Technology at the University of Hertfordshire, where he lectures in design and manufacturing with a focus on automotive engineering. He also leads the UH Racing Autonomous–Automated Driving System (ADS) team, playing a pivotal role in the IMechE Formula Student AI competition. His research interests include Driver Monitoring Systems (DMS), ADAS, and Connected Autonomous Vehicles (CAVs).

Julien Le Duigou is Full Professor at the Mechanical Engineering Department of the Université de Technologie de Compiègne (UTC), France, and Researcher at Roberval Research Institute of Mechanics Energy and Electricity. He is currently Head of integrated production and logistics specialty for engineering degree.

He obtained his MSc in Mechatronics and a Mechanical Engineering degree in 2006, a PhD in Mechanical Engineering from Ecole Centrale in 2010, and an accreditation to supervise research (HDR) from UTC in 2017. He has been Administrator of AFIS (French INCOSE Chapter on System Engineering) since 2018 and Scientific Advisor of IFIP WG5.1 on Product Lifecycle Management since 2012.

His recent research interests include product/process digital thread, decision-aided systems for production, and agile manufacturing systems. He published more than 100 papers in refereed international journals, books, and conferences on these subjects.

Christopher Milne is Head of Information Assurance and Governance at the University of St Andrews. Although born in Glasgow, he considers the Northeast of Scotland home and resides in Brought Ferry.

Chris read Librarianship and Information Studies at the Robert Gordon University (RGU), (1994), before reading for a Postgraduate Diploma in Information Analysis (1995), and a Masters in Records Management, Northumbria University (2007).

Chris has had a varied career, predominately in higher education, as academic librarian, He has also published in the fields of information literacy (2004) and records management—advocating the use of records management information retrieval philosophies as a means of improving access to information from corporate Websites (2007) and (2010).

In his current role at the University of St Andrews, he is responsible information governance and records management across the institution. Chris has a business focused view of legislation and regulation—in that it should aid institutions in their mission and development and not hinder them via obsessive compliance and the introduction of unnecessary and/or artificial barriers.

From 2018 to 2024, Chris was Chair of the Scottish Higher Education Information Practitioners' Group and through that work supported Universities Scotland and the Scottish Government in initiatives for universities to fairly, lawfully, collect and process personal data concerning criminality (UK General Data Protection Regulation, Article 10) to support student welfare and safety initiatives surrounding gender-based violence.

John Mitchell graduated from the University of Leeds with a Bachelor's and Master's degree in Mechatronics and Robotics in 2020. He is currently in his final year of studying for a PhD with the School of Mechanical Engineering at the University of Leeds in the application of sensor systems and artificial intelligence towards remote gait analysis. He also participates in the teaching and demonstrating of the laboratory sessions for the Master's-level module 'Mechatronics and Medical Robotics' at the University of Leeds, which teaches 3D modelling and machine learning for the development of healthcare and medical robotic systems.

Stanislao Patalano received the master's degree in mechanical engineering cum laude at the University of Naples Federico II (UNINA) in 1997 and a PhD degree in machine design and construction in 2001. From 1999 he was first Researcher and then Associate Professor of Design Methods for Industrial Engineering. Since 2018, he is Full Professor of Design Methods for Industrial Engineering. From 2011 to 2022, he was responsible for the research activities of the Dept. of Industrial Engineering within PON Projects DIGIPAT (Digital Pattern Product Development), ICOSAF (Integrated Collaborative Systems for Smart Factories), and H2020 Project ENERMAN (ENERgy-efficient manufacturing system MANagement).

His research interests include CAD methodologies, knowledge-based engineering and digital prototyping, variational analysis and tolerance design, object-oriented programming and simulation, mechatronic system design.

Olivia Penas received his MSc in Material Science Engineering in 1999, PhD in Physics of Materials (Smart Space composites) in 2002 from INSA Lyon (France), and HDR degree (French habilitation to supervise PhD) in 2019. She was Head of the Research Technical Support Teams of Supmeca in 2006, then Deputy Director of LISMMA Laboratory in 2013, and Research Deputy Director of Supmeca (Paris) since 2015.

Her main research topics are systems engineering (MBSE), agility, knowledge management/ontologies, digital twin, design metrics, mechatronic design, and multiphysics conceptual design.

Régis Plateaux obtained a Master of advanced studies in automated manufacturing from ENS Cachan, France (today ENS Paris-Saclay). He obtained a PhD degree of Industrial Engineering in 2011 from Ecole Centrale Paris, France (today CentraleSupélec), related to the guaranty of the consistency and the continuity during mechatronic design process by the means of analogy, topology, and algebraic topology.

Associate Professor at ISAE-Supméca, France, he is Head of the Mechatronic and Complex System (MSC) department. His teaching activities address tools, languages, and methodologies for the collaborative and agile design of complex systems, on Model-Based Systems Engineering (MBSE), the digital twin concept, and multiphysics modelling.

His main research focusses on: model-based system engineering (MBSE), agile design, modelling and simulation of complex, mechatronic and cyber and cyber-physical systems and their digital twins, traceability of designs and simulations, coherence of topologies and interactions and knowledge management.

Robert Rayner is Senior Lecturer at the University of Hertfordshire, UK, specializing in mechatronic systems, robotics, control theory, and autonomous vehicles. He teaches and conducts research extensively in these fields. Dr. Rayner has held academic positions at several leading UK institutions, including the University of Bath, where he completed his PhD in high-speed servomechanical system design, Imperial College London, and Kingston University London, before joining the University of Hertfordshire in 2020. He also served as Deputy Head of mathematics at a private school in St Albans, Hertfordshire.

Dr. Rayner's industry experience includes working with renowned companies such as Renishaw, Cummins Power Generation, Molins ITCM, and Mercedes-Benz. He is also Part-Qualified Patent Attorney. In addition to his academic and industry roles, he is highly involved in engineering education, having served as Education Officer for the Institution of Mechanical Engineers (IMechE) Greater London Region for several years, and currently holds multiple leadership positions within the organization.

Rob Richardson is Professor of Robotics, at the University of Leeds, where he also serves as Director of the Real Robotics Lab within the School of Mechanical Engineering. Additionally, he is Principal Investigator (PI) and Director of the National Facility for Innovative Robotic systems—an investment by UK Research and Innovation (UKRI) focused on advancing research in future robotic technologies. From 2019 to 2023 he held the role of Executive Chair for the EPSRC UK-RAS network, a strategic collaboration involving 37 UK Universities within the UK. He currently serves as Member of the executive committee of the UK Robotics Growth Partnership, dedicated to improving the UK through the integration of robotics and autonomous systems. As Innovation Director for Acuity Robotics, a spin-out company from the University of Leeds, he is actively working towards real-world impact in the field of civil inspection tasks. His diverse research interests span applied robotics including areas such as robotics for civil infrastructure inspection and repair, the making of smart bodies for smart robots, and robotics for 3D printing applications. His robotic platforms operate across various environments, including the air, ground, water, and underground.

Pablo Rocamora-Garcia is Biomedical Engineering Graduate (2023) with a Master's in Artificial Intelligence from the University of Alicante. His research focuses on computer vision and machine learning, specifically in clinical image and data analysis. His Bachelor's Thesis, which developed a deep neural network for segmenting diffusion MRI data, was published at the Iberian Conference on Pattern Recognition and Image Analysis. He has completed additional courses in deep learning and data mining technologies and participated in knowledge transfer programmes such as GENNERA and the T4EU Entrepreneurship School. Pablo received the Impulso Award for Digital Transformation and a UA award for innovative Bachelor's Thesis projects. He continues his research, publishing in specialized scientific journals.

Roland Rosen studied mathematics with a focus on economics at the Technical University of Munich and has been working in research and development at Siemens AG since 1989. In various roles as Project, Group, and Technology Field Manager, he has worked on different application fields. His technological perspective is shaped by simulation and digital twins. For example, development methodologies for mechatronic systems, and the use of simulation in the operation and service of complex technical systems, production systems and applications in the Industry 4.0 environment. Currently, he is Project Manager of the consortium Factory-X in the German initiative Manufacturing-X.

Mutaz Ryalat was born in Amman, Jordan. He received a BSc degree in Mechatronics Engineering from the University of Jordan, Amman, Jordan, in 2005, an MSc in Mechatronics Engineering from Loughborough University, Loughborough, UK, in 2007, and a PhD in Nonlinear and Robust Control from the University of Southampton, Southampton, UK, in 2015.

He is currently Associate Professor in the Mechatronics Engineering Department at the German Jordanian University, Amman, Jordan. His research interests are focused on nonlinear and robust control of mechanical/ electromechanical systems, Hamiltonian systems, mechatronics system design, robotic systems, and Industry 4.0. and AI. He serves as Active Reviewer for a number of journals and conferences in the area of mechatronics engineering.

John Ryan is Chartered Mechanical Engineer with over 25 years of project experience. He is Former Lead Design Engineer and Senior Project Manager for Laing O'Rourke, Network Rail, Tideway, and Crossrail. His experience and background in delivering projects on Victorian era infrastructure with over one-hundred-year design life gives a unique view on what lifecycle management means.

He is CEO of SymTerra, a communication platform established in 2021 and now being used to manage over £50bn worth of infrastructure assets across the UK and Europe. SymTerra has been backed by some of the largest global Venture Capitalists and in October 2023 was recognized as one of Bloomberg's top 25 UK startups.

Marcelo Saval-Calvo is currently Associate Professor at the University of Alicante. He obtained a degree in computer engineering from the University of Alicante in 2010 and a Master's degree in 2011. He received his PhD in Information Technology and Computing from the same university in 2015, funded with a public grant from the Generalitat Valenciana. His research focuses on 3D registration and reconstruction; analysis of deformable elements; neuroimaging; autonomous vehicles sensorization; and behavioural analysis using artificial intelligence techniques. He has an active collaboration with the University of Edinburgh that started in 2014 carrying out part of his thesis research; then in 2017 he went as Post-doctoral Researcher for a year and was Visitor for 4 months as Senior Researcher in 2021.

He has 25 publications in journals, most of them indexed in the JCR (IEEE Access, PeerJ, PAA, CVIU, ASOC, ESWA) in quartiles Q1 and Q2; 3 book chapters; and 19 prestigious international conferences (3DV, IROS, WCCI, IJCNN, IWANN, IWINAC). Also, Marcelo has been Reviewer for several international journals and conferences and Member of several evaluation panels. He has participated in a total of 10 research projects funded in competitive public calls, 2 research projects funded by the University of Alicante, being Principal Investigator (PI) of one of them (GRE16-28), 3 projects from regional calls, being PI of one of them (GV/2020/056), a national project in as Senior Collaborator, and 3 projects funded by the Spanish government, being PI of PID20023-149562OB-I00.

Matthew P. Shuttleworth is Research Fellow in Advanced Manufacturing at the University of Leeds in the UK. He gained his MEng in Aeronautical Engineering from Loughborough University, UK, in 2016 and his PhD in the additive manufacture of three-dimensional, conformal, and flexible electronics at the University of Leeds in 2020. Since then, he has diversified into biomedical engineering, where he developed bespoke measurement systems for monitoring dual-mobility hip implants without line of sight during wear testing and more recently into manufacturing for robotics with a focus on automated manufacture and assembly of small-scale swam robots. In his latest venture, Matthew has co-founded Hydra Manufacturing, a UK-based spin-out company from the University of Leeds, developing machinery for manufacturing components from difficult-to-process advanced materials such as technical ceramics.

Edward Simpson spent seven years in the army with a range of challenging appointments across the globe including service for the United Nations before entering Higher Education. Eddie still maintains strong links with the Armed Forces as an Active Veteran. As Chartered Quantity Surveyor and Chartered Builder, during 30 years of HE lecturing and research, Eddie has been Programme Leader on undergraduate and postgraduate courses covering construction, environment, and sustainable engineering and design. Eddie's experience and reputation led to appointments such as Higher Education representative (Scotland) for the built environment and Chairman of Validation Panels and Qualification Development Specialist. His professional qualifications and recognitions cover a broad range of skills and experience beyond construction including Logistics Manager, Chartered Environmentalist, and Senior Fellow of the Higher Education Academy.

Eddie's recent research interests and recent publications include issues covering higher education, procurement in the context of MoD projects, Stoic philosophy and its alignment with environmental issues, and sustainability. Recent publications include the following indicative issues: innovation and failure in mechatronics design education, corporate environmental sustainability, and applying Stoic philosophy to environmental decision-making.

Andrew Spiers was a fellow of the Royal Society of Biology, Andrew graduated from the University of Auckland in New Zealand, where he remained to complete his Masters and PhD degrees. Before joining the University of Abertay where he was Reader in molecular and environmental microbiology, Andrew held positions at the Department of Biochemistry and the Department of Plant Sciences at the University of Oxford. Though his main research interests lay in the adaptation of bacteria to soil and plant environments, Andrew also had significant interests in the modelling of complex systems of all types and of the issues associated with the implementation and operation of such models. In his teaching, Andrew used these interests to develop courses for engineering students looking in particular at issues of sustainability and the problems associated with the design and operation of complex systems and environments. Sadly, Andrew died unexpectedly before the book was completed.

Maarten Steinbuch holds Chair of Systems and Control at Eindhoven University of Technology (TU/e) (0.6 FTE) where he is Distinguished University Professor. He is mostly known for his work in the field of advanced motion control, as well as in robotics for high precision surgery. He is Prolific Blogger and Key Opinion Leader on the influence of new technologies on society. He received the MSc degree (cum laude, 1984) and PhD degree (1989) from Delft University of Technology. From 1987 to 1999, he was with Philips Eindhoven.

Since 1999 he has been Full Professor of Systems and Control at Eindhoven University of Technology. He was Editor-in-Chief of IFAC Mechatronics (2009–2015) and is Chairman of various supervisory and advisory boards and (Co)founder of seven companies. He was Scientific Director of the TU/e High Tech Systems Center 2014–2020. Since 2018 he has been Scientific Director of Eindhoven Engine.

In 2003, 2005, 2008, and 2015 he obtained the '*Best-Teacher*' award of the Dept. of Mech. Eng. In 2008 and 2014 his research group obtained the QANU excellence rating [5555]. In 2015 he received the KIVI Academic Society Award, and in 2016, he was awarded as Simon Stevin Meester 2016, the highest Dutch award for Scientific Technological research.

Maarten Steinbuch has numerous ISI (web of science) highly-cited papers, more than 350 publications in ISI web of science, more than 4,500 citations (H-index 39), and more than 13,000 citations in Google Scholar (H-index 60). He is IEEE Fellow and Fellow of the Netherlands Academy of Engineering.

Martin Törngren is Professor in Embedded Control Systems at the Mechatronics unit at KTH Royal Institute of Technology. He has a particular interest in trustworthiness and dependability of autonomous cyber-physical systems (CPS), including complexity management and architectural design. In related efforts at KTH with partners, he is driving the development of an open research testbed for CPS interacting with a smart infrastructure including with applications in connected automated vehicles.

He has authored/co-authored more than 200 peer reviewed publications and part of developing and leading both graduate and continued education courses, as well as the development of new/renewed master programmes. He has so far supervised 21 graduated PhD students.

Networking and multidisciplinary research have been characteristic throughout his career. From 1999 to 2004, he served as Chairman of the Swedish real-time systems association, and he has represented KTH as Core Partner in the EU networks of excellence in Embedded systems design, Artist2 and ArtistDesign, and in the corresponding industrial associations (Artemis, Inside). He is moreover Principal Initiator of the Innovative Centre for Embedded Systems (www.ices.kth.se), launched in 2008 (serving as director until 2020), and is Director of the TECoSA Swedish national competence centre on Trustworthy Edge Computing Systems and Applications (www.tecosa.center.kth.se) in close collaboration with industry. He spent a 10-month sabbatical in 2011/12 at UC Berkeley and a 2-month sabbatical in the spring of 2018 at the Stevens Institute of Technology (Hoboken, New Jersey). This was followed by a 2-month sabbatical again at UC Berkeley in the fall of 2018.

Ferdinando Vitolo is Assistant Professor of Design Methods for Industrial Engineering at the University of Naples Federico II. He obtained his Master's degree in Mechanical Engineering (cum laude) from the University of Naples Federico II in 2012. From 2012 to 2014 he collaborated with the University of Naples Federico II in the National Project: '*PON 01_1268, Digital Pattern Product Development;*

a pattern-driven approach for industrial product design, simulation and manufacturing'. After his PhD in Industrial Engineering, in 2017 he won a postdoctoral fellowship for activity research on '*Human-robot collaboration: modelling and design of workplace and device*'.

His main research interests are focused on MBSE and KBE, mechanical design, GD&T, advanced CAD modelling methodologies, and mechatronic systems, which is reflected in his courses on technical drawing, concept design of new vehicles, and digital modelling of systems.

Adrian-Catalin Voicu graduated from the Valahia University of Targoviste, the Faculty of Materials Engineering, Mechatronics and Robotics, with the specialization of Mechatronics, in 2009 and received his Master's degree in Industrial equipment and installations, in 2011. He obtained his PhD degree in Mechanical Engineering, from Doctoral School of Engineering Sciences—Electrical Engineering, Materials Engineering and Mechanical Engineering, Valahia University of Targoviste, in 2016, with the theme: '*Studies and research on 3D measurement of complex parts for automotive industry by multiple laser scanning*'.

In 2009, he joined the Smart Measurement Mechatronics Department of the National Institute of Research and Development in Mechatronics and Measurement Technique in Bucharest, Romania, currently being, Senior Researcher 3rd degree of the Integrated Control Mechatronic Systems Laboratory. He has 15 years of experience in mechatronics and mechanical design, integrated dimensional control, measurement and control equipment for the automotive industry, 3D scanning, sensors, and transducers.

Dr. Eng. Voicu has participated in several courses, seminars, and conferences in the above specific fields, coordinated various research projects, and worked as Member of the research teams of numerous national and international projects.

He has a series of 16 scientific papers published in World of Science and BDI indexed journals, as well as 17 scientific works presented and published in the volumes of national and international specialized scientific events, being Co-Author of a scientific book.

Steven A. Watt is currently Chief Information Officer at the University of St Andrews in Scotland. He is European Engineer, Chartered Engineer, Chartered IT Professional, Fellow of the British Computer Society, and Certified Information Security Systems Professional (CISSP), with over 30 years' experience working within the IT industry. As CIO he is responsible for the strategic development of IT services at the University. For the last five years, key focuses have been cyber security and digital transformation which has positioned the University in its response to the pandemic and in responding to this existential threat.

He has also worked in the private sector for Low & Bonar plc in both Europe and North America. In this role he was responsible for the coordination and delivery of corporate IT provision at around 52 locations which included the provisioning and management of wide area networking, telecommunications, groupware and systems management, working with outsourced partners centralizing the core ICT support for all sites to the UK. He has also worked as Consultant across a range of industries including healthcare and manufacturing.

During Steve's career, he has spent over twelve years on the boards of both a University and Further Education College, giving him a broad insight into challenges faced across the sector, not just at St Andrews. He has also served as Chair of the Higher Education Information Directors Scotland group. He is an Active Participant in a number of UK IT groups.

In taking the University's corporate social responsibility very seriously, Steve led a project to consolidate equipment from over 50 locations to a single green-energy data centre. For the third time in a row, the data centre has achieved the highest level of award for its performance and environmental design—the GOLD award under the Certified Energy Efficient Datacentre Award (CEEDA) Scheme, and is also Participant in the European Code of Conduct for data centres. The IT team has won many UK industry-standard awards for its approach, design, development, and operation of this facility.

In 2016, Steve formed a small IT security team, which, from a standing start, provides a comprehensive IT security system, achieving Cyber Essentials certification in March 2018 and Cyber Essentials Plus in April 2019. This team was named joint winner (along with Police Scotland) of the award '*Outstanding Cyber Team*' at the Scottish Cyber Awards in October 2018, and in 2022, one of his young IT Security Analysts was named '*Cyber Rising Star 2023*'. He is Active Researcher and Practitioner in the area of cyber security and privacy.

Steve is passionate about inspiring the next generation of IT professionals and since 2012 has recruited over thirty-five Modern Apprentices into his IT teams. Since then, the opportunities, commitment, and support provided to his IT Apprentices have been acknowledged by Skills Development Scotland, Microsoft, and the Chartered Institute for IT, the BCS. Steve's team has received numerous awards for their work in this area, including 'Microsoft Apprentice Employer of the Year'. He also provides mentoring for a range of IT professionals across the UK. Steve was listed in the Computing influencers top 100 in 2023.

George R. S. Weir holds academic qualifications from three Scottish universities: a Master of Arts degree from the University of Glasgow, PhD from the University of Edinburgh, and a postgraduate diploma from the University of Strathclyde (where he taught Computer Science for over thirty years). As Senior Academic, in addition to supervising many successful PhD students, he has authored more than two hundred publications in the areas of human computer interaction, education, applied language technology, information security, cybercrime, and digital forensics. Recently, he has worked closely with colleagues in the UK, Canada, and Saudi Arabia on the application of textual analysis to content classification. He is Adjunct Professor at the School of Criminology, Simon Fraser University, in Canada and Visiting Lecturer in Information Security at the University of Colombo School of Computer Science, in Sri Lanka. George is Fellow of the Sir Winston Churchill Memorial Trust, Senior Member of IEEE, Associate Member of the Association of Certified Fraud Examiners, and Fellow of the Higher Education Academy. Having been elected in 2022 as

Local Government Councillor in South Ayrshire, he continues his academic research activities in parallel, under the auspices of Mellow Research, based in Ayr, Scotland.

Roger J. White is Founder and President of RC2 Inc., a research consultancy based in Ontario Canada with over 35 years R&D, marketing, and entrepreneurship experience in optronic sensing, robotics, and AI. Focus has been largely on the development of novel analytical systems for use in real-time macromolecular characterization, bio-process monitoring, and protein crystallography.

He completed a PhD in Materials Science from Bath University in 1985, then worked at Thorn-EMI CRL on thin film optics, and distributed laser sensing for use in military and civilian applications, including battlefield intelligence systems and smart homes. Moved into technical marketing and product management for Millipore Waters, a US market leader in HPLC analytical separation systems, before joining a specialist personal robotics startup in London as European Product Manager for Laboratory Automation and Assisted Living, working on a collaborative research project to develop low-cost paraplegic rehabilitation systems.

He established a specialist analytical instrumentation distribution, contract analysis, and R&D company based in North Wales in 1992, managing and participating in a range of collaborative research projects funded by UK & EU bodies while developing a compact protein aggregation analyser using guided-wave light scattering. In 2006, the UK instrumentation business was sold to a Boston MA startup and RC2 Inc. was founded in Ontario Canada as an analytical instrumentation R&D consultancy.

Projects undertaken to date include the spin-out of an innovative low-cost micro-volume protein analyser and a multi-channel fibre-optic UV detector capable of monitoring 12 process separations in parallel. Current focus is on the development of microfluidic sensor arrays for use in rapid diagnostic systems used in wide spectrum screening and in-home AI well-being monitoring systems for combatting cognitive decline in people living alone.

Chapter 1
Mechatronic Futures Revisited—An Introduction

Peter Hehenberger and David Bradley

Abstract Since the concept of mechatronics was first proposed, it has been subject to a continuous process of change as it adapts to developments in technology. Nevertheless, the underlying concepts of integration at the component, device and system levels associated with a transfer of functionality from the physical to the software domain remain essentially the same. Here, in the introduction to this revised edition of Mechatronic Futures, written some 7 years after the initial version, some of the issues and challenges now facing mechatronics engineers and designers are considered to give context ahead of the more detailed analysis provided by the individual chapters. These chapters include both updated versions, in areas such as home systems, healthcare, manufacturing technologies and the Internet of Things, plus new material covering areas such as artificial intelligence, cybersecurity, issues of sustainability, system complexity and lifecycle monitoring which have evolved and developed in recent years and which impact significantly on the mechatronic concept. Many of these new ideas overlap with existing issues, and while this introduction sets out to provide an overall context, they are of necessity and, importantly, reflected in the individual chapters in different ways and with differing emphases. The changes and challenges identified through the various chapters also impact on the ways in which mechatronics education is developing and the resulting links into project management. The final chapters of the book reflect on these issues. Finally, it is becoming apparent that mechatronics engineers and designers are increasingly required to take into consideration not only the technical issues that lie at the core of mechatronics but also social, and indeed political, factors, as for instance in respect of individual privacy, into their thinking. The introduction seeks to point out some of these issues and concerns.

P. Hehenberger (✉)
University of Applied Sciences Upper Austria, Wels, Austria
e-mail: peter.hehenberger@fh-wels.at

D. Bradley
Abertay University, Dundee, UK

P. Hehenberger and D. Bradley (eds.), *Mechatronic Futures*,
https://doi.org/10.1007/978-3-031-83571-1_1

1.1 Background

The period of over 50 years since the concept of a mechatronic system was introduced by Tetsuro Mori [1, 2] to express the growing impact that the availability of electronic components was having on the control and operation of inherently mechanical systems has been and continues to be, a period of significant and rapid technological change. In particular, there has been a shift in emphasis within systems from hardware to firmware and software, leading to the introduction of a wide range of consumer products structured around the use of 'smart' devices, many of which remain essentially mechatronic in nature in that they bring together a core of mechanical engineering with increasingly sophisticated electronics and software. When combined with enhanced local and remote communications, this has led to the evolution of systems based around the ability of 'smart' objects to communicate with each other and hence to effectively self-configure according to context.

1.1.1 Motivations for This Edition

The period since the original Mechatronics Futures book was published in 2016 has seen significant changes in terms of both technology and applications, as for instance in relation to artificial intelligence and cybersecurity, which relate not only to the design and operation of mechatronic systems, but also to the way in which they are used. For instance, Applied AI is now a major component of many mechatronic systems which can in turn act as an information source for other forms of Big Data and AI-based analysis, resulting in potential social and ethical issues to which designers and developers of mechatronic systems need to be aware. Such systems are also liable to hacking attacks, hence issues such as cybersecurity need to be accommodated.

Other factors considered in determining potential content included consideration of reports such as those produced by the World Economic Forum on the future nature of manufacturing and jobs, and the associated impact on the ways in which mechatronics was and indeed is, to be deployed as part of and in support of developments and changes in these and related areas [3, 4].

1.1.2 Megatrends and Their Impact on Mechatronics

A particular driver for the revised edition was the concept of Megatrends as put forward by the Zukunftsinstitut. These Megatrends are considered as drivers of change and hence impact on the ways in which mechatronic systems are designed and operated. The 12 proposed Megatrends are [5]:

Connectivity—This relates to the ways in which the digital infrastructure and associated communications technologies are impacting on the ways in which we live and work, bringing with them new lifestyle and behaviour patterns as well as impacting upon economic and business models.

Impact areas for Mechatronics include Artificial Intelligence, Big Data & The Cloud, Complexity, Design, Human-Centred Systems, Industry, Internet of Things, Lifecycle Management & Transport.

Gender Shift—Traditional gender roles assigned to men and women in society are changing, resulting in changes to societal norms as well as to economic models.

Impact areas for Mechatronics include Artificial Intelligence, Big Data & The Cloud, Human-Centred Systems, Internet of Things & Lifecycle Management.

Globalisation—This refers and relates to the ways in which societies are coming together, supporting the exchange of ideas, talents and goods.

Impact areas for Mechatronics include Big Data & The Cloud, Complexity, Design, Human-Centred Systems, Industry, Internet of Things, Lifecycle Management & Transport.

Health—The importance of health as a fundamental value has increasingly been recognised, and the provision of and access to appropriate and effective levels of healthcare impacts on quality of life.

Impact areas for Mechatronics include Artificial Intelligence, Big Data & The Cloud, & Human-Centred Systems.

Individualisation—This is driven by an increase in personal freedoms; it defines the relationship between the individual and society.

Impact areas for Mechatronics include Artificial Intelligence, Big Data & The Cloud, Complexity, Design, Human-Centred Systems & Transport.

Knowledge Culture—The world is becoming smarter: global education levels are higher than ever and are continuing to grow almost everywhere. Fuelled by the Megatrend Connectivity, our understanding of the world and the way we interact with information is changing. Education is becoming more digital. Cooperative and decentralised structures for knowledge generation are spreading, and our understanding of knowledge, its creation and dissemination, is increasing.

Impact areas for Mechatronics include Artificial Intelligence, Big Data & The Cloud, Complexity, Design, Human-Centred Systems, Internet of Things, Lifecycle Management & Transport.

Mobility—This relates to the mobility of the individual in respect of both the physical and information domains and is a driver for change in the form of new products and services as well as developments in transportation.

Impact areas for Mechatronics include Artificial Intelligence, Big Data & The Cloud, Design, Human-Centred Systems, Internet of Things & Transport.

Neo-Ecology—This relates as to how issues of sustainability impact on and change individuals, societies, economies and organisations.

> Impact areas for Mechatronics include Artificial Intelligence, Big Data & The Cloud, Complexity, Design, Human-Centred Systems, Industry, Internet of Things, Lifecycle Management & Transport.

New Work—How will change impact the future of work, and what will that work look like?

> Impact areas for Mechatronics include Design, Human-Centred Systems, Industry, Lifecycle Management & Transport.

Security—The nature of both collective and individual security and the responsibility for ensuring this is evolving rapidly and is likely to continue to do so.

> Impact areas for Mechatronics include Artificial Intelligence, Big Data & The Cloud, Complexity, Design, Human-Centred Systems, Internet of Things & Transport.

Silver Society—Societies are ageing, presenting significant challenges but also opening up new social and cultural opportunities for new vitality. To support this demographic transformation, new social and economic conditions along with a new attitude to ageing are needed.

> Impact areas for Mechatronics include Artificial Intelligence, Big Data & The Cloud, Complexity, Design, Human-Centred Systems, Internet of Things & Transport.

Urbanisation—As more people live in cities, these need to evolve to meet the needs of the population as well as act as the hubs of a globalised economy.

> Impact areas for Mechatronics include Artificial Intelligence, Big Data & The Cloud, Complexity, Design, Human-Centred Systems, Internet of Things, Lifecycle Management & Transport.

1.1.3 Long-Term Thinking

The lead time for the development of new and novel technologies to the point where they become commercially viable can be several decades while the implementation time for large projects can also be of the order of a decade, placing additional stress on the design and project management. Thus, while the first DARPS Grand Challenge of 2004 [6] paved the way for autonomous vehicles, as can be seen from Fig. 1.1, there are still some years to go before anticipated full adoption [7].

Further, many systems are achieving extended lifetimes through a process of refurbishment and refitting. Perhaps an extreme example of this is the Boeing B52 *Stratofortress*, the last airframe for which, of a total of 742, was built in 1963. Yet thanks to a programme of refitting avionics and engines remains in service and is

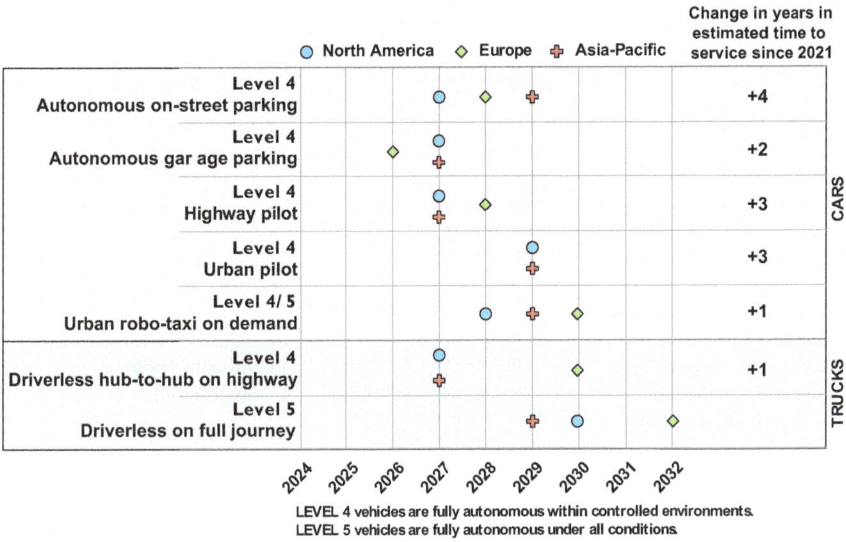

Fig. 1.1 Estimated timeline for adoption of Level 4 and Level 5 autonomous vehicles (after [7])

intended to do so for another 20 years. This concept of life extension requires the designer to futureproof systems by designing in the potential for several upgrades during in-service life [8, 9]. Notwithstanding the remark often attributed to Nils Bohr that:

> '*It is difficult to make predictions, especially about the future.*'[1]

Table 1.1 attempts to identify some of the issues associated with this approach.

1.2 (Some) Challenges Around for and Mechatronics

Mechatronics has always faced challenges in terms of the way in which the concept was conceived and implemented. This was particularly so in respect of the growth in capacity and capability of computing technology which has taken place over the last 50 years, giving mechatronics engineers a range of new and highly capable tools with which to work. More recently, the growth of Information Technology has resulted in the development of concepts such as cyberphysical systems, the Internet of Things and The Cloud, all of which have impacted on the nature of mechatronics, moving it from a device-oriented concept to a systems-oriented concept.

These developments in the ability to collect, process and utilise data along with the enhanced technical capacity of devices have also resulted in ideas and concepts which were in place 20 or more years ago becoming genuine applications in areas

[1] As well as by others including Yogi Berra!

Table 1.1 Potential technical development in the short, medium and long terms

Now	Short to medium term 2025 to 2035	Long term 2035 to 2050
Available technologies • Big data analytics (in some areas) • Cloud technologies • Cyberphysical systems • Digital twin (limited capability) • Industry 4.0 (developing) • Internet of things • Networked sensors • '*Smart*' technologies (relatively limited and basic[2])	Available technologies • Big data analytics (updated and refined) • Digital twins (second generation) • Enhanced sensor technologies • Enhanced communications (5G +) • Industry 5.0 • Smart technologies for domestic and industrial energy management[3]	Available technologies • Big data analytics (established) • Digital twins (established) • Enhanced sensor technologies • Enhanced communications (5G +) • Industry 5.0 (established) • Smart technologies (established)[4]

such as transport, manufacturing and healthcare. This places further demands on mechatronic systems designers and engineers both in terms of translating concepts into applications and in determining future directions of travel for mechatronics.

1.2.1 Artificial Intelligence

Perhaps the greatest challenge to mechatronic engineers and designers to emerge in recent years is that of AI which is impacting on the way in which systems operate at all levels. Consider for instance the situation of Fig. 1.2 with vehicle systems as the exemplar. Here, at the lowest level, the role of mechatronics is concentrated around those components, systems and sub-systems which lie at the core of any system to provide the primary sensing and actuation functions, as for instance in engine management. Low-level intelligence may be incorporated for functions found at this level such as data validation and actuator control.

These systems are then integrated on a functional basis to create larger sub-systems, as for instance a vehicle's stability control system. Again, intelligence may be incorporated to manage the functionality of this larger system and to support communication between systems. These two lower levels may be further integrated with the addition of additional AI features to create a cyberphysical system along the lines shown in Fig. 1.3. Here, the embedded processor elements can offer a range of functions across a range of systems, as for instance in the case of adaptive cruise

[2] The term '*smart*' as used here generally implies the capability to communicate via the internet rather than indicating any form, however limited, of embedded intelligence and decision-making capability.

[3] Potentially incorporating embedded intelligence and decision-making capability.

[4] Incorporating embedded intelligence and decision-making capability.

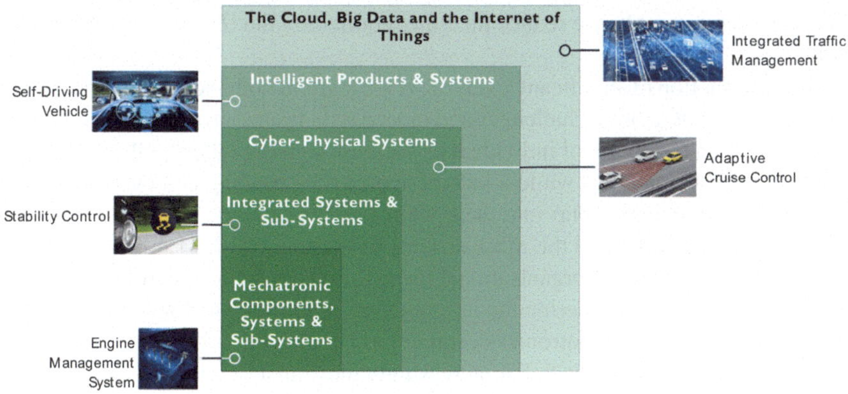

Fig. 1.2 An AI hierarchy in mechatronics

control where speed, lane control, vehicle tracking and other features are brought together with the role of the cyberelement being to integrate these functions within the safe operation of the vehicle.

Further integration of features, again with added AI elements, adds additional capability to create advanced intelligent products, illustrated here by a self-driving vehicle. Such products assumed added responsibilities in areas such as safety along with an ability to react to complex, and unlearned, situations and environments.

Finally, the top layer sees the use of AI to manage a wide range of features across large numbers of component elements. Thus, the Internet of Vehicles can be used to manage traffic flows, to find and assign parking spaces and, through a link to a network of Digital Twins, to gather performance-related data which can be used both to upgrade current systems and to feed into the design of next-generation systems.

In addition to these system-level functions, AI is also expected to contribute to the design process by enabling the use of '*virtual experiments*' to assess the effectiveness

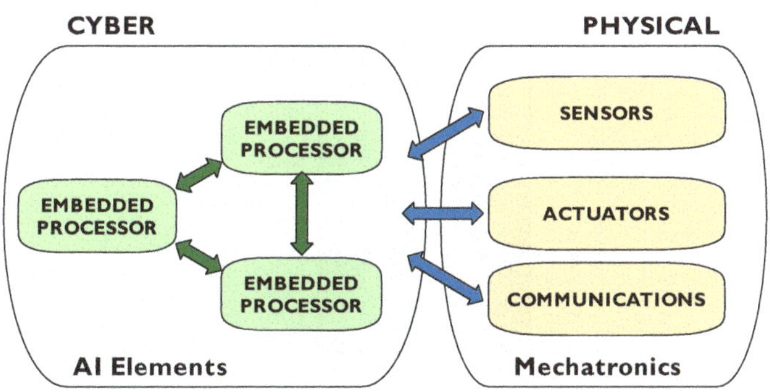

Fig. 1.3 A cyberphysical system

of design options and hence reduce the time required by the '*Design-Test-Evaluate*' cycle.

The introduction of significant levels of AI throughout both the design process and applications is not without challenges and indeed risk. Issues have already been identified in relation to aspects of individual privacy through the, often well-intentioned, application of AI analytics while a lack of appropriate training and evaluation data can introduce unintended bias into the system.

Further, and referring to the mechatronics ecosystem of Fig. 1.4, it is clear that there also are a number of organisational layers of which the mechatronics designer must be aware when considering the introduction of AI-based technologies. Historically, the focus of the mechatronics designer has been on the core of this ecosystem, the intelligent or smart product, but increasingly they must be aware of the existence of these other layers, if not necessarily of the details of their relationships either with the product or with each other. Figure 1.5 and Table 1.2 then indicate some of the issues associated with the introduction of AI within a mechatronic environment.

This must all be undertaken while taking into account of the need to manage the introduction of AI systems at all levels. In its recent report *Governing AI for Humanity* the United Nations High Level Advisory Body on Artificial Intelligence establishes some of the potential opportunities and risks associated with the introduction of AI while establishing the need for standards to govern its introduction and use (Fig. 1.6). Again, these will all impact on the mechatronic environment [10–17].

Fig. 1.4 A mechatronics ecosystem

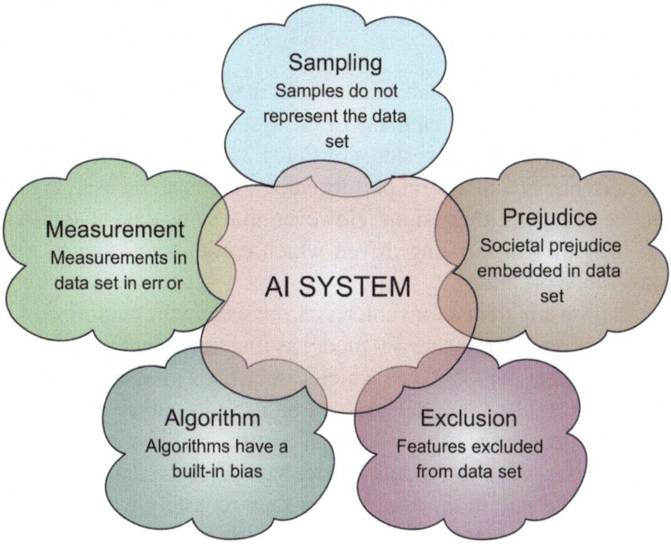

Fig. 1.5 AI issues

Table 1.2 Issues in AI design

Ethical design	Unethical design
White box structures • Operation of AI is transparent to the user • Decisions are explained and able to be easily assessed	Black box structures • Operation of AI is hidden from the user • Decisions are not explained and cannot be readily assessed
Privacy is protected • Personal data is held privately and securely • Individuals retain control over who has access to their data and can define how it is to be used	Privacy is not protected • Individuals do not have control over their personal data • Hence do not know how and by whom it is being accessed and used
Algorithmic fairness • Bias is designed out at every stage of the process from data collection to application	Algorithmic bias • Algorithms contain a built-in bias which distorts the outcomes from the system

Fig. 1.6 Growth of standards governing the role and use of AI (after [17])

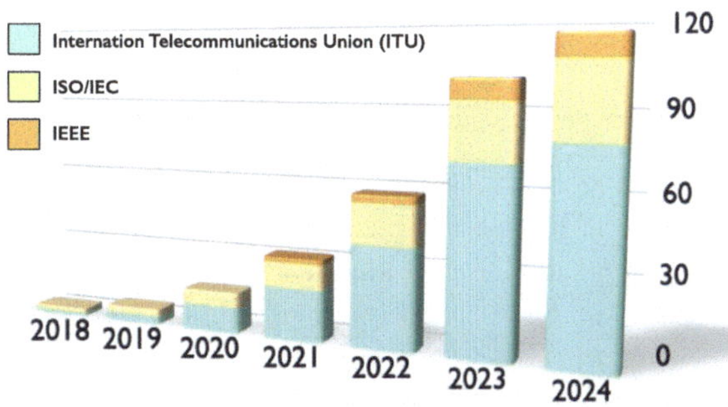

1.2.2 Privacy

Consider again Fig. 1.2 which illustrates the ways in which mechatronics, particularly in association with AI, impacts on many of the systems found in the home and elsewhere. In respect of the vehicle systems of Fig. 1.2, the data collected both enables operation and provides feedback on performance which can be used to support both upgrades and next-generation designs. However, that same data also provides information as to how the car is being driven which could, for instance, impact upon individual insurance premiums.

A similar relationship to that for vehicles can be seen in the evolution of healthcare provision to the current Healthcare 4.0 model as illustrated by Fig. 1.7 where a wide range of devices collect personal data which can be used to assess their current state of health. However, these same devices can be used to gather data which may be capable of being used against the individual and their ability to access healthcare and treatment. This has led to models where, say, personal data collected within a person's home is held within the home rather than remotely with access controlled by the individual in agreement with their medical team.

This conflict between the ability, particularly of mechatronic systems and components, to access personal data has led to the introduction of the concepts of what has been referred to as *Privacy by Design* and the introduction of legislative frameworks such as the General Data Protection Regulation in Europe [18, 19]. Both of which have a structure similar to that of Fig. 1.8 at their core.

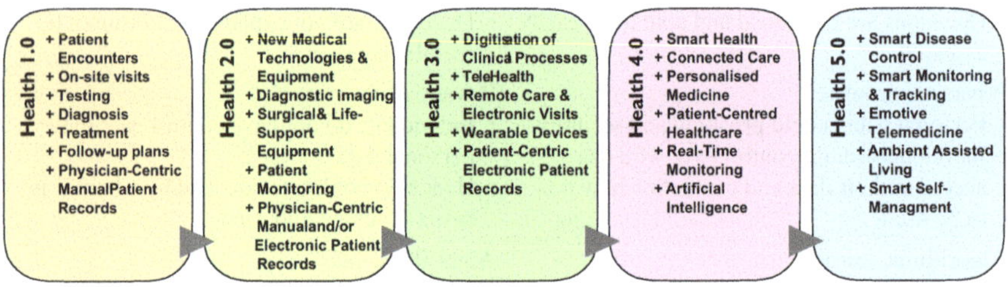

Fig. 1.7 Evolution of healthcare 4.0 and beyond

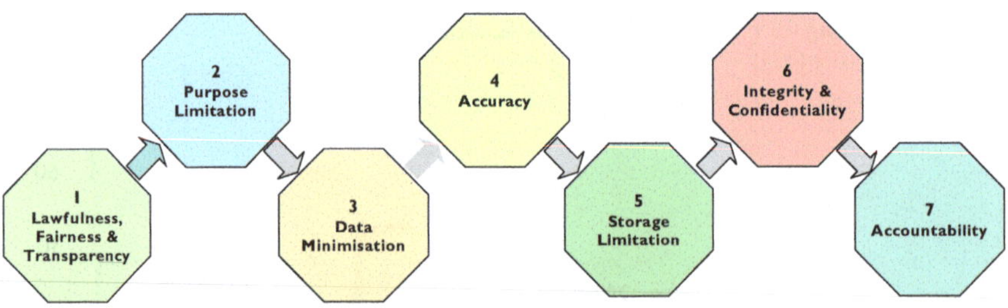

Fig. 1.8 Conceptual structure of the general data protection regulation and of privacy by design

Though the approach advocated by privacy by design has been criticised as being too general in its concept, it has nevertheless provided a useful starting point from which to develop technical, social and legal arguments in support of data privacy and the privacy of the individual across the entirety of the data lifecycle. In the context of mechatronics, the *Privacy by Design* approach has tended to focus on individual organisations or a specific technology whereas systems-level integration may, in mechatronic terms at least, require the bringing together and interoperability of a range of different technologies designed, developed and operated by a range of organisations, hence implying a shift away from organisational to infrastructural design [20–25].

In 2018, the European Data Protection Supervisor Giovanni Buttarelli argued [24] that:

> '*While privacy by design has made significant progress in legal, technological and conceptual development, it is still far from unfolding its full potential for the protection of the fundamental rights of individuals…*'[5]

And went on to call upon the European Commission:

- To ensure strong privacy protection, including privacy by design, in e privacy regulation.
- To support privacy when adapting or creating legal frameworks which influence the design of technology, by increasing incentives and substantiating obligations, including appropriate liability rules, to integrate privacy by design in products and services.
- To foster the roll-out and adoption of privacy by design approaches and Privacy Enhancing Technologies (PETs) in the European Union (EU) and at the member states' level through appropriate implementing measures and policy initiatives.
- To ensure continuous availability of competence and resources for research and analysis on privacy engineering and privacy enhancing technologies at EU level, either by maintaining the current capacity and tasks for ENISA,[6] or by allocating appropriate resources to other entities.
- To support the development of new practices and business models through the research and technology development instruments of the EU.
- To support policy initiatives for EU institutions and national public administrations to lead by example and to integrate appropriate privacy by design requirements in public procurement.
- To support an inventory and observatory of the '*state of the art*' of privacy engineering and PETs and their advancement, and to raise awareness on the subject with citizens and economic and political actors.

[5] In paragraph 22 of the Opinion.

[6] European Union Agency for Cybersecurity.

1.2.3 Cybersecurity

In respect of a computer system, a vulnerability is taken as referring to any defect or failing in that system that compromises its security. To secure any computer or computer-based system such as mechatronics, it is necessary to understand the nature of the attacks that can be used, and Table 1.3 therefore identifies some of the more common forms of attack used [26–28]. The UK Government's *Cyber Security Breaches Survey 2023* [29] obtained responses from 2,263 businesses ranging in size from micro-businesses with between 1 and 9 employees and large businesses with more than 250 employees along with 1,174 high and very-high-income charities.[7] Of these, 32% of businesses and 24% of charities reported breaches or attacks over the previous 12 months, the most disruptive being as set out in Table 1.4.

In the UK, the National Cyber Security Centre [30] has set out five principles for the design of cybersecure systems as being:

Establish the context before designing a system—To achieve a secure design it is necessary to fully define and understand the context in which the system is to operate, the fundamental requirements of the system and its operational limitations.

Make compromise difficult—Apply and deploy concepts and techniques which make it harder for attackers to compromise data or systems.

Make disruption difficult—For high-value or critical services that rely on technology for delivery and implementation, it is essential that the relevant technology is always available.

Make compromise detection easier—Despite all precautions, there is always a chance that a system will be subjected to new or unknown forms of attack. Compromise indicators provide a means of detection for such attacks.

Reduce the impact of compromise—Design systems such as to naturally minimise the severity of any compromise.

They go on to suggest that this can be achieved in part through the deployment, of techniques and methods such as those set out in Table 1.5.

Some Mechatronic Systems at Risk

Aviation—Aircraft depend on highly complex and integrated computer systems to manage and maintain them in flight, including data collection and communications to support the operation of remote Digital Twins monitoring the operation of individual system elements such as the airframe and engines. The aviation industry is also extremely reliant for its operation on complex interconnected computer systems that support a range of functions from air traffic control to ticketing [31–33].

Consumer Devices—Consumer devices such as computers, tablets, smartphones, tablet computers, smartwatches and other mobile devices contain a wide range of internal systems such as accelerometers, cameras, GPS receivers and microphones,

[7] High income corresponding to > £500,000 and very-high income to > £5 million.

Table 1.3 Some common forms of cyberattack

Type of attack	Notes
Backdoor	In a computer system a backdoor refers to a means of access which bypasses normal authentication procedures and security protocols, allowing access to the core of the system
Denial-of-service attack	Denial-of-service attacks (DoS) aim to make a computer network or resource unavailable to its intended users, for instance by overwhelming the system with automatically generated spurious requests for access
Direct-access attack	A direct-access attack is associated with an unauthorised user gaining access to the system, enabling them to access stored data and to modify core software
Eavesdropping	Eavesdropping typically involves the tracking of unsecured communications between network hosts
HTML smuggling	An HTML smuggling attack embeds malicious code inside an HTML file or web page. When the HTML is opened by a user, this code is activated to place malware onto the target system
Malware	Malware, or '*Malicious Software*', refers to code deliberately intended to disrupt or harm a computer system and its users. Types of malware include: • Viruses • Worms—these are a form of self-replicating malware that spreads through a computer system or network without need for user interaction • Trojan Horses—these are programs that aim to fool users into installing them, perhaps by pretending to provide a service • Spyware is used gather information from an infected computer, as for instance in the form of a *keylogger*s which records keyboard inputs and keystrokes
Man-in-the-middle attack	A Man-in-the-Middle (MITM) attack involves an attacker attempting to intercept, monitor or modify communications between users by simulating user identities
Multi-vector, polymorphic attack	These are a complex form of attack based on a combination of other forms of attack including, for instance, combinations of viruses, worms and trojans
Phishing	Phishing involves fooling users into disclosing information, for instance through the use of false emails, messages and phone calls. They often direct users to a fake website which then asks for personal information such as login details and passwords to be entered Spear-phishing attacks target specific individuals
Social engineering	This form of attack is intended to get a user to disclose personal information or grant physical access by impersonating trusted contact or source
Spoofing	Spoofing a variation on phishing in that the attacker pretends to be a valid contact of the user in order to gain access to information or resources
Tampering	Tampering involves the intentional but unauthorised modification of a computer system

Table 1.4 Most disruptive type of breach or attack as reported by the UK Government *Cyber Security Breaches Survey 2023* [29]

	Businesses (%)	Charities (%)
Phishing attacks	59	64
Impersonating an organisation in emails or online	35	41
Hacking or attempted hacking of online bank accounts	15	9
Takeovers of organisation or user accounts	15	13
Viruses, spyware or malware (excluding ransomware)	14	11
Denial-of-service attacks	8	11
Unauthorised accessing of files or networks by outsiders	3	1
Ransomware	2	5
Unauthorised accessing of files or networks by staff	1	3
Unauthorised listening into video conferences or instant messaging	0	0
Other breaches or attacks	4	6

any or all of which may be associated with the collection of personal data. These devices are also increasingly linked to a widening range of home automation devices such as '*smart*' thermostats and video doorbells which can be used to provide an indicator of occupancy [34, 35].[8]

Internet of Things—The Internet of Things (IoT) is comprised of the interconnection of a range of physical devices and objects including building systems and smart home technologies[11], eHealth technologies, personal devices and transport systems that use network connectivity to collect, distribute and exchange data [36–39].

Medical Devices and Healthcare—Medical devices including diagnostic equipment and implanted devices including pacemakers and insulin pumps have all been subject to cyberattack, creating significant risks to patients. Hospitals have also suffered data breaches leading to the loss of sensitive data, including patient records which are often held on in-house networks [40–44]

Utilities—Telecommunications networks, power grids and power plants all contain significant mechatronic elements vulnerable to cyberattacks which can disrupt such networks, rendering them unavailable to users [45–48].

Vehicles—Vehicles contain significant numbers of mechatronics elements related to applications such as engine management, braking systems, cruise control and advanced driver-assistance applications. Additionally, vehicles may increasingly be able to communicate both with other vehicles and with external systems via the Internet of Vehicles. Self-driving cars will be even more complex [49–51].

[8] The term '*smart*' as used here implies the capability to communicate via the internet rather than indicating any form, however limited, of embedded intelligence and decision-making capability.

Table 1.5 Ten steps to cybersecurity mitigation (after [30])

1	Risk management	Cybersecurity risk assessment
2	Engagement & training	Staff training and awareness-raising activities
3	Asset management	Identify critical assets
4	Architecture & configuration	Have at least three of the following in place: • Up-to-date malware protection • Network firewall • Restricted IT admin and access rights • Security controls on organisation-owned devices, • Only allow access via organisation-owned devices • Separate Wi-Fi networks for staff and visitors • Specific rules for personal data storage and transfer • A Virtual Private Network (VPN)
5	Vulnerability management	Apply software security updates within 14 days
6	Identity & access management	Require two-factor authentication when people access the organisation's network, or for applications they use
7	Data security	Cloud backup or other forms of backup
8	Logging & monitoring	At least one of the following: • Use of tools for security monitoring • Monitoring of user activity
9	Incident management	Have a formal incident response plan or at least three of: • Written guidance on who to notify of breaches • Roles or responsibilities assigned to specific individuals during or after an incident • External communications and public engagement plans • Guidance around when to report incidents externally
10	Supply chain security	Monitor risks from suppliers or wider supply chain

1.2.4 Complex Systems

To understand the way in which mechatronic systems have become embedded within increasingly complex systems environments consider Figs. 1.9 and 1.10 which show the evolution of vehicle systems along with an outline of the range of systems that might be found on a current vehicle. From these figures, it is readily apparent that mechatronics has been closely involved in the development of the great majority of the systems currently found on board vehicles.

The first mechatronic systems used in support of Vehicle Dynamic Stabilisation such as antilock brakes and traction control replaced largely mechanical systems in these roles but were essentially stand-alone in nature. As vehicle systems continued to evolve, in particular with the introduction of automated vehicles, electric and hybrid technologies (Table 1.6), this has introduced further layers of complexity at the system level, requiring the mechatronics engineer to be aware of the role that their system, subsystem or component has in relation to the whole.

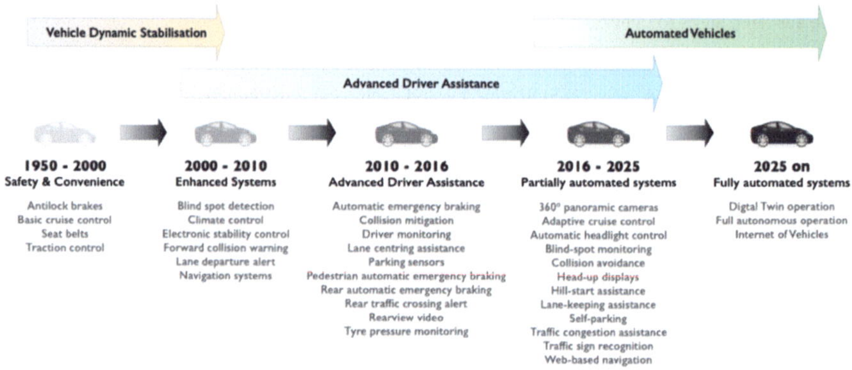

Fig. 1.9 Evolution of vehicle systems (see also Fig. 1.2)

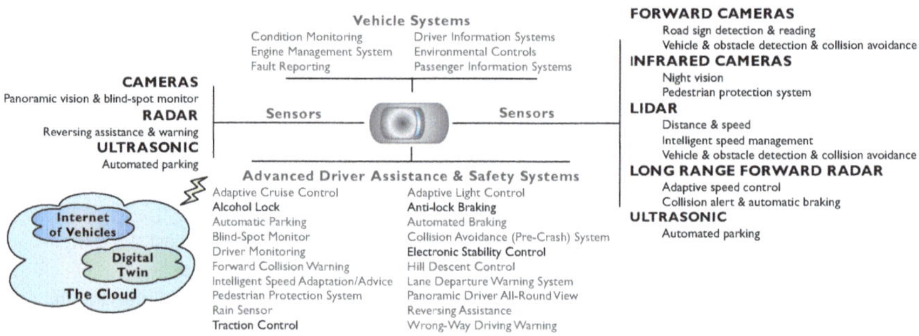

Fig. 1.10 Current vehicle systems

Table 1.6 Types of hybrid vehicles

Dual hybrids	Contain two different energy recovery systems
Mild hybrids	Mild hybrids utilise a limited range hybrid features, for instance some engine-assist or regenerative braking and usually cannot operate in a purely electric mode
Full-hybrids	Sometimes referred to as strong hybrids they can run on either the engine, the batteries or a combination of the two
Plug-in hybrid	These are full hybrids which can be plugged into an electrical outlet to be charged. They also tend to provide a greater battery capacity
Dual hybrids	These contain two different energy recovery systems and mild-hybrid, full-hybrid and plug-in hybrid forms are available

While vehicle systems have been used here to illustrate the growing complexity of systems structured around mechatronics, similar arguments could also be made for a wide range of other systems including manufacturing, healthcare, building

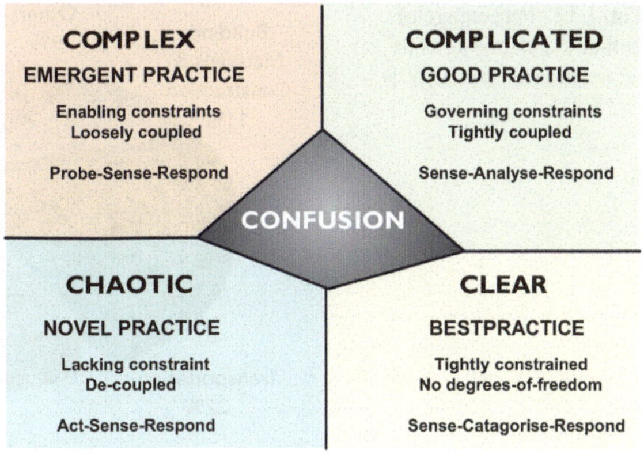

Fig. 1.11 *Cynefin* framework

management and domestic systems. In each case, the mechatronic elements form a key part of the system and hence of its operation.

The recognition of the growth of complexity has led to the introduction of methods of analysis and interpretation that aim to help the design engineer understand the relationships between the various elements of the system. One such approach is that of the *Cynefin* framework developed at IBM in 1999 by Dave Snowden as an aid to decision-making. This approach breaks the problem down into a series of domains which can then be analysed in their turn [52–58]. Using the current terminology[9] these domains and their characteristics (Fig. 1.11) are:

Clear—This represents the *'known knowns'* for which there are rules or best practice in place. This is essentially a stable situation is stable with clear relationships. In this case, the process is one of *'sense–categorize–respond'* by first establishing the facts and then categorising the possible responses using the appropriate rules or best practice.

Complicated—This contains the '*known unknowns*' for there is available a range of answers accessible by analysis or prior knowledge and expertise. Here the approach is one of 's*ense–analyze–respond*' in which the facts are first established and analysed to determine the appropriate actions.

Complex—This contains the '*unknown unknowns*' when there are no right answers and cause and effect can only be deduced by a process of *'probe–sense–respond'*, typically based on the use of experimental and test procedures.

Chaotic—Here, cause and effect are unclear and not readily subject to analysis and experiment. The approach is then one of '*act–sense–respond*' with the aim of transforming the chaos into complexity.

---

[9] The domain names have changed over the years though the context remains generally the same. See www.youtube.com/watch?v=N7oz366X0-8 (accessed August 2024).

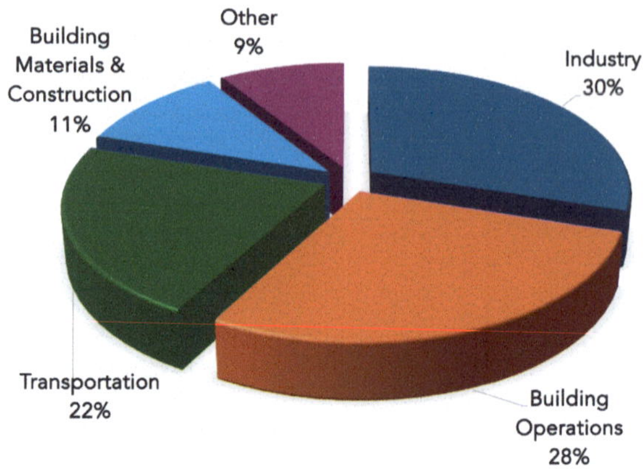

Fig. 1.12 Percentage of global CO2 emissions by sector in 2019 (after [60])

Confusion—This refers to those situations which cannot be readily encompassed by the other domains. Often requires the breaking down of the problem into its constituent parts which can each then be allocated to an appropriate domain.

1.2.5 Sustainability

It is estimated that in 2023 global energy-related CO2 emissions were of the order of 37.4 billion tonnes (Gt), an increase of 410 million tonnes (Mt) or 1.1% from 2022, with human activities being the major contributor. This rate of increase is however being mitigated to a degree by the increased use of clean technologies [59, 60]. Figure 1.12 then suggests the percentage of emissions by sector.

It is also increasingly the case, given that quality and price are equivalent, that environmental impact becomes a critical factor to be considered in respect of the purchase of a product or service [61, 62]. However, for many consumers, though purchasing decisions may be influenced by a consideration of environmental impact, it may not matter such consumers are unwilling to pay more for reduced environmental impact. The consideration of environmental implications throughout the design process however has the potential to lead to the development of innovative, competitive and sustainable design outcomes.

In this context, approaches such as that of the Circular Economy and the use of lifecycle monitoring all have a role to play in the design of future mechatronic products and systems in the move towards more sustainable economic models. What is also clear is that such moves cannot and indeed must not be carried out in isolation and that there is a need to support the transition towards more sustainable products and processes through the appropriate use of drivers such as standards and legislation. Further, the design of mechatronics courses must also reflect the increasingly important, and challenging, role that sustainability needs to play in the development of these new, and sustainable, products and processes.

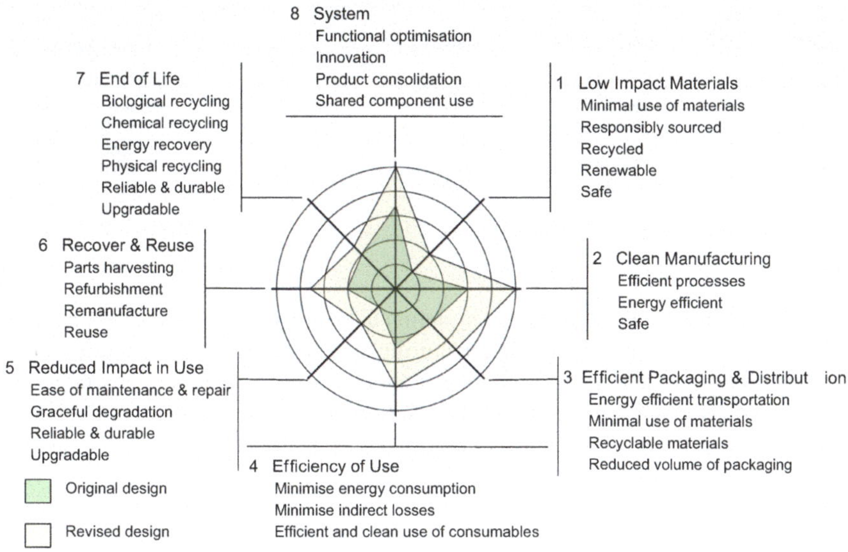

Fig. 1.13 EcoDesign wheel

EcoDesign

EcoDesign is a holistic approach to product design that pays special attention to the environmental impact of a product over the entirety of lifecycle. Particular areas for consideration include:

- Consumption of resources—energy, materials, water and land.
- Emissions, including noise emissions, impacting on the environment and on human health,
- Waste production and management with all consumables, materials and parts being inventoried and accounted for across the entire lifecycle of the product.

Here, the concept of the EcoDesign wheel of Fig. 1.13 provides a useful tool both for assessing a new product in respect of its environmental impact and in comparison with older or alternative products [63–73].

In terms of mechatronics, the potential for mechatronic components, assemblies and systems to impact on the environmental impact of a product or process is significant and increasingly a factor in their design and implementation.

1.3 Mechatronics Design

The importance of design and development methodologies to support the whole life cycle of technical systems such as mechatronic systems, cyberphysical systems and others has become increasingly significant in recent years. The main drivers for these

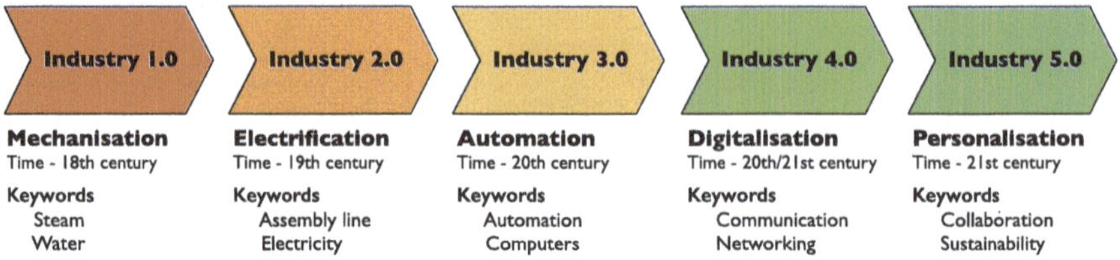

Fig. 1.14 Industry 1.0–industry 5.0

developments are the digitalisation of systems such as lies at the heart of the concepts of Industry 5.0 [74] and Smart Manufacturing which along with the guiding principles of processes such as Agile Management and Agile Production integrate with aspects of sustainability such as Green Production and the Circular Economy to drive forward the sustainability agenda as suggested by Fig. 1.14. Enabling technologies for Industry 5.0 are then [75]:

- Human-centric solutions and human–machine-interaction technologies that interconnect and combine the strengths of humans and machines.
- Bio-inspired technologies and smart materials that enable materials with embedded sensors and enhanced features while being recyclable.
- Real-time-based digital twins and simulation to model entire systems.
- Cybersafe data transmission, storage and analysis technologies that are able to handle data and system interoperability.
- Artificial intelligence, e.g., to detect causalities in complex, dynamic systems, leading to actionable intelligence.
- Technologies for energy efficiency and trustworthy autonomy as the above-named technologies will require large amounts of energy.

The VDI 2206-guideline establishes a framework for the holistic approach to Mechatronic systems, a field integrating mechanical, electrical, control and computer engineering. It recognises the increasing prevalence and importance of mechatronics across various industries, including automotive, aerospace, consumer goods and manufacturing. The standard aims to provide a structured methodology for the development and assessment of these intricate systems, guiding engineers and practitioners in achieving optimal design and performance.

The V-model [76] of Fig. 1.15 addresses a step-by-step approach with the conceptualisation, modelling, prototyping, testing and implementation stages. Furthermore, the standard promotes a transparent and systematic documentation process, enabling effective communication and collaboration across multi-disciplinary teams.

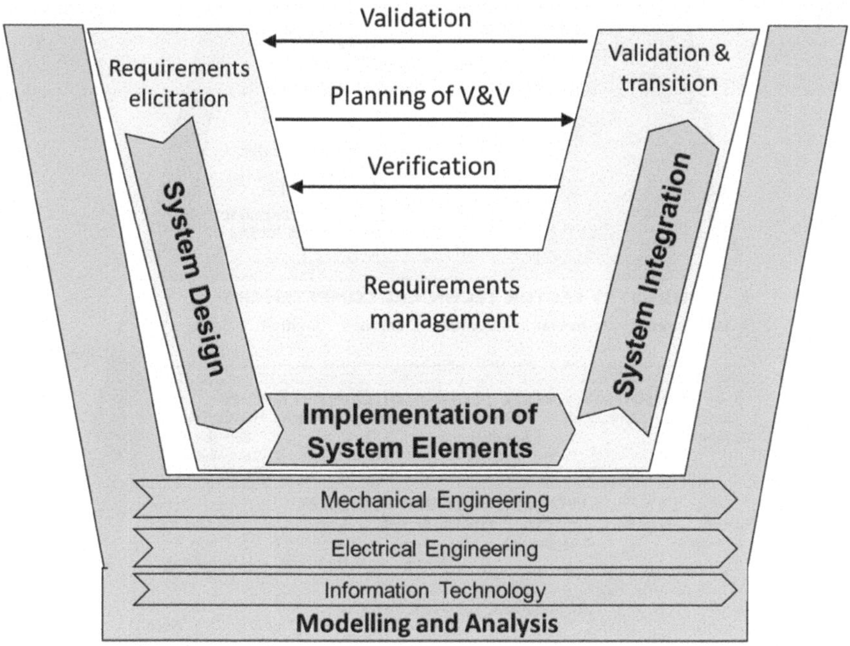

Fig. 1.15 V-model for design of mechatronic systems [76]

1.4 Education

The underlying challenge for mechatronics education and educators remains essentially the same as it has been since courses in this area began to be developed in. the 1980s, namely how to achieve an effective balance between the level of detail associated with traditional and conventional engineering programmes and the cross and inter-disciplinary approaches required by mechatronics. However, to this underlying issue has been added the need to accommodate the increasingly important systems aspect of mechatronics along with the added complexity that is associated with the continuing growth of Cloud and Internet of Things-based products and systems.

Thus, while acknowledging the need for mechatronics-oriented education to remove compartmentalisation of approach from between disciplines and focus on integration, other aspects of educational programmes and structures such as the requirements of governments and professional bodies can result in conflicting objectives and requirements. This can then translate into areas of product development and design where failures to communicate cross and inter-disciplinary needs and requirements can result in delays and indeed in product failure.

However, the situation is further complicated by the fact that the way in which mechatronics is viewed and perceived also impacts upon the way in which mechatronics courses are designed and implemented, which in turn has resulted in numerous differing and occasionally contradictory, definitions of mechatronics used to support

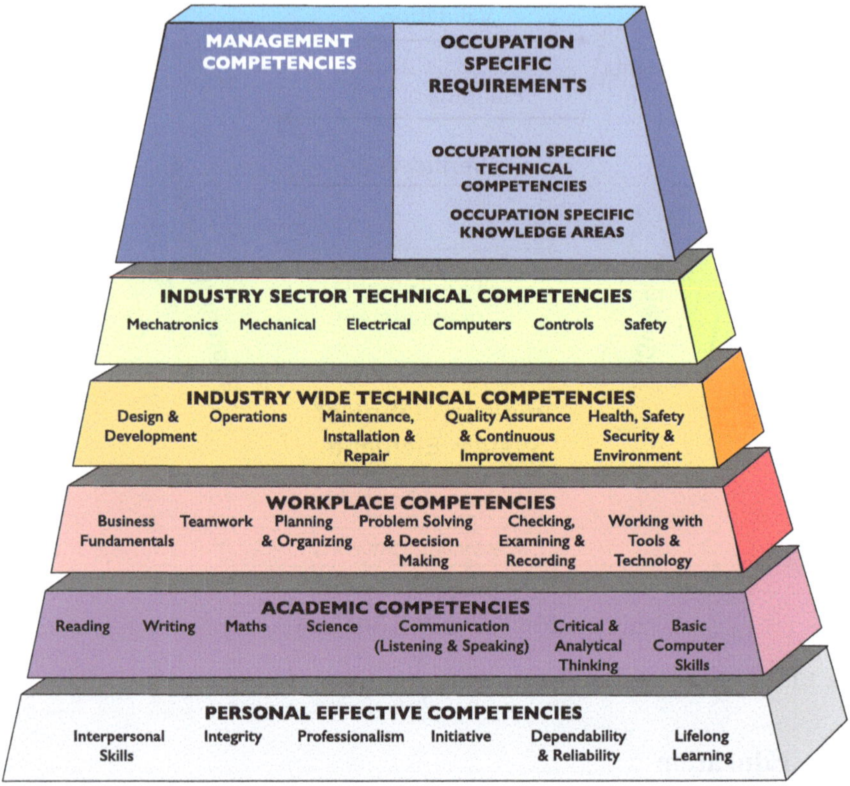

Fig. 1.16 Mechatronic competences (after [77])

the course structure. For instance, consider the diagram of Fig. 1.16. This is based on the US Department of Labor competency model [77] as adapted for mechatronics, the elements of which attempt to define what is expected of individuals in industry.

While models such as this may well be of use in helping to establish basic content, factors such as the skill and expertise of the teaching staff, the local industry structures and needs and research areas all need to be accommodated. As indeed do new technologies and concepts such as AI and Digital Twins.

Finally, there is the potential to extend the conventional concept of mechatronics as being something which is developed and taught at undergraduate and postgraduate levels to younger school-age groups through the deployment and use of a variety of advanced teaching aids and support tools [12, 78–83].

1.5 Structure of the Book

The structure of the revised edition is as illustrated in Fig. 1.17. Following on from the Foreword and the Introduction which sets out the background to the revised the contents are divided into a series of sections as follows:

Attributes & Challenges—Here, issues such as the future challenges facing mechatronics designers and engineers are discussed along with issues such as the social and ethical issues associated with the use of artificial intelligence, the challenges of dealing with complexity and decision-making throughout the design process before looking at issues of cybersecurity and sustainability in the context of mechatronics.

Mechatronics Design—This section looks at and considers various aspects of the mechatronics design process and places them in the context of developments of mechatronics technologies.

Digital Technologies—These are at the heart of mechatronics and the section considers areas such as the application of artificial intelligence in the operation of mechatronic components, devices and systems, lifecycle monitoring and the use of Digital Twins together with the continuing development of the Internet of Things.

Hard Technologies—These form the physical core of mechatronics and the section considers their use in respect of a number of mechatronic environments ranging from biometrics and biomechanics, production technologies and manufacturing and electronics and nanotechnology.

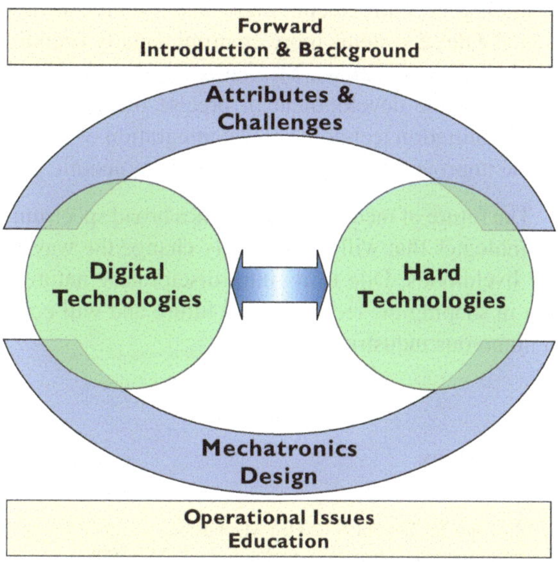

Fig. 1.17 Structure of the revised edition

Operational Issues and Education—This final section brings together issues of project management and the training and education of future mechatronic engineers.

1.6 Observations

The future of mechatronics promises to be a fascinating development in which the integration of mechanics, electronics and computer science will continue to lead to new innovations, more efficient systems and ground-breaking applications. In addition, aspects such as security, privacy, artificial intelligence and so forth are playing an increasingly important role. Some forward-looking aspects associated with these and other developments are discussed in the book and include:

- Intelligent and autonomous systems: In the future, mechatronic systems will increasingly be able to act autonomously. By combining advanced sensor technology, artificial intelligence and powerful actuators, intelligent mechatronic systems will be able to perform complex tasks autonomously.
- Industry 4.0/5.0 and digitalisation: Mechatronics will play a key role in the context of Industry 4.0/5.0 and further digitalisation. The integration of Internet of Things (IoT) technologies, data analysis and predictive maintenance will have a significant impact on the efficiency, flexibility and quality of future manufacturing processes.
- Sustainability and resource efficiency: Future mechatronic systems will be increasingly focused on sustainability and energy efficiency. Innovative materials, intelligent control technologies and resource-saving manufacturing processes will make these systems more environmentally friendly.
- Medical applications: Mechatronic systems will play a central role in medicine, be it in the development of precise medical devices, robot-assisted surgery or rehabilitation technology. The integration of mechatronics will thus contribute to the improvement of diagnostic and therapeutic processes.

The future of mechatronics offers a broad spectrum of innovative applications and technologies that will permanently change the way we produce, interact and shape our livelihoods. Due to its inter-disciplinary nature, mechatronics will play a key role in shaping the technological future and will continue to open up new horizons in numerous industries.

References

1. Kozono H, Yokota S (2008) Mechatronics coined by Tetsuro Mori, Proc. 12th International conference mechatronics technology, ICMT 2008
2. Dixit US, Hazarika M, Davim JP (2017) History of mechatronics, A brief history of mechanical engineering, Dixit US, Hazarika M & Davim JP (Eds), Springer, pp 147–164

3. World Economic Forum (2023) The future of jobs report 2023 @ www.weforum.org/reports/the-future-of-jobs-report-2023/. Accessed September 2024
4. World Economic Forum (2023) Global lighthouse network: shaping the next chapter of the fourth industrial revolution @ www.weforum.org/whitepapers/global-lighthouse-network-shaping-the-next-chapter-of-the-fourth-industrial-revolution? Accessed September 2024
5. www.zukunftsinstitut.de/zukunftsthemen/dossier/megatrends-en
6. Behringer R, Sundareswaran S, Gregory B, Elsley R, Addison B, Guthmiller W, Daily R, Bevly D (2004) The DARPA grand challenge-development of an autonomous vehicle, IEEE Intelligent Vehicles Symp., pp 226–231
7. Chiao C, Deichmann J, Heineke K, Kelkar A, Kellner M, Scarinci E, Tolstinev D (2024) Autonomous vehicles moving forward: perspectives from industry leaders, McKinsey Center for Future Mobility @ www.mckinsey.com/features/mckinsey-center-for-future-mobility/our-insights/autonomous-vehicles-moving-forward-perspectives-from-industry-leaders. Accessed September 2024
8. Krznaric R (2021) The good ancestor: how to think long term in a short term world, WH Allen
9. Shrivastav S & Joshi R (2021) Design thinking for long-term product planning. Proceedings Intl. conference research into design (ICoRD 21), Vol 3, Design for Tomorrow, pp 387–398
10. Arinez JF, Chang Q, Gao RX, Xu C, Zhang J (2020) Artificial intelligence in advanced manufacturing: current status and future outlook. J Manuf Sci Eng 142(11):110804
11. Aguilar J, Garces-Jimenez A, R-moreno MD, García R (2021) A systematic literature review on the use of artificial intelligence in energy self-management in smart buildings. Renew Sustain Energy Rev 1(151):1115305
12. Cintra Faria AC, Barbalho SCM (2023) Mechatronics: a study on its scientific constitution and association with innovative products. Appl Syst Innov 6(4):72–113
13. McLean S, Read GJ, Thompson J, Baber C, Stanton NA, Salmon PM (2023) The risks associated with artificial general intelligence: a systematic review. J Exp Theor Artif Intell 35(5):649–663
14. Salmon PM, Baber C, Burns C, Carden T, Cooke N, Cummings M, Hancock P, McLean S, Read GJ, Stanton NA (2023) Managing the risks of artificial general intelligence: a human factors and ergonomics perspective. Human Factors Ergon Manuf Serv Ind 33(5):366–378
15. Sharifani K, Amini M (2023) Machine learning and deep learning: a review of methods and applications. World Inform Technol Eng J 10(07):3897–3904
16. Wach K, Duong CD, Ejdys J, Kazlauskaitė R, Korzynski P, Mazurek G, Paliszkiewicz J, Ziemba E (2023) The dark side of generative artificial intelligence: a critical analysis of controversies and risks of ChatGPT. Entrepreneurial Bus Econ Rev 11(2):7–30
17. United Nations AI Advisory Body (2024) Governing AI for humanity: final report, United Nations
18. ISO (2023) ISO 31700–1:2023 Consumer protection—privacy by design for consumer goods and services Part 1: High-level requirements @ www.iso.org/obp/ui/#iso:std:iso:31700:dis:ed-1:v1:en. Accessed September 2024
19. www.gdpr.eu/what-is-gdpr/. Accessed September 2024
20. Nissenbaum H (1998) Protecting privacy in an information age: the problem of privacy in public. Law Philos 17(5):559–596
21. Rubinstein IS, Good N (2013) Privacy by design: a counterfactual analysis of google and Facebook privacy incidents. Berkeley Technol Law J 28:1333–1413
22. Felzmann H, Fosch-Villaronga E, Lutz C, Tamò-Larrieux A (2020) Towards transparency by design for artificial intelligence. Sci Eng Ethics 26(6):3333–3361
23. Jezova D (2020) Principle of privacy by design and privacy by default, Reg Law Rev, 127–139
24. Kostova B, Gürses S, Troncoso C (2020) Privacy engineering meets software engineering. On the challenges of engineering privacy by design, arXiv preprint arXiv:2007.08613
25. Aljeraisy A, Barati M, Rana O, Perera C (2021) Privacy laws and privacy by design schemes for the internet of things: a developer's perspective. ACM Comput Surv 54(5):1–38
26. Kaur J, Ramkumar KR (2022) The recent trends in cyber security: a review. J King Saud Univ Comput Inform Sci 34(8):5766–5781

27. Admass WS, Munaye YY, Diro A (2023) Cyber security: state of the art, challenges and future directions, Cyber Secur Appl, 100031
28. WEF Global Cybersecurity Outlook (2024), January 2024 @ www.weforum.org/docs/WEF_Global_Cybersecurity_Outlook_2024.pdf. Accessed August 2024
29. UK Government (2023) Cyber security breaches survey 2023 @ www.gov.uk/government/statistics/cyber-security-breaches-survey-2023/cyber-security-breaches-survey-2023. Accessed August 2024
30. www.ncsc.gov.uk/collection/cyber-security-design-principles/cyber-security-design-principles. Accessed August 2024
31. Schafer DJ (2022) Cyber threats in united states aviation: a review of inherent blind spots. J Homeland Nat Secur Perspect 7(1):51–69
32. Ukwandu E, Ben-Farah MA, Hindy H, Bures M, Atkinson R, Tachtatzis C, Andonovic I, Bellekens X (2022) Cyber-security challenges in aviation industry: a review of current and future trends. Information 13(3):146–167
33. Asbaş C, Tuzlukaya ŞE (2024) Aviation cybersecurity: shielding against cyberthreats in the air, Harnessing digital innovation for air transportation (ed Yilmaz AA) IGI Global, pp 42–55
34. www.nordvpn.com/blog/ring-doorbell-hack/?srsltid=AfmBOooFFKavjjwUF57pnChemD2OVNranU2blqfw8kem-6sMQt8-fjclfoundation.mozilla.org/en/privacynotincluded/ring-video-doorbell/. Accessed September 2024
35. Vasilescu NG, Pocatilu P, nea M (2023) IoT security challenges for smart homes, education, research and business technologies: Proc. 21st International conference on informatics in economy, IE 2022, pp 41-49
36. Lee I (2020) Internet of Things (IoT) cybersecurity: literature review and IoT cyber risk management. Future Internet 12(9):157–177
37. Djenna A, Harous S, Saidouni DE (2021) Internet of things meet internet of threats: new concern cyber security issues of critical cyber infrastructure. Appl Sci 11(10):4580
38. AlSalem TS, Almaiah MA, Lutfi A (2023) Cybersecurity risk analysis in the IoT: a systematic review. Electronics 12(18):3958
39. Amoo OO, Osasona F, Atadoga A, Ayinla BS, Farayola OA, Abrahams TO (2024) Cybersecurity threats in the age of IoT: a review of protective measures. Int J Sci Res Arch 11(1):1304–1310
40. Coventry L, Branley D (2018) Cybersecurity in healthcare: a narrative review of trends, threats and ways forward. Maturitas 113:48–52
41. Al-Qarni EA (2023) Cybersecurity in Healthcare: a review of recent attacks and mitigation strategies. Intl J Adv Comput Sci Appl 14(5):135–140
42. Javaid M, Haleem A, Singh RP, Suman R (2023) Towards insighting cybersecurity for healthcare domains: a comprehensive review of recent practices and trends. Cyber Secur Appl 1:100016
43. Kelly B, Quinn C, Lawlor A, Killeen R, Burrell J (2023) Cybersecurity in healthcare, Trends of artificial intelligence & big data for E-Health, pp 213–231
44. 2023 HIMSS Healthcare Cybersecurity Survey @ www.himss.org/resources/himss-healthcare-cybersecurity-survey. Accessed August 2024
45. Rass S, Schauer S, König S, Zhu Q (2020) Cyber-security in critical infrastructures, Vol. 297, Springer International Publishing
46. Sun CC, Cardenas DJS, Hahn A, Liu CC (2020) Intrusion detection for cybersecurity of smart meters. IEEE Trans Smart Grid 12(1):612–622
47. Lehto M (2022) Cyber-attacks against critical infrastructure, Cyber security: critical infrastructure protection, Eds Lehto M & Neittaanmäki P, Springer International Publishing, pp 3–42
48. Michalec O, Shreeve B, Rashid A (2023) Who will keep the lights on? Expertise and inclusion in cyber security visions of future energy systems. Energy Res Soc Sci 106:103327
49. Kim K, Kim JS, Jeong S, Park JH, Kim HK (2021) Cybersecurity for autonomous vehicles: review of attacks and defense. Comput Secur 103:102150

50. Takahashi J (2018) An overview of cyber security for connected vehicles. IEICE Trans Inf Syst 101(11):2561–2575
51. Ji B, Zhang X, Mumtaz S, Han C, Li C, Wen H, Wang D (2020) Survey on the internet of vehicles: network architectures and applications. IEEE Commun Stan Mag 4(1):34–41
52. Kurtz CF, Snowden DJ (2003) The new dynamics of strategy: sense-making in a complex and complicated world. IBM Syst J 42(3):462–483
53. Snowden DJ, Boone ME (2007) A leader's framework for decision making. Harv Bus Rev 85(11):68–76
54. Gorzeń-Mitka I, Okręglicka M (2014) Improving decision making in complexity environment. Proced Econ Finan 16:402–409
55. Nachbagauer A (2021) Managing complexity in projects: extending the cynefin framework. Project Leadersh Soc 2:100017
56. Beale D, Dazzi F, Tryfonas T (2023) Unweaving the definitions of complexity. IEEE Trans Syst Man & d Cybern Syst 54(2):682–692
57. Maree L, Swart M, Huma GJP (2023) Exploring the integration of complexity leadership theory and the Cynefin framework, 34th Southern Africa Institute of Management Scientists Conf. (SAIMS), pp 275–285
58. Vidal GH, Coronado-Hernández JR, Minnaard C (2023) Measuring manufacturing system complexity: a literature review. J Intell Manuf 34(7):2865–2888
59. International Energy Agency (2022) Global energy review: CO2 Emissions in 2023 @ iea.blob.core.windows.net/assets/33e2badc-b839–4c18–84ce-f6387b3c008f/CO2Emissionsin2023.pdf
60. www.communityenergy.ca/projects/embodied-emissions/. Accessed September 2024
61. Deloitte (2023) Gen Z and Millenial Survey @ www.deloitte.com/global/en/issues/work/content/genzmillenialsurvey.html. Accessed September 2024
62. European Commission (2020) Attitudes of Europeans towards the environment, Directorate-general for environment @ data.europa.eu/doi/https://doi.org/10.2779/902489. Accessed September 2024
63. Tu JC, Hsu FL (1999) The ecodesign strategy on product research and development from the life-cycle design, Proceedings 1st IEEE international symposium on environmentally conscious design and inverse manufacturing, pp 351–356
64. Favi C (2013) Toward eco-design: an integrated lifecycle engineering system to develop sustainable mechatronic products and services, PhD Thesis, Università Politecnica delle Marche Scuola
65. Ferreira AG, Neves DB, Moreira RM, Filho ER (2017) Innovation opportunities focusing design for sustainability: an ecodesign strategy wheel application. Prod Manag Dev 15(2):113–123
66. Nishant R, Kennedy M, Corbett J (2020) Artificial intelligence for sustainability: challenges, opportunities, and a research agenda. Intl J Inform Manag 53:102104
67. Merschak S, Diallo TM, Hehenberger P (2022) Life cycle sustainability assessment as a decision-making tool for the design of mechatronic systems. Int J Prod Lifecycle Manag 14(2–3):142–173
68. Van Doorsselaer K (2022) The role of ecodesign in the circular economy. Elsevier, Circular Economy and Sustainability, pp 189–205
69. Shu K (2023) Transforming construction: the growing role of mechatronics in building our infrastructure. Sci Technol Eng Chem Environ Prot 1(4):7
70. Hąbek P, Palacz M, Saeed F (2024) Embedding sustainability into mechanical engineering master programs—a case study of the top technical Universities in Europe. Sustainability 16(2):941–961
71. Has M (2024) Sustainable products: life cycle assessment, risk management, supply chains, ecodesign. Walter de Gruyter GmbH & Co KG
72. Onu P, Adesoji AA, Emmanuel A (2024) A mini review on the role of mechatronics in addressing energy and environmental issues, IEEE International conference on science, engineering and business for driving sustainable development goals, pp 1–5

73. Peter O, Pradhan A & Mbohwa C (2024) Smart Mechatronics: Innovating Sustainable Solutions for the Future, Proceedings 5th African international conference industrial engineering & operations management, pp 1037–1047
74. Industry 5.0 (2021) European commission, directorate-general for research and innovation directorate f—prosperity, Unit F.5—Industry 5.0
75. Enabling Technologies for Industry 5.0: Results of a workshop with Europe's technology leaders (2020) European commission, directorate-general for research and innovation, Directorate F—Prosperity, Unit F5—Industry 5.0
76. www.vdi.de/fileadmin/pages/vdi_de/redakteure/richtlinien/dateien/VDI_2206_V-model_en.pdf. Accessed September 2024
77. Ennis MR (2008) Competency models: a review of the literature and the role of the employment and training administration (ETA), Office of Policy Development and Research Employment and Training Administration US Department of Labor @ https://www.dol.gov/sites/dolgov/files/ETA/publications/Competency%20Models%20-%20A%20Review%20of%20Literature%20and%20the%20Role%20of%20the%20Employment%20and%20Training%20Administration.pdf. Accessed August 2024
78. Karkoub M, Abdulla S (2020) Transformative learning experiences in mechanical engineering through mechatronics from high school to college. Intl J Mech Eng Educ 48(1):3–31
79. Phan MH, Ngo HQT (2020) A multidisciplinary mechatronics program: From project-based learning to a community-based approach on an open platform. Electronics 9(6):954–997
80. Guc F, Viola J, Chen Y (2021) Digital twins enabled remote laboratory learning experience for mechatronics education, 1st IEE International conference on digital twins and parallel intelligence, pp 242–245
81. Habib MK, Nagata F, Watanabe K (2021) Mechatronics: experiential learning and the stimulation of thinking skills. Educ Sci 11(2):46–67
82. Rayner R, Kerwin K, Valentine N (2022) Robot-assisted teaching—the future of education?, EcoMechatronics: challenges for evolution, development and sustainability, Hehenberger P & Bradley D (eds), Springer International, pp 329–357
83. Tantawi K, Nasab A, Potter L, Tantawi O, Wilson N, Henrie A, Sirinterlikci A, Kaplanoglu E (2023) Mechatronics education-current and future trends, 11th IEEE International symposium digital forensics & security, pp 1–5
84. Buttarelli G (2018) Preliminary opinion on privacy by design, European Data Protection Supervisor Opinion 5/2018 @ www.edps.europa.eu/sites/default/files/publication/18-05-31_preliminary_opinion_on_privacy_by_design_en_0.pdf. Accessed August 2024

Chapter 2
Mechatronics Drives Moore's Law

Maarten Steinbuch

Abstract The field of mechatronics developed itself in the last decades alongside with the rise of the semiconductor industry. The combination of smart controls, electronic and mechanical design not only enabled smart manufacturing processes, but also was a key ingredient for the exponential development of computational power of integrated circuits. The machines used for the production of ICs are at the edge of motion precision performance possibilities. This requires the best from the mechatronics field. The systems engineering and systems thinking aspects become even more prominent as the complexity increases and other disciplines are as relevant, such as physics, optics and also AI and computer science. The impact for academic research can be observed as the need for science teams and pressure on the speed of innovation. This can only be achieved with partnerships between industry and academia and releases a new (4th) generation of university.

Keywords Motion · Wafer scanner · Mechatronics · Robotics · Precision · University · Systems

2.1 How It Started

The field of mechatronics started in the seventies when mechanical systems needed more accurate controlled motions. This forced both industry and academia to explore sensors and electronic-assisted feedback, while using mostly electrical drives instead of, for instance, mechanical camshafts in production facilities. This introduction of feedback-controlled motion formed the basis for the need to enable mechanical engineers and electronic engineers to work better together and to understand each other's language. Note that in those days control engineering departments were mostly part of the electrical development or research departments of industry and academia. Various initiatives were also undertaken to develop a common language or methodology. Some institutes pushed Mechatronics forward as being a new discipline.

M. Steinbuch (✉)
Eindhoven University of Technology, Eindhoven, The Netherlands
e-mail: m.steinbuch@tue.nl

P. Hehenberger and D. Bradley (eds.), *Mechatronic Futures*,
https://doi.org/10.1007/978-3-031-83571-1_2

In industry, the design teams were typically forced to really discuss at the specification level deeper insights from within their specific disciplinary knowledge. Computer-assisted design and simulation tools really boosted the field in the late eighties and nineties. An example of the project-oriented mechatronics way of working has been the development of optical storage devices [1]. In those years also ASML was started as a spin-out of Philips Electronics. The mechatronics challenges were there even more pronounced as the complexity of those early day wafer steppers was unknown up to then. Later the development of wafer scanners even posed greater challenges in mechatronics thinking. Teams of mechanical designers, using their finite element programs, and electronics and control specialists, with their specific simulation tools, co-developed mechanisms with very tight specifications on manufacturability, cost and dynamics.

In that same time frame of the end of the eighties, in many industries and academia, mechanical engineers started more and more to also address dynamics and control, and control groups started to emerge also in mechanical engineering departments, all of which signalled a move away from mono-disciplinary approaches. Within Philips Electronics it was a challenge in the late eighties and early nineties to convince others about this mechatronics thinking, and even more the systems engineering aspect for complex motion machines. For this reason, the cartoon shown in Fig. 2.1 was often used at many occasions and management discussions [1].

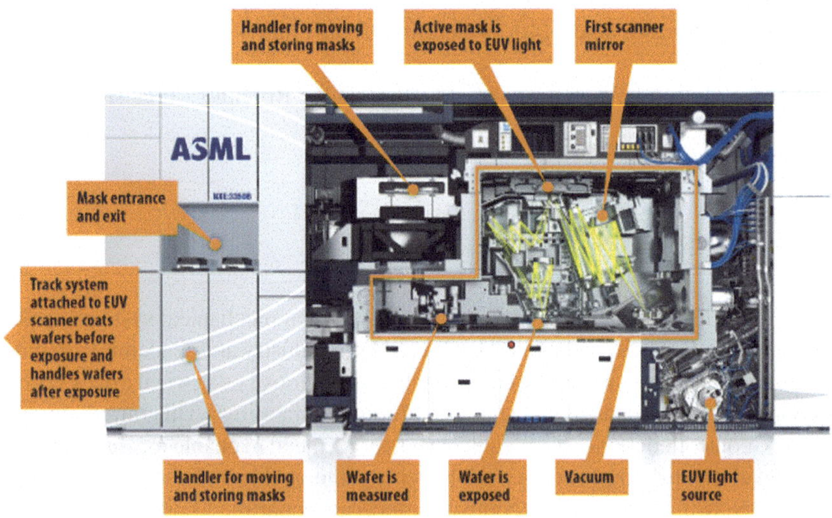

Fig. 2.1 A modern ASML wafer scanner

2.2 Computer-Controlled Devices

The rapid development of the personal computer enabled the better use of simulation and design tools, hence improving the overall design process and quality of exchange of design ideas in an early phase. However, and equally relevant, the PC-enabled digitized computer-controlled mechatronic systems testing and implementation. This required addressing the role of computer science engineering and showed the need to include the software discipline, but to a still rather limited extent. This also led to include more and more in the field of systems engineering as a way of working in industry on more complex products and high tech systems. However, thinking about the 'common' language, or at least to understand each other better, clearly is far less trivial between the hardware and software domains, than within the hardware domain itself.

From a research perspective the questions start at the discrete time level, i.e. how to use the computer to implement control functions such that the performance previously done with analogue implementation was maintained as much as possible. However, soon the higher level supervisory control modes were taken into the mechatronics field, and this forced research to make the switch towards the much more difficult questions of discrete event systems, facing continuous time dynamics in the mechanical system. This has led to the research field of hybrid systems within the systems and control discipline. This part forms the natural interface between the hardware (the '*old*' mechatronics) and the software (computer science) field. Nowadays, the rapid developments in AI and data-based methods further enlarge the scope of mechatronics.

2.3 Applications

The performance improvements due to mechatronic thinking have been profound and are broadly acknowledged. Applications of mechatronics can be found in many products and production environments. Although in the early days, the control of electric motors was an often seen application, mechatronics thinking also is used in the design of hydraulic systems, piezo driving actuators, the modelling and control of production equipment, scientific equipment, opto-mechatronics, automotive mechatronics, optical communication devices, etc.

Overseeing the inflow of submitted mechatronics papers over the last few years, more application papers are submitted on medical devices, on high precision systems, drones (UAV), automotive and robotics. The papers on scientific achievements on modelling languages and tools have reduced, meaning probably that appropriate tooling is now more common. The same seems to be true for papers on education in mechatronics. This was a hot topic in the late nineties, where good examples were found including experimental work for the students.

There are not so many discussion papers anymore about what could be called the mechatronic design method, because it is by now maybe clear that part of the innovation done in mechatronics in practice has more to do with helping disciplines to communicate, preferably via the use of shared models or quantified simulations. The scientific methods addressed in mechatronic journal submission are mostly seen in the systems and control area, where the mechatronic application is often used as a validation or simply as a showcase. An emerging field is the use of optimization algorithms, not only for finding optimal control laws, but more and more also for component design, up to system topology optimization as a new design tool [2], which is similar to the use of digital twin modelling and simulation packages in order to automate the whole design flow. The core of the mechatronics submissions and community still is Mechanical Engineering, Electrical Engineering and the area of Systems and Control. The interrelation with Computer Science and Physics is still rather limited, but especially for Computer science this is going to shift to coming years with the automated design packages and AI techniques.

2.4 Multi-Physics

High-end mechatronic systems such as wafer scanners such as that of Fig. 2.1 for optical lithography or electron scanning probes and in space applications and scientific instrumentation have an error budget that is getting closer to being a flat distribution over the various sources.

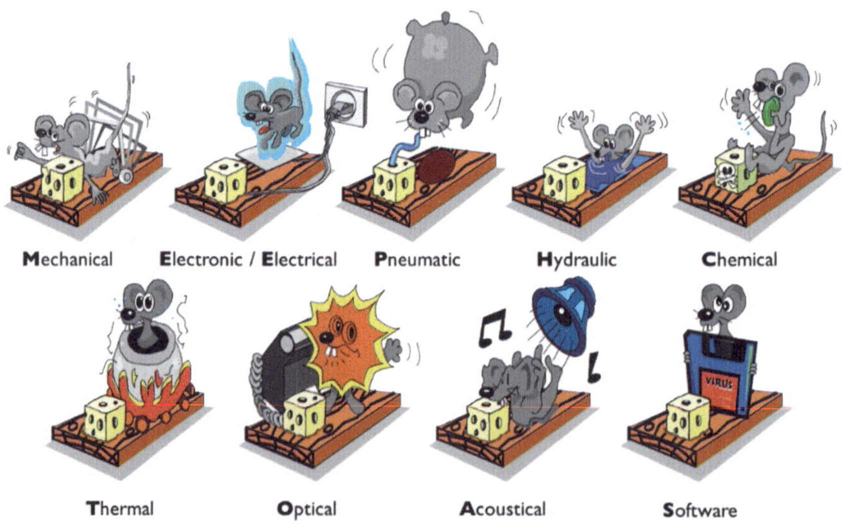

Fig. 2.2 Cartoon emphasizing the many mono-disciplinary solutions for a given problem [1]

For instance, for modern wafer scanners thermal and cooling-fluids-induced vibrations now are as significant as mechanical modal vibrations excited by the actuators. This has to do with the extreme conditions and requirements; moving an 80 kg mass with accelerations more than 20 g, and achieving accuracies below nanometres with mKelvin temperature variation [3]. This means that the '*normal*' mechatronics and its motion control systems now start to have a dynamic interaction with the thermal and fluid control dynamics. The overall performance assessment and design improvements now start to cover not only mechanical and electrical/electronic and software disciplines, but also physics issues like thermal and fluid partial differential equation-based modelling. And what will be the impact for mechatronics design thinking when we include the possibilities of additive manufacturing? If a 3D industrial metal or ceramics printer can be used to freely shape our mechanisms, how to arrive at an overall optimal design? New developments are also for instance the use of superconduction motion platforms, see Fig. 2.3 as an example [4].

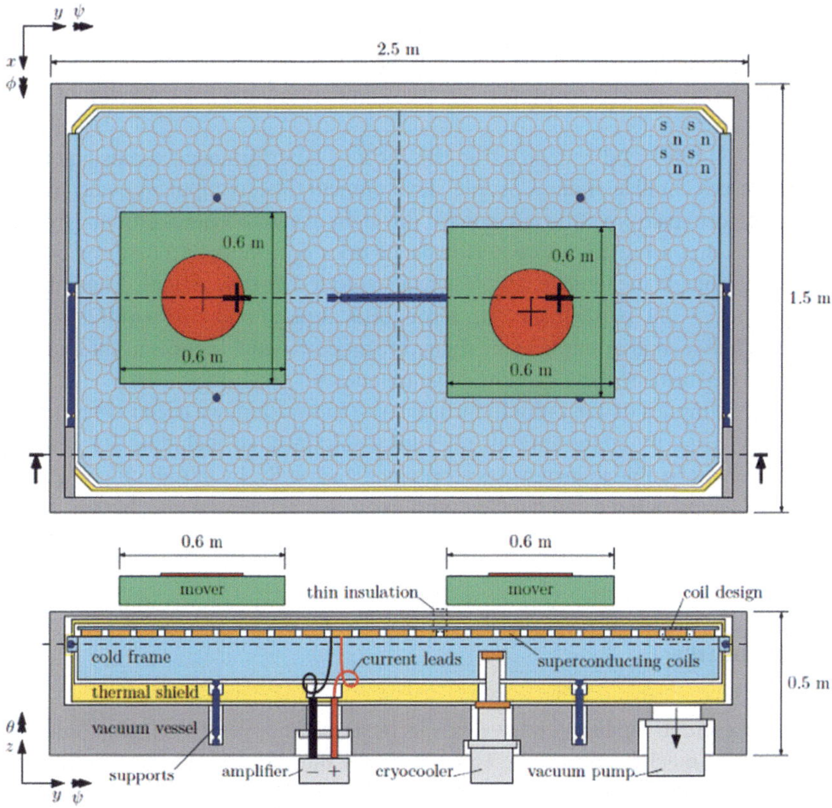

Fig. 2.3 Schematic representation of superconducting planar motor for lithography application. Two movers (green) carrying the wafer (red) are magnetically levitated at room temperature [4]

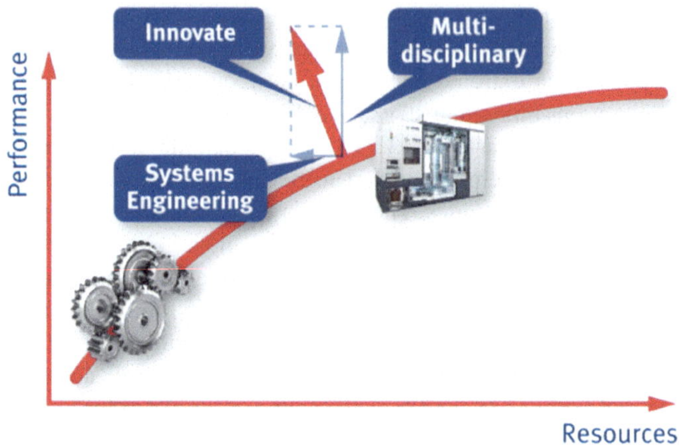

Fig. 2.4 Performance complexity (resources) trade-off

The performance trade-off can now only be lifted to the next level if we are able to handle this complexity by proper systems engineering and the inclusion of more disciplines.

In Fig. 2.4, this trend is depicted in the form of a performance versus resources plot. Resources could be money, people, development time, computer power, energy, etc. The performance typically is accuracy, throughput and robustness/reliability. The curve shows that achieving more performance does cost more and more resources, until not feasible. In the figure examples are also plotted; firstly, a simple transmission gear system has low performance (in terms of accuracy) and also requires limited resources. The second example in the figure is a modern wafer scanner as the example of extreme performance and needing huge resources.

The curve implies that in order to further boost innovation, we need to incorporate two means. First, by addressing all relevant disciplines, so including for instance physics, we will be able to increase performance. Secondly, by introducing a systems engineering approach we can handle complexity in a better way and, hence, go left on the resources axis.

2.5 Robotics

Almost opposite to the high-end systems as described above, the robotics field also influences the mechatronics area. Here, it is not the multi-physics discipline that is required, but the computer science field to cope with unstructured and changing environments. In robotics, the developments are directed towards vision, mapping and localization, so understanding the environment ('*world modelling*') but also the field of Artificial Intelligence (AI)—which has already been a promise for decades, but could evolve rapidly in coming years. Both areas are currently in an accelerating phase because of the upcoming autonomous vehicles. The disruption seen in the automotive industry is huge, both in the area of power trains (i.e. electric drives and transmissions), and the use of computer science, as for instance the sensors in a modern car, including the rapid developments in autonomous functions implemented in passenger cars as well as in commercial vehicles. This in fact is all about mechatronics, AI, controls!

The field of robotics, including autonomous cars, could be treated as a separate research area, next to mechatronics, but for instance the speed requirements of industrial robots or the accuracy requirements of surgical robots such as the *Preceyes* robot of Fig. 2.5 necessitate the inclusion of the description of the dynamic behaviour of the robots. The change from rigid body modelling towards flexible systems, then directly makes it in the heart of mechatronics. The same holds for the systems engineering thinking and the system topology optimization, which is also similar in hybrid power trains for vehicles. So where does mechatronics end and robotics start?

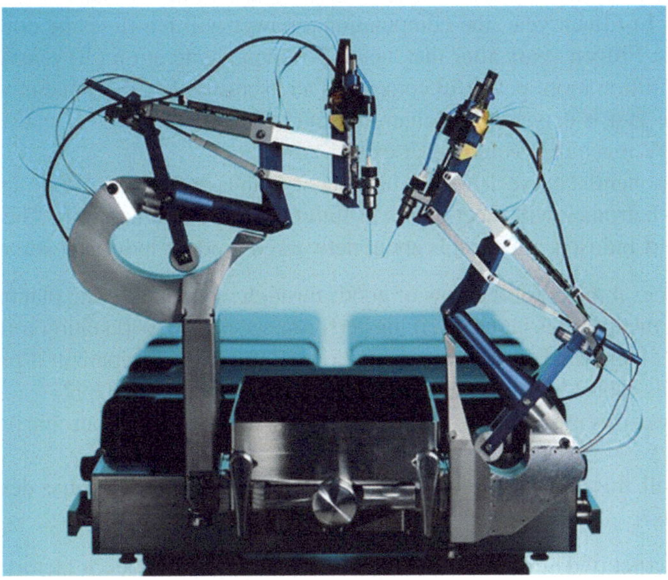

Fig. 2.5 *Preceyes* eye surgery robot [5]

2.6 Cyberphysical Systems, Moore's Law and Smart Industry

The shift from decentralized mechatronic systems towards networked connected systems is known as the field of cyber-physical systems, referring to the field of cybernetics. The research questions are how to guarantee stability and performance during or after packet (information) loss, and how to deal with variable delays. The domain is even further away from the hardware of mechatronics, but is developing so rapidly, that we should ask the question of how to embrace to potential of network-controlled systems, for instance in the field of remote condition monitoring and servicing. In the next decade, the explosion of the Internet of Things (IoT) further necessitates finding the answers to this question [6].

In fact, the ASML machines enabled the law in the past years and enabled it to follow Moore's law for the decades to come.

Figure 2.6 shows the development of calculation power per unit of money as a function of years. As an approximate straight line, it predicts an exponential rate of change. In fifteen year, the computational power will ten times be doubled, i.e. 1000 times. Fifteen years after that again. So in one generation (30 years), we will have one million more powerful computers and sensors. For the same money. This remarkable fact is driven by mechatronics and systems thinking. Because without ASML EUV machines would never exist.

This exponential curve drives the Internet of Things as well as AI. One application where mechatronics will meet IoT is in future of our manufacturing. The Industry 4.0 or Smart Industry attention is about networked modern industrial automation.

- What does it mean for the flow of goods through a manufacturing plant if knowledge of the logistics is shared, if the performance of one workstation is optimized as part of the total logistics or operation, if service and repair in a production facility is robust because workstations are flexible and can adapt?
- What does this imply for the industrial robotics and smart mechatronic production devices?
- How will this impact the design requirement of our mechatronic devices and products?

The Internet of Things and AI will not only change the modern factory. Sensors will be everywhere. This means it will be entering our households and equipment used at home, as well as our cars. When wearable electronics are pushed further, and we are surrounded by sensors, we only need the step towards actuation to be able to close the loop and by that enter the world of mechatronics again [6]!

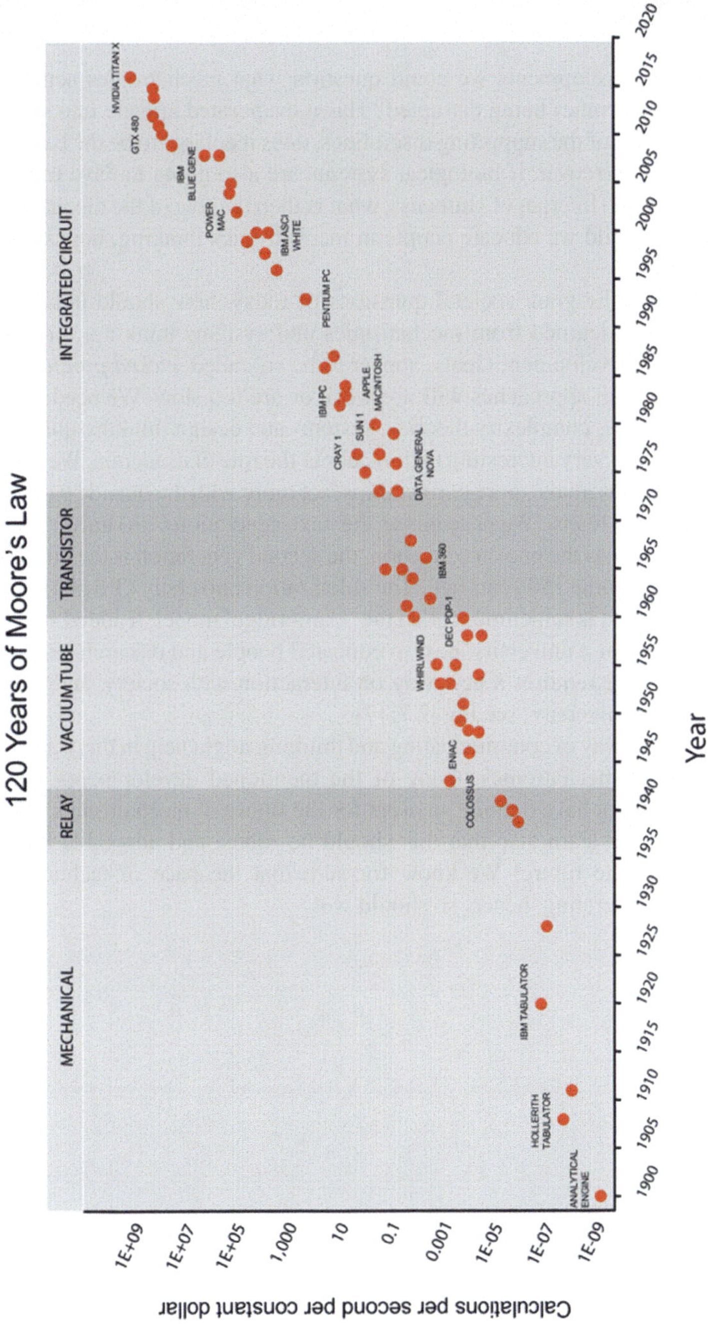

Fig. 2.6 Moore's law

2.7 Towards Systems Integration and a Next-Gen University

Overseeing these developments we could question what mechatronics actually is or will be. Is mechatronics being disrupted? Has it evaporated already into systems engineering, is it part of the supporting disciplines, does it enlarge to be the backbone of cyber physics? Moreover, if biological systems are also going to have technical devices implemented (Internet of Humans), what is then the role of the mechatronics discipline? How should we educate people in mechatronics thinking, how small or how broad?

Moreover, given the grant societal questions of today, how should universities embrace the lessons learned from mechatronics and systems thinking. Indeed all of the 17 Societal Development Goals, appear to be so-called *wicked problems*, for which classical design approaches will not work or are too slow. We need at least outside-in approaches, complexity theories, system- and design thinking qualities.

In this respect, it is very interesting to think about the role of academia. We need to open up as universities, make more partnerships, act more with the outside world and work more in science teams. We pose this as the next-generation (4th) university [7]. The first generation was the one on education, the second generation is the university of research, starting some 150 years ago. The valorization university (3rd gen) started about 30 years ago and is sometimes called the 'Cambridge' model. It added start-ups and IP to the output of a university, next to educated people and research results.

However, our time requires a new way on interaction with society, and we call this the fourth-gen university; see Fig. 2.7. [7].

Our mechatronic way of communicating and thinking might help in this transition.

In this book on mechatronics, many of the mentioned developments will be addressed. We will not have definite answers for the future of mechatronics, nor for its education, but we learn also that this should be robust and adaptable because we cannot predict the future! We know for sure that the pace of technological development is accelerating, hence, so should we!

	3rd Generation	4th Generation
Objective	Education, research & know-how exploitation	Education, open innovation (research)
Role	Creating value	Enabling value creation
Method	Inter-disciplinary science	Multi-actor innovation
Human capital development	Professionals, scientists & entrepreneurs	Professionals, scientists entrepreneurs, artists, customers, ecosystemparticipants
Orientation	Global	Ecosystem
Language	English	English
Organization	Institutes & centers	Innovation spaces
Management	Professional management	Disruptors

Fig. 2.7 Third- and new fourth-gen university [7]

References

1. Munnig Schmidt R, Schitter G, van Eijk J (2011) The design of high performance mechatronics. Delft University Press, Netherlands
2. Silvas E, Hofman T, Serebrenik A, Steinbuch M (2015) Functional and cost-based automatic generator for hybrid vehicles topologies. IEEE/ASME Trans Mechatron 20(4):1561–1572
3. Butler H (2011) Position control in lithographic equipment : an enabler for current-day chip manufacturing. IEEE Control Syst Mag 31(5):28–47
4. Koolmees HB (2020) A superconducting magnet plate for a planar motor application, PhD Thesis Eindhoven Univ Technol, ISBN: 978–90–386–5175–0
5. Meenink HCM (2011) Vitreo-retinal eye surgery robot: sustainable precision. PhD Thesis Eindhoven Univ Technol, https://doi.org/10.6100/IR717725
6. Bradley D, Russell D, Ferguson I, Isaacs J, MacLeod A, White R (2015) The internet of things—the future or the end of mechatronics. Mechatronics 27:57–74
7. Steinbuch M, Oomen T, Vermeulen H (2022) Motion control, mechatronics design, and moore's law. IEEJ J Ind Appl 11(2):245–255. https://doi.org/10.1541/ieejjia.21006010

Part I
Attributes and Challenges

Chapter 3
Ethical Design Choices for AI Systems—Machine Autonomy or Human Alliance?

Roger J. White

Abstract Artificial intelligence has proven over the years to be an essential adjunct to human endeavour through the automation of routine tasks better suited to computational solutions, and recent advances in this field have greatly enhanced this potential. By stripping away the media hype around 'intelligent machines' that will either release humanity to live in wealth, peace and harmony or become the slaves of super-intelligent robots, the truth is that it is a morally neutral technology that can be used to either assist or impede mankind's future progress. This chapter discusses a range of ethical issues that result from this simple truth that need to be addressed in mechatronic systems design. This constraint in the implementation of advanced machine automation systems is often overlooked as new solutions such as **Gen AI** are launched into the marketplace, driven by the financial imperative to commercialise recent advances by a small number of investor-driven corporations with the encouragement of governments interested in potential economic and security benefits. After years where a 'free Internet' masked the true intentions behind cloud data collection and blinded people to the dangers this created, the technology is now bumping up against the real world where mathematical logic is only one factor in defining an equitable, progressive and innovative future for a planet beset with a range of existential issues, such world war and climate change. It should be clear that true machine autonomy is not achievable in the medium term without some level of human population re-education to make it act in a predictable and machine logical way for AI software to operate at a much higher level of accuracy than is plausible currently, mechatronic design must focus on systems that augment rather than replace human intelligence, playing to strengths not forcing compliance.

R. J. White (✉)
RC2 Inc, Kingston, ON, Canada
e-mail: rjw_rc2@icloud.com

P. Hehenberger and D. Bradley (eds.), *Mechatronic Futures*,
https://doi.org/10.1007/978-3-031-83571-1_3

3.1 Introduction

While there has been tremendous media attention in the past few years to the progress being made in the field of AI/deep machine learning, leading some to predict that conscious machines may be just around the corner, it is worth pausing to look more closely at the wider world reality. Are these claims of a revolution simply incremental improvements on ideas first proposed in the 1950s but untestable at the time on the cumbersome computer hardware available, or truly paradigm shifting innovations that open a pathway to fully autonomous machines in the near future? This chapter will attempt to answer this question based on direct research experience dating back to the previous AI boom of the 1980s when laser/fibre optic distributed sensors coupled with personal computers were developed that could start to generate complex data sets in areas as diverse as autonomous vehicle control, smart homes, personal medical record storage and bio-chemical process control.

Stripping away all the recent media hype around '*intelligent machines*' designed to create commercialisation interest, the truth is that **Artificial General Intelligence** remains a distant dream of utopian technologists that may never become a reality; indeed, until an agreed definition of human intelligence/consciousness can be produced it is impossible to start assessing how to create it in machines despite what tech. investors try to claim. Having said that, **Narrow AI** applications have been solving many key problems in chemistry, medicine, sustainable transportation and technical search for many years, and the recent advances in Large Language Model (LLM) development show the potential to expand and enhance these areas of applications into many other areas of industry, creating opportunities for fresh innovation and new, sustainable technologies. However, such developments also come with real-world risks around data security & privacy, misinformation and fraud that have to be guarded against to prevent misuse.

Recognising that an AI does not have the capacity to think for itself, and only does what a human programmes it to do, there is a need to construct an ethics framework around developers and users of this technology, and the aim of this chapter is to provide a mechatronics systems designer with pointers to highlight the main issues that need to be addressed prior to any system rollout. This is based on prior experiences and the latest (as of mid-2024) legislative proposals from Europe and the US to address how to begin regulating the industry in the wake of rapidly growing public concern about the current lack of accountability shown by the relatively small number of corporations that drive innovations in this market.

When discussing how to create sustainable, ethical AI solutions, the starting point must be that it is a morally neutral technology (c.f. nuclear physics). Whether it is a force for good capable of sustainably enriching humanity without destroying representative democracy, or a weapon of mass destruction and authoritarian control, is determined through human agency. It was a human that programmed HAL9000,[1] and it could have been rebooted or switched off very easily unless a programmer had decided otherwise.

---

[1] The supposedly sentient computer that went rogue in the 1968 film 2001: A Space Odyssey.

This chapter will raise a range of common issues that result from this simple truth encountered when designing AI systems. With the advent of powerful new Large Language Models (LLM), such as ChatGPT, these need to be addressed as a matter of urgency since the push to commercialise this technology in an unregulated fashion intensifies. This is being driven by a small number of powerful corporations assisted by governments interested in potential security benefits and economic growth that the current '*Big Data*' approach can provide in opposition to a wider public concern about the sociological impact of this latest envisaged technological gold rush.

After years during which a '*free Internet*' model has masked a range of unforeseen and unintended consequences associated with cloud data collection arising from the realisation by some business leaders of the potential value contained within the individuals' data collected, the AI world is now bumping up against a real world where mathematical logic is only one factor in defining the needs of an expanding human population facing climate change, increasing inequality, political polarisation and the demographic challenges of a rapidly ageing population. It is not possible here to cover all of the issues that need to be considered in great detail as so many interrelated disciplines are involved, while the field of ethical AI is also currently in a highly fluid state. A bibliography of selected texts has therefore been included at the end of the chapter, along with references to two podcasts, for the interested reader to investigate more fully the subjects discussed.

3.2 Security

The development and rapid growth in the use of cloud servers for amassing vast databases of information on the web through the creation of social media, online shopping, search and general media digitisation and streaming, has created a need for methods to protect the valuable information captured. This is an area of the Internet's rapid expansion to support AI developments that has been ignored from the early days of social media commercialisation and have only now started to become important as cybercrime, mass hacking and privacy breaches have become dominant issues and a significant drag on world economies. In large part, this problem is due to the initial philosophy of the World Wide Web where free access to information and communications between cultures was seen as an unmitigated benefit for the growth of liberal democracies, where the mantra of '*only the guilty have something to hide*' was used to disguise the loss of privacy that resulted. The initial emphasis was on the free exchange of information as a way to democratise global media and communications, with no real thought given to how the necessary infrastructure investment being created by the bubble of venture capital raised would be monetised prior to the Dotcom crash.

As soon as it became apparent that new business models would need to be developed to ensure existing corporate, financial and political interests could bring the technology back under control and start making money, the importance of data harvesting as a potential route to commercial success with AI as well as a source of revenue through advertising became apparent to a few key players actively researching how to develop their online businesses (Google, Meta, Amazon, Microsoft), as well as Silicon Valley investors such as Elon Musk funding start-ups such as OpenAI.

This led to the creation of what has become known as '*surveillance capitalism*', initially structured around online shopping and social media, where the aim was to convert users into raw data sources that could be exploited commercially through micro-targeted advertising using data harvested from users. This kick-started the data centre boom, creating the model of '*free*' websites and low-cost shopping being paid for by the user through the supply of private information to the corporations running the sites, turning users from clients into data sources and social media and search sites into advertising agencies. A key element of the business strategy that developed was to work to loosen regulations imposed by national governments to maximise the commercial advantage data harvesting created, allowing for rapid growth in infrastructure and profitability at the expense of personal privacy and national tax revenues.

In recent years, there has been a growing public awareness of this loss of privacy, driven by various data breaches, as for instance the Facebook-Cambridge Analytica scandal in 2016 when attempts were made by the latter to influence the Brexit referendum in the UK and the Presidential election in the US through micro-targeting of online political advertising. This has increasingly focussed attention on how to ensure that an individual's data stored online is protected and only accessible to those having a legitimate need to do so, and then only for a fully authorised purpose, for instance through the use of biometrics.

The nature of most common data storage, with multiple copies of similar personal data sets for an individual spread across multiple financial, government and e-commerce providers makes security a lot harder as agreement has to be obtained at a global level on common protocols around encryption and data access. Particularly as AI makes it relatively easy for a bad actor to compile a complete ID for an individual through the analysis of these multiple partial data sets located across the web.

3.2.1 Data Structures

One of the major problems with Internet cloud storage systems is that they have been built on what was originally intended to be a free access, open network for unrestricted communication between people around the world. Following the Dotcom crash at the turn of the century, Web 2.0 was built on this framework as a commercial data mining infrastructure that led corporations such as Google to move from being an unprofitable search engine to the most successful advertising agency in the world, while social media start-ups such as Facebook were able to monetise their users through advertising, promoting at the same time the belief that there was no longer any such thing as privacy and only someone with something to hide would be worried about this.

This exponential growth in social data capture also created opportunities for bad actors to exploit the generally weak security and open structure of the Internet through everything from misleading advertising and echo chamber creation, through to state-controlled hacking, financial scams, extortion and cyber warfare. Health services have been crippled for weeks at a time around the world by extortionists [1], power grids have been disrupted and retailers have lost billions of dollars through disruptions to their online sales operations.

As the use of AI in online services of all types becomes more and more embedded into everyday life, so the risks associated with the concentration of data storage in the hands of a few corporations, coupled with the fragile nature of the web infrastructure, make the world far more vulnerable to the negative aspects of the current configurations. In a world where autonomous vehicles are being trialled on public roads, and over-stretched power, health and justice infrastructures are increasingly looking to AI to optimise efficiency to enable systems to operate at 90% + utilisation rates, it would be very easy for a hostile power, should it choose to do so, to create a social meltdown in another country by disrupting the networks controlling vehicles, power generation, public body IT systems and so forth.

To avoid this doomsday scenario and make it easier for individuals to control their own data while at the same time participating in developing a much safer AI-assisted economy, researchers are looking at ways of breaking down data into smaller ring-fenced parcels stored as locally as practicable so that the highest level of personal information (e.g. health and financial) is held on a server controlled exclusively by an individual, while anonymised scoring algorithms can take aspects of this data authorised by the user and use it in population level training activities. Emergency channels to medical centres, banks, etc., could then be set as needed by the individual. The key here is to better secure key data from hacking attacks to reduce the incidence of ID theft and reduce the market value of personal data by the devolution of storage. At the same time, this should make systems more sustainable as power consumption would be decreased, and from an AI perspective, the theoretical background to an AI system would more closely mimic human intelligence which comprises individual processors (brain) communicating socially (network) with other processors to share aspects (data score outputs) of their own experiences in the creation of shared intelligence.

3.2.2 Encryption Standards

With an increased focus on data security in a largely unregulated cloud server environment, there has been an increasing interest in, and emphasis on, data encryption as a way to prevent hacking in support of user privacy requirements. Indeed, Apple Inc uses encryption as the foundation of its marketing strategy of providing protected hardware systems despite increasing governmental pressures around the world to add backdoors for national security services. Blockchain encryption is starting to become a standard technology for communication apps as it is deemed at the present time to be unbreakable, but currently, the level and type of encryption used are largely decided at the corporate level for purely commercial reasons that bear little real concern for the privacy of individual customers.

Moves are being made to try and promulgate legally enforceable standards to back up privacy protections, particularly in the EU following on from the pioneering work they had undertaken with General Data Protection Regulations (GDPR), but movement in other jurisdictions in North America and the Far East towards a joint approach has been delayed through a combination of commercial and national security pressures, coupled with a general growth in nationalism.

3.3 Privacy

In the modern digital environment, it has become increasingly difficult to protect an individual's privacy when the Internet is based on a free-use model that treats site users as a raw material source for their advertising and AI business units. More and more agencies are being allowed access to databases normally reserved for government use—everything from police records through credit files to health records—that impinges on the human right enshrined in European law to a private life. As a result, access to mortgages or business finance is restricted for many people, while employment opportunities are closed off because of some context-free record on an individual's digital file. This is taken to extremes in authoritarian countries where population control is underway through the use of social credit ratings based on the information pulled together online.

While clearly bad for an individual, this unregulated use of personal data has negative effects on business innovation, crime and inequality as context-free decisions on potential life choices become restricted by past mistakes or family circumstances. To combat this move towards a loss of societal free will, regulation on private data will need to be enacted and enforced as discussed below, but some fundamental changes to the design of algorithmic systems will also be needed.

3.3.1 Transparency

Before an individual can challenge the use of their personal data that has been collected for a different purpose in another context, they have to be made aware of what is held and by whom. This in the current environment of the Internet where any individual online has their digital exhaust hoovered up and sold in obscure digital auctions to advertisers, ratings agencies and so on. Increasingly, government agencies are broadening access to confidential data held on individuals in the name of security and business efficiency, a trend that will only accelerate as they get caught up in the dash for AI.

Looking specifically at AI, an additional issue to address is how to make the algorithms used more transparent, so that as well as having input into what data is being used, an individual can understand how decisions made by an autonomous system are made, allowing context to be added in by human assessors to reduce the chances of discrimination and safety issues with commercial autonomous systems.

Regulatory measures are starting to be put in place that will give citizens of mainly European countries rights to request access to data held on them, but the applicability and enforcement of early measures still tip the balance in favour of data holders, and it is still up to an individual to make themselves aware of who has collected their personal data. In this context, Apple Inc do provide information as to the levels of private data access that any app downloaded on the Apple Store, but this is not backed up with enforceable regulations.

3.3.2 Error Correction

Once an individual has been made aware of data held online through enhanced transparency, the next step will be to put in place mechanisms that allow for any errors in the information, either factual or contextual, to be rapidly corrected and decisions based on this misinformation revised. This is particularly important in areas such as credit scoring, insurance, debt collection and recruitment where there is an increasing body of information showing that the finance industry in particular, having gained enhanced access to previously secured government data, base life-changing decisions on the basis of '*objective*' scoring algorithms that use low statistical scores (70%–80%) to claim a predictive correlation to outcomes.

3.4 Regulation

A major factor in controlling the way that AI is used to minimise potential harm is to create a regulatory framework with sufficient flexibility and enforceability to ensure that democratically controlled governmental bodies rather than a small cabal of technology corporations are in control of how this powerful, potentially life-enhancing technology is developed moving forward. This will not be an easy task due to the transnational nature of the corporate interests involved, and the opaque nature of the modern data economy, but some initial steps are now being taken using both existing legal mechanisms and new regulatory concepts that need to be considered when designing AI in mechatronic systems, as this rapidly evolving field will have serious implications for the structural elements needed in a system that takes advantage of the obvious benefits derived from AI.

3.4.1 IP Ownership

Large Language Models such as ChatGPT burst on the public imagination in 2022 when over-dramatized claims of human-like behaviour were suggested by some key technology influencers as being the dawn of General AI that could make humans redundant in many sectors of business, artists that would be outshone by computers, and musicians no longer necessary. Within a matter of months, this hype turned into a different conversation as influential musicians, authors and artists started to combine and file legal actions claiming copyright infringement against the corporations involved in the commercial development of Generative AI. These suits are all based on the unauthorised use of digital information stored in, and disseminated through, the cloud and claim royalties for this use that will be almost impossible to calculate accurately.

On the flip side, AI users have started filing patent and copyright applications for '*inventions*' and artistic works based on their novelty, despite the fact that the training models used are all based on existing, published innovations. To date, this seemingly absurd paradox has not been tested in law as cases rejected for patents have focussed on the legal constraint that only human innovations can be protected, but should corporations be successful in their attempts to change this restriction, the question of how to prove something is uniquely inventive when generated by AI will need to be addressed.

The whole question of IP ownership will be extremely influential on the future direction of AI in enterprise applications since any ultimate decision that training model developers will need to gain a licence for the materials used and pay royalties will fundamentally alter the economic model driving Gen AI, as well as the obvious restrictions such as need will place on the creation and deployment of LLMs in general.

3.4.2 Legal Liability

A major consideration that, in the 1980s AI research & investment boom, was one of the insurmountable obstacles in the way of using autonomous vehicles on highways is that of legal liability. Most readily understood in the context of self-driving cars, the problem the insurance industry and highway authorities had was '*in the event of an accident, who is responsible for the vehicle*?' Since the vehicle has no human driver in fully autonomous mode, should liability fall on the dealership that sold the car, the manufacturer of the vehicle, or the supplier of the AI software used to control it? From an insurance perspective, third-party claims could no longer be charged to a driver deemed liable for the accident since none would exist, so where would the guaranteed income from mandatory insurance regulations around the world come from? Since the insurance and automotive industries are multi-national, these questions would need to be answered following global agreements on regulatory changes—a difficult enough prospect in the 1980s when transnational cooperation was very much in fashion, but far harder in these more nationalistic times.

Until answers to the question of how to define global regulations that control legal liability for robotic systems, it is difficult to see how a successful commercial rollout of fully autonomous vehicles can be undertaken without a level of authoritarian control to force either corporations or individuals to be financially and legally liable for the actions of autonomous systems they own or manufacture.

3.4.3 General Data Protection Regulation

One of the earliest attempts to draft regulations to control online data collection and movement was the EU General Data Protection Regulation (GDPR) 2016/679, agreed in April 2016 and fully enacted across the EU states in May 2018 [2]. Largely dealing with data privacy, the regulation provides users of digital data with a legal framework for ensuring data is managed in a fashion that protects the privacy and ownership rights of an individual's personal data, including the duty to obtain informed consent from a subject before processing data.

In the time since its introduction, while it only applies to European markets, there has been a fairly widespread acceptance of the constraints imposed by the GDPR amongst most tech corporations after some initial opposition from US-based corporations who objected to a foreign jurisdiction curtailing their freedom to operate. With the launch of generative AI products into the market, defining whose data has been used in any given product is impossible when global data storage is controlled by a few US mega-corporations who have been free to mine their data archives for training data. Some legal actions in this area are pending in the EU and they, along with other countries, are now moving forward with further legislation to enhance regulatory control in their respective jurisdictions.

3.4.4 Next Steps

With recent concerns about social media and misinformation online and how it affects societal cohesion and the democratic process, regulatory focus has turned to how to bring this area of the Internet under some form of governmental control. The urgency to enact such legislation increased with the launch of ChatGPT in 2023 when it became clear how AI can be used to assist bad actors in the dissemination of propaganda through AI-generated fake news articles and deepfake audio/video clips.

Collaborative bodies from national governments and the tech. industry have started to be created with the aim of defining what regulation is needed for AI, and how it will be enacted and enforced. A number of legal cases around existing IP copyright legislation in the US have also been launched in 2024 that will determine who is responsible for licencing materials used in generative AI training, while a recent federal case filed by the US Dept of Justice has resulted in a ruling that Google operates as a monopoly and may need to be broken up as a result. As of summer 2024, this rapidly evolving process of industry/regulator collaboration and legal confrontation is hard to interpret with tens of media articles published daily pointing in a multiplicity of directions. As the pace of announcements from AI developers tails off (early signs are that this is starting to happen) and the regulatory landscape begins to crystallise, greater clarity on the best way forward should be achieved.

3.5 Discrimination

Since both the data sets and factors of interest used in training models are selected by human operatives, it is inevitable that the experiences of individuals concerned will colour the assumptions they use when selecting and tagging data. As an example, analysts select historical criminal record data without accounting for previous levels of systemic racism in a police force when training algorithms to predict localities where high levels of crime are to be expected. This can either be unconscious as a result of having always lived in rural areas where the majority population simply have no interaction with racial minorities, or, as for instance in a highly charged partisan political environment, a belief that high levels of crime are a function of being part of a racial minority so something the AI is expected to predict.

Whatever the source of the human discrimination that is projected onto the assumptions used when training a model, it is essential to build in processes that try to minimise the effect this has on the model predictions as well as testing protocols to detect bias as the model is rolled out for public use.

3.5.1 Human Experience

Since a machine, unlike a human, does not possess the functionality to either think or emote, it is incapable of discriminating when predicting outcomes from probabilistic multi-variate data analysis. Indeed, a main concern with using AI in real-world situations is that machines are seen as objective because all they do is calculate numbers. What is rarely understood is that the human subjectiveness a machine is supposed to remove from decision-making is baked in during the algorithm definition and training phase of product development.

While analysts can be selected and trained to avoid discrimination up to a point, it should never be assumed that it has been eliminated prior to rollout. Rather, products need to be designed to be used as an assistant with human arbiters making the final decisions, i.e. the technology should be treated as a filter for the rapid screening of documents, flagging potential targets to a human who can then take more time to fully assess the information it contains. Marketing AI as a replacement for humans in areas such as job application screening to save money would, in the long run, reduce productivity, innovation and profits through a standardisation of the workforce leading to group-think.

3.5.2 Hardware Sensitivity

One way of reducing the effect of discrimination within an AI system would be to increase the number of data inputs used in the neural network so that the influence of each individual assumption on a final prediction can be reduced. This leads to higher initial model training costs and a need for higher-performance computer hardware, but cost savings in operation would balance this over time and discrimination could be minimised.

3.5.3 Algorithmic Underpinning

An aspect of AI that needs to be looked in far more detail is how net layouts are selected for any given task, and the weights applied and ultimately used in calculating outputs. Many models used today make predictions using probabilities at the 1-sigma (68.2%) level, while the norm in experimental environments for assuming causation would be 4-sigma (99.99%) or higher. By bringing weights more in line with statistical norms, predictions will be improved and the effect of input discrimination at worst more obvious to a human operator, and thus easier to counter. By creating models that have much better-defined outputs, '*fuzziness*' will also be reduced, making the models far less opaque.

3.6 Bias

A major issue with machine learning-based AI is always going to be the quality of data used to train a model for whatever the purpose proposed. Simply scraping enormous amounts of data from the Internet and throwing it at a deep learning neural network—as increasingly happens today with LLMs—embeds historical biases into the predictive outputs from a model. This has been highlighted in many critical application areas in recent years, ranging from predicting criminal behaviour or financial risk, through CCTV-linked facial recognition systems to autonomous vehicles and battlefield recognition.

Current cloud data scraping methodologies make the elimination of bias impossible, not least because the historic human interactions built into the data are themselves biased by the differences in how individuals perceive the world around them.

However, it is essential to be aware of the sources of bias when designing an application so that its minimisation can be incorporated into the model design at as early a stage as practicable. When this is done correctly through the careful selection of training sets to provide a statistically balanced data archive that accounts for the factors described below, then AI can provide an excellent tool for automating routine analysis tasks currently performed sub-optimally by humans, finding crucial correlations in a fraction of the time that a human would be capable of (if at all) to better inform the analyst in the interpretation of complex arrays of data.

3.6.1 Skewed Historical Data

This is probably the most obvious source of bias in data sources identified in recent years, particularly in areas related to facial recognition and security/crime prevention applications. Despite this, many police jurisdictions in the US have been rolling out algorithmic methods for predicting both locations where crimes are more likely and personality profiles of potential offenders. Training the models on historical crime statistics where segregation and racism have been a driving force for more than a century creates outputs that assume skin colour is a major predictive correlation with criminality, leading to a range of issues, from higher insurance premiums in racially tagged '*high crime*' areas, through inflated false positive rates for minority racial groups in cctv facial recognition surveillance, to miscarriages of justice and longer jail sentences based on a judicial belief in the objectivity of computer evidence.

Scientific publications have also been used to train LLMs, with an increasing use of these models to assist researchers with writing the never-ending stream of papers demanded for career progression. However, the exponential increase in papers and scientific journals [3] seen in recent years creates the conditions for another concerning form of bias. In the period 1900–2000, – 1,000,000 articles were published, while since 2000 there have been – 1,5000,000. The major scientific advances (relativity, quantum mechanics, genetics, computing) all occurred in the twentieth century, yet the majority of papers online come from the 21st where the pressure to publish has led to a growth in pay-to-play articles that are either inconsequential, re-submissions of previously published work, or deliberately slanted to be controversial in the hope of driving citations.

In an increasing number of cases, data has been manipulated to ensure a predetermined thesis is confirmed. At the same time, there has been a drop in published studies that try to replicate surprising findings as funding for such work has been reduced, so a key element of the scientific method has been omitted from the process. Issues around the quality of peer review have also arisen in recent years, while paper retraction rates, even from prestigious journals, have brought the whole system into question.

LLM training sets cannot take digital publication quality into account very easily, as algorithms have no way to discriminate between good and bad writing, so the net result is that they train far more on the mass of recent publications, including those that have been retracted but still visible on the Internet. As was seen with papers relating to the COVID pandemic a vast array of deliberately misleading, or premature data releases from studies subsequently found not to be reproducible, created real-world controversies around the measures taken to attempt to control the disease that still, in 2024, create political unrest and a lack of trust in public health initiatives that have led to deaths and serious injury.

3.6.2 *Internet Access*

A subtler form of bias that has to be taken into account, particularly with the advent of LLM models, is the inequality of access to the Internet created as a result of geography. While it is true to say that true unlimited data, high-speed fibre optic access (1 Gb/sec +) is currently available in the majority of large towns and cities in the developed world, rural areas in most parts of the world still only have access to data capped, low speed (>100 Mb/sec) DSL, satellite or wireless—the latter two being very susceptible to signal degradation due to wet weather and/or signal obstructions in the landscape. Indeed, a significant percentage of households in the US, around 20% according to a 2022 report from the US Department of Commerce [4], have no home Internet access at all.

These disparities skew the data available online for LLM training dramatically towards wealthier, more educated urban populations, excluding the life experiences from the training data of a significant minority of the global population and making it more difficult for these already disadvantaged groups to fully participate in the modern digital socio-economic structure being assembled.

3.6.3 *Inequality of Opportunity*

As more and more reliance comes to be placed on the use of automated AI tools in recruitment, education and even the management of public health, the already existing inequalities identified over the years created by systemic racism, sexual/gender discrimination and so on will not, as promoted by advocates for the use of '*objective*' screening tools to remove '*subjective*' human bias, make processes fairer. Instead, as outlined previously, human biases will be baked-in to the processes that define the quality of life for an individual, making it increasingly difficult for an individual to alter their life goals. With recruitment, the assumptions built into the algorithmic forecasts of job performance will most closely match wealthier urban dwellers since their career profile has been used predominantly to train the unregulated commercial algorithms used. In addition, this sub-set of the population will be more aware of how a resume needs to be worded to get through screening and onto the desk of a human assessor.

As well as being inherently unfair to '*non-standard*' applications, this bias works against the interests of potential employers since any work team requires a mix of varied backgrounds and experiences to provide an innovative and efficient approach to problem-solving. By filtering out those with unusual career paths, the trend towards 'group-think' in corporations will increase at the expense of innovation.

While AI is demonstrably of great benefit in many areas of diagnostic medicine, performing far better than human operators in areas such as early diagnosis of life-threatening diseases, the biases inherent in current training data sets can cause problems in this seemingly objective area of complex data analysis. However, socio-economic factors, environment, ethnicity and sex/gender of a patient are key factors in diagnosing diseases and predicting likely treatments, so the bias in historical data towards white, heterosexual males of sufficient wealth to be able to afford medical treatment creates difficulties when attempting to predict cause and outcome in other populations. As a result, the efficacy of algorithmic diagnosis is lower in populations that fall outside the historically dominant male population, reducing overall positive outcomes.

3.7 Sustainability

3.7.1 Future Cloud Server Requirements

Before making any bold predictions about how quickly generative AI models will lead to the Holy Grail of Artificial General Intelligence (AGI), consideration needs to be given as to whether the current approach using LLMs can be sustainable in the short-medium term. In 2020, before the explosive growth in LLM development, the data centres underpinning the Internet data economy accounted for 1% of all global power consumption [5], while water consumption and land use were also highlighted as causes for environmental concern. More recent research published online [6], looking at the increased need for data centres and IT systems to process data for LLMs such as ChatGPT, show that this will exceed 10GW by early 2025. It was also estimated that demand globally for data centres used in AI will increase by up to 10 × every six months. Already in 2024, power outages in several West Coast states in the USA due to this demand have been reported in the general media as an under-invested power infrastructure struggles to keep up with the commercial growth of generative AI, and the Internet is now regarded as having a larger carbon footprint than the airline industry. Climate change concerns, coupled with the overloaded electricity generation/distribution infrastructure, are a factor in limiting the growth of generative AI in the short term.

Aside from the sustainability issues highlighted in this section, before anyone starts to herald the arrival of super-human artificial intelligence that will supplant humanity, it will need to be able to demonstrate a similar power requirement to the human brain, which requires ~ 100 W to function.

3.7.2 Training Model Climate Drift

An area of concern that seems to be largely ignored in the literature is the effect on LLMs trained on data collected under one set of historic climatic conditions when applied to predicting behaviour as the climate changes with time. Applications ranging from automated agriculture, through autonomous transportation, to a range of health conditions all need to take this into account as, key to all AI, is that the training data used is a match for the task that it is to be used in.

3.8 Areas of Opportunity

There are a lot of negative issues still to be resolved with AI as the hype associated with LLMs starts to switch from over-excitement/fear towards scepticism as the overblown claims made since the launch of ChatGPT 4.0 in 2023 are found to be unachievable. Political decisions will hugely influence how this technology is rolled out in coming years and whether it will be a power for more individual freedom and equality or increased surveillance and control by those in control of an increasingly unequal society. To be positive, these dilemmas were also faced in the early twentieth century following the Industrial Revolution and ultimately sense prevailed and governments acted to reduce inequality and regulate society to make life easier for the general population.

To finish on an optimistic note, two areas where AI can be (and to a certain extent already is being) applied as an assistive tool to improve the health and well-being of people from all backgrounds are outlined below with suggestions of how to overcome current obstacles to their implementation.

3.8.1 In-Home Health Support

A lot of work has gone into developing automated care solutions for the elderly and infirm living at home to ease the high costs and lack of infrastructure currently encountered when handling an ageing population. Many companies have developed connected sensors to check a range of medical conditions operating through a mobile phone and video systems for remote consultations with a doctor became popular during the COVID-19 pandemic.

Far less attention has been given to the use of AI to maintain health in people living alone who, while perfectly healthy physically, do not receive a range of normal social stimuli that help to maintain good physical and mental health. Loneliness and lack of regular exercise can create conditions that accelerate issues associated with ageing such as reduced mobility, social isolation and potentially dementia. Age-related diseases such as Glaucoma can also go undetected until treatment is not an option. The net result is that in-home care provision or relocation to a residential home becomes necessary at an earlier age than would have been the case if normal activity levels had been maintained, while degenerative diseases are more costly to treat than if detected at an earlier stage.

What is needed in these circumstances is a holistic data collection system in the home that can monitor in detail how an individual moves around the house and garden throughout the year, patterns of visitors, meal times etc. and use this data to look for changes in behaviour over time that may indicate the start of a decline into dependency. Where necessary, medical conditions can be monitored, and emergency alerts made to a local medical centre, etc. The same system would also provide verbal prompts to get the client to contact relatives, friends etc., maintain a '*concerned friend*' dialogue to ease loneliness and ensure routines are maintained in the first instance. Clearly a lot of highly personal data would be collected, so the AI unit[2] would need to be capable of processing everything on board and then connecting overnight while recharging with an archive server that is not connected to the Internet unless instructed to upload data to a medical centre or elsewhere.

Since there are many elderly unfamiliar with the use of mobile phones and computers, or have no/poor Internet connections in rural areas, the AI unit would need to be capable of communication through landline telephony and restrict data processing to the on-board unit rather than rely on online servers.

3.8.2 Sustainable Transportation

Despite all the bold claims made by people such as Elon Musk that fully autonomous vehicles will be on the road by the end of 2024 (previously 2023!), the reality is far more complex and the current situation is not very different to that experienced in the 1980s when industry predictions were that the self-drive cars then being track tested would be available to the general public within 5 years. In both cases, it is true that the technology worked reliably and did find some application in warehouse and airport locations, but could not cope with sharing the road with humans programmed to behave in a different logic universe to machines. At the same time, a range of non-technical obstacles also became apparent, so much so that a failure to agree on international safety regulations for road vehicles shut down progress.

The current situation, while not so constrained by International regulatory negotiations as research has been concentrated in Silicon Valley and combined with EV development efforts, is still subject to the same issues. This was highlighted in recent years through a number of crashes involving Tesla cars being driven in the self-drive mode that had been installed as a test module, and some road deaths resulting from other test vehicles colliding with pedestrians. Software modifications to make the cars avoid all unexpected human movements encountered led to autonomous taxis tested in San Francisco being pulled from the road after initial tests showed the cars

[2] Envisaged by the author as a robotic pet such as *K9* from Doctor Who!

travelled so slowly because of the ubiquity of humans moving around them that in some instances, passengers were getting out and walking to their destination arriving ahead of the taxi.

This issue illustrates the wider problem when mixing autonomous and human-piloted vehicles in the same road space; humans are highly unpredictable using machine logic and, for the AI to learn human avoidance, it would first need to be trained by getting involved in real-world collisions and causing death and injury that even the most libertarian government body will not permit. The alternative approach that has been deployed instead is a safety-first approach, which works well at saving lives but creates efficiency problems as seen with the San Francisco taxi trials. Also, humans are fast learners and once they are aware that an autonomous vehicle will stop if you step out in front of them there will be increasing traffic congestion in cities as pedestrians simply cross the road when they feel like it in the knowledge that the oncoming traffic will stop.

Alternative approaches as proposed in the 1980s may be the way forward in the short-medium term. Rather than mixing self-drive with normal cars, the proposal was to dedicate a lane on multi-lane highways to autonomous '*road trains*' so that long-distance traffic could travel at high speed with minimal inter-car spacing to increase traffic density. Vehicles in the A-Lane could also recharge from overhead cables as they travelled, ensuring faster journey times. Once a vehicle needed to leave the highway, it would revert to semi-autonomous control with the driver relying on guidance and safety systems only (as is the reality today). The A-Lane network could also have been extended by incorporating unused rail tracks. Cities could also be adapted such that only autonomous vehicles are allowed inside an outer perimeter road.

3.9 Concluding Remarks

The current structure of the AI industry centred on unregulated data capture from Internet users using '*free*' cloud-based services, while it has enabled a few early adopters in the technology sector to fund the rapid development of a range of new autonomous systems, is starting to encounter resistance on a number of fronts that will lead in the short-medium term to a decision point on how to move the whole field of AI forward.

For continuing progress on the current development pathway, true machine autonomy is not likely to be achievable without some enhanced level of human population re-education to act in a logical way that can be predicted by algorithms at a much higher level of accuracy than is plausible currently.

The alternative approach would be to continue to focus on where AI can provide benefits by enhancing specific applications such as search, image recognition and smart process control where the logic and speed of algorithms can create better data in a much shorter space of time to a human operator who can then use the superior deductive skills of the brain to make decisions based on that data.

While the current trajectory is favoured by the small pool of technology corporations and investors driving the current surge of interest, there is increasing evidence that the approach of '*scrape more data and build faster computers*' is starting to stall as people become more aware of the privacy and security issues this creates, data creators are starting to demand payment and control over the IP used to create LLMs, and simple sustainability limitations are encountered.

Even if these can be overcome, crucial elements of human development would not be served as predictions based on patterns from historical data contain a range of biases while predicting the future from history removes true inspiration and lead innovation from science and the arts. Machine logic is purely deterministic and based on the information and assumptions programmed into it, whereas human intellect is an evolutionary process capable of random lateral jumps (inspiration) that are pseudo-quantum mechanical in nature—how would machine learning mimic the leaps of imagination that took Einstein to Relativity & Quantum Mechanics when all available data at the time that could have been used to train with suggested Newtonian physics was a complete theory of the universe?

Acknowledgements This chapter has been based on ideas developed and issues encountered, working over the years on a range of collaborative projects with research colleagues in the UK, Canada, the European Union and the USA involved in sensor development and complex data analysis. In particular, I would like to thank Professor David Bradley, Eddie Goodbrand and Brendan McGlynn for the numerous brainstorming sessions they have contributed to as I have developed my ideas on how best to tackle the complex issues surrounding the use of machine intelligence for the benefit of all humanity, rather than as a driver of increasing inequality and social discord.

References

1. HIMSS Healthcare Cybersecurity Survey (2023) @ www.himss.org/resources/himss-healthcare-cybersecurity-survey. Accessed August 2024
2. EU General Data Protection Regulation 2016/679 @ eur-lex.europa.eu/legal-content/EN/TXT/PDF/?uri=CELEX:32016R0679. Accessed September 2024
3. Thelwall M, Pardeep S (2022) Scopus 1900–2020: growth in articles, abstracts, countries, fields, and journals. Quant Sci Stud 3(1):37–50
4. Cao M, Goldberg R (2022) Switched off: why are one in five US households not online?, National Telecommunications & Information Administration Bulletin @ www.ntia.gov/blog/2022/switched-why-are-one-five-us-households-not-online. Accessed August 2024
5. Obringer R, Rachunok B, Maia-Silva D, Arbabzadeh M, Nateghi R, Madani K (2021) The overlooked environmental footprint of increasing internet use. Resour Conserv Recycl 167(105389):1–4
6. Patel D, Nishball D, Ontiveros JE (2024) AI Datacenter energy dilemma—race for AI Datacenter Space, @ www.semianalysis.com/p/ai-datacenter-energy-dilemma-race. Accessed August 2024

Bibliography

Amoore L (2020) Cloud ethics, Duke University Press

Arthur WB (2009) The nature of technology. Simon and Schuster
Autor DH, Mindell DA, Reynolds E (2022) The work of the future. Princeton University Press
Bhaskar M, Suleyman M (2023) The coming wave, Crown
Broussard M (2019) Artificial unintelligence. MIT Press
Christian B (2020) The alignment problem. Norton
Cobb M (2020) The idea of the brain. Basic Books
Conway EM, Oreskes N (2023) The big myth. Bloomsbury Publishing
Crawford K (2021) Atlas of AI. Yale University Press
Daub A (2020) What tech calls thinking, FSG Originals × Logic
Davis E, Marcus G (2019) Rebooting AI. Pantheon Books
De Langhe B, Puntoni S (2024) Decision-driven analytics, Wharton School Press
Du Sautoy M (2019) The creativity code. Belknap Press
Doctorow C (2019) Radicalized, Head of Zeus
Dubber MD, Pasquale F, Das S (2020) The Oxford Handbook of Ethics of AI, OUP
Eubanks, V (2019) Automating inequality, Picador
Fosch-Villaronga E (2019) Robots, Healthcare, and the Law, Routledge Research
Frey CB (2019) The technology trap. Princeton University Press
Harford T (2021) The data detectives, Riverhead Books
Hoepman J-H (2021) Privacy is hard and seven other myths. MIT Press
Huttenlocher D, Kissinger HA, Schmidt E (2021) Cambridge analytica, Little Brown & Co
Johnson S (2011) Where good ideas come from, Riverhead Books
Kahneman D, Sibony O, Sunstein CR (2021) Noise, William Collins
Kearns M, Roth A (2019) The ethical algorithm, Oxford University Press USA
Kelly K (2011) What technology wants. Penguin Books
Lanier J (2014) Who owns the future. Simon and Schuster
Lanier J (2018) Dawn of the new everything, Picador
Lanier J (2019) 10 Arguments for deleting your social media accounts right now, Picador
Lepore J (2020) If then, Liveright Publishing Corp
Levy K (2022) Data driven. Princeton University Press
Mackenzie D, Pearl J (2018) The book of why. Basic Books
Madi, MAC (2021) **The dark side of nudges**, Routledge Focus
Mitchell KJ (2023) Free agents. Princeton University Press
Morgan SL, Winship C (2014) Counterfactuals & Causal Inference. Press, Cambridge Uni
Muller JZ (2018) The tyranny of metrics. Princeton University Press
Murgia M (2024) Code dependent, Henry Holt & Co
O'Neil C (2017) Weapons of math destruction, Broadway Books
Pasquale F (2016) The black box society. Harvard University Press
Pasquale F (2020) New laws of robotics. Harvard University Press
Perez CC (2019) Invisible women, Abrams Press
Prescott T (2024) The psychology of artificial intelligence. Routledge
Richards N (2022) Why privacy matters. Oxford University Press
Russell S (2019) Human compatible, Viking
Sumpter D (2021) The ten equations that rule the world, Flatiron Books
Sunstein CR, Thaler RH (2021) Nudge. Penguin Books
Tenner E (2018) The efficiency paradox, Knopf Books
Toon N (2024) How AI thinks, Transworld Publishers
Tzanou M (Ed.) (2021) Health data privacy under the GDPR, Routledge Research
Veliz C (2021) Privacy is power, Bantam Press
Vigen T (2015) Spurious correlations, Hachette Books
Vincent J (2022) Beyond measure, WW Norton & Company
Wolfram S (2023) What is chat GPT doing …, Wolfram Media
Wooldridge M (2020) The road to conscious machines, Pelican Books
Wylie C, Halstead G (2020) Mindf*ck: inside Cambridge Analytica's Plot to Break the World, Profile Books
Yates K (2020) The math of life & death. Scribner
Zuboff S (2018) The age of surveillance capitalism, Public Affairs

Podcasts

Roose K, Newton C (2022) Hard Fork, The New York Times (March 2024)
Safi M (2024) Black Box, The Guardian (February 2024)

Chapter 4
Complexity and Decision-Making

Daniel Gilmour, Edward Simpson, Andrew Spiers, David Blackwood, and David Bradley

Abstract Systems are becoming increasingly complex both in terms of their structure and operation, impacting in turn on the ways in which design and operational decisions are made. The chapter therefore sets out to examine the nature of complexity from the perspective of Mechatronics with particular reference to the design and decision-making processes that are involved. The chapter concludes by considering the role of models in these processes along with the potential impact of artificial intelligence.

4.1 Introduction

A dictionary definition of complexity is:

> *The state of having many parts and being difficult to understand or find an answer to.*

Which unfortunately in the context of technology provides little in the way of clarity as to what might be considered a complex system from the perspective of either the systems designer or operator, or indeed with respect to mechatronics. In his 1998 paper, Corning [1] suggested that complexity is associated with the following attributes:

- A complex phenomenon consisting of many parts.
- Many relationships/interactions among the parts.
- Parts produce combined effects that are not easily predicted and may often be novel.

In 2018, Mayfield et al. [2] suggested that:

> *Engineered systems can be complicated or complex. A complicated system has many aspects or components to consider. It is nonetheless amenable to rational planning, design and execution, despite being on a large scale perhaps. A complex system has many interrelated*

D. Gilmour (✉) · E. Simpson · A. Spiers · D. Blackwood · D. Bradley
Abertay University, Dundee, UK
e-mail: d.gilmour@abertay.ac.uk

P. Hehenberger and D. Bradley (eds.), *Mechatronic Futures*,
https://doi.org/10.1007/978-3-031-83571-1_4

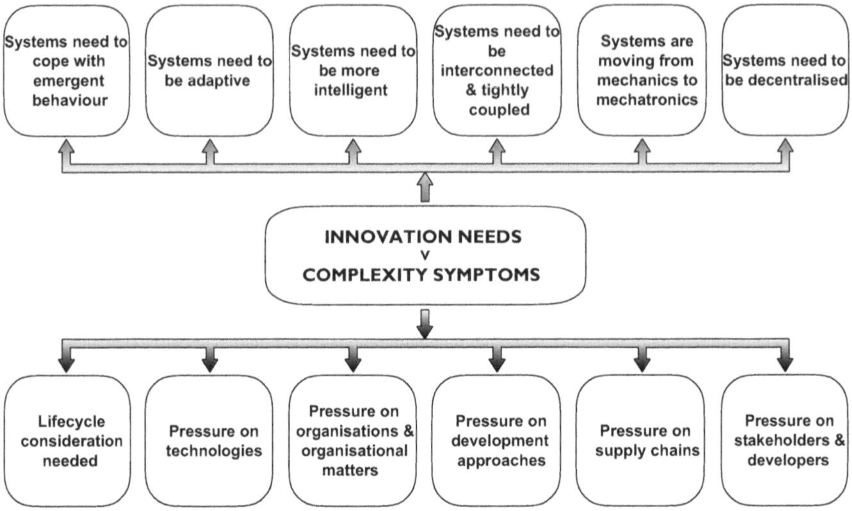

Fig. 4.1 Relationship between innovation and complexity (after [3])

> *aspects or components to consider, each of which has its own (at least partly) autonomous or independent behaviour, which is influenced by interactions with other elements. The* ('emergent') *properties and behaviour of a complex system are not* (fully) *predictable, confounding many classical system engineering techniques and methodologies.*

And that:

> *Components of a complex system can themselves be independent systems, whose development, operation and governance may be subject to different authorities; there may be no* 'top-down' *overall control. These component systems can, therefore, have separate and independent life cycles, and typically, there is no single top-level* 'chief system engineer' *whose authority all the component systems are subject to.*

Then in their 2023 paper, Morales, Nizamis and Bonnema [3] argued that the relationships between innovation and complexity were along the lines of those suggested by Fig. 4.1.

This chapter investigates the nature of complexity regarding engineering systems in general, and mechatronics in particular, and hence considers how this impacts sub-systems such as the design process and decision-making.

4.2 Mechatronics and Complexity

Historically, mechatronics focused on the implementation of devices, sub-assemblies and components in which sensors and actuators are integrated under some form of computer control. These mechatronic elements then formed the basic building

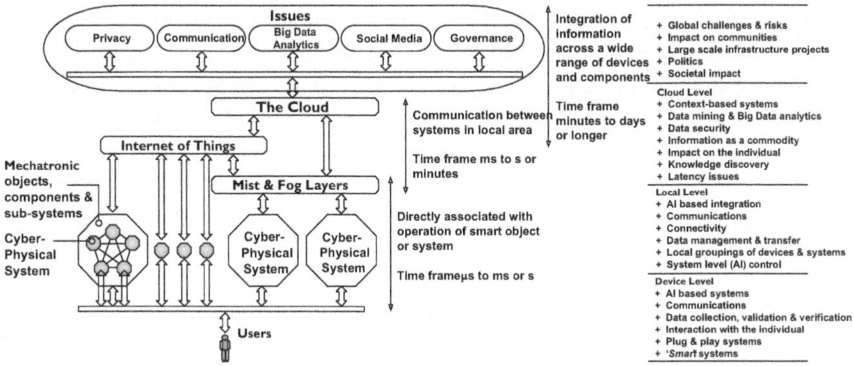

Fig. 4.2 System structure incorporating the cloud, internet of things, mist & fog layers and cyber-physical system elements

blocks for applications in areas ranging from manufacturing to construction, transport and domestic environments. More recently, the development and implementation of Cyber-Physical Systems (CPSs) within which mechatronics elements are integrated through advanced computing strategies, including artificial intelligence (AI), together with the introduction of the Internet of Things (IoT) and Cloud Computing suggests that a more systems-oriented view of mechatronics now needs to be put in place to accommodate these developments as suggested by Fig. 4.2.

From the perspective of the design team responsible for system configuration and operation, it is quite conceivable that they may not be fully capable of understanding, or realising, exactly how systems such as those represented by Fig. 4.2 might function or operate across all possible circumstances. This is particularly the case where they may be dependent upon embedded AI constructs for the whole or part of their operation, as for instance in the form of an embedded Cyber-Physical System (CPS), which itself may constructed from a range of mechatronic components, assemblies and sub-systems. Further, the fact that the exact configuration of the system at any time will be dependent on the functional and operational requirements at that time creates difficulties in the interpretation of the role of specific system elements. Thus, while the functional requirements for a component, sub-assembly or sub-system may well be capable of being defined, along with their role concerning system operation, the exact in-service conditions may not be known to the designer.

From the perspective of the user, the primary area of interaction with such a system is through the medium of a range of '*smart*',[1] and often mechatronic, elements, which themselves are increasingly incorporating AI structures. It is also these same levels

[1] The term '*smart*' is increasingly used to refer to devices which are capable of communicating with each other by WiFi or other media and which can be utilized as part of an Internet of Things construct. However, here and in subsequent discussion it is also taken as being more focused on devices which incorporate within themselves a (albeit generally limited) degree of decision-making capability.

which are often primarily associated with the collection and onward distribution of system data, including personal data.

Users are therefore faced with a growing range of challenges in respect of their ability to assimilate and use the information associated with the operation of the system, and hence to make choices in relation to that operation. Further, they may be dealing with systems which are capable of semi-autonomous, or indeed autonomous, self-(re)configuration in relation to changes in operational context, drawing both information and components from The Cloud via the Internet of Things (IoT) as required [4, 5].

Consequently, systems are becoming increasingly complex [6–8] with multiple components being integrated at several levels as illustrated by Fig. 4.2. Here, the lowest level constitutes the **Device Level** and is largely made up of mechatronic and other forms of smart objects, components and sub-systems, which may themselves be integrated to form a Cyber-Physical System. Communication, and hence integration, between individual elements is facilitated by the introduction of the Mist & Fog layers [9–11] to handle data integration and transfer. The **Device Level** is also where the primary interaction with the user takes place along with the collection of operational data [12–14].

The next level in the structure of Fig. 4.2, the **Local Level**, is then associated with increased interconnection and data transfer between devices through the medium of the IoT. In many instances this will not be a fixed system configuration but will see system elements coming in and dropping out in relation to both need and context. Finally in Fig. 4.2, the **Cloud Level** provides further layers of integration along with a range of functions such as Big Data Analytics and the combination of data from across multiple sources to support knowledge discovery [15, 16]. Table 4.1 then provides examples of how these levels come together across manufacturing, vehicles and the built environment.

Thus, it can be seen that the role of mechatronics is becoming increasingly complex, no longer simply that of providing the components, sub-systems and systems necessary for the operation of the system but also that of managing and supporting the system across its full range of potential operating conditions.

4.3 The Engineering Design Process

Figure 4.3 provides a representation of the engineering design process [17–19] which attempts to show the connections between the various elements of the supply chain and manufacturing processes. It must also be taken into consideration that a great many products undergo a process of near-continuous modification and upgrade over their lifetime to take into account both developments in technology and user feedback, resulting in a cyclical model for continuous development.

Table 4.1 Operating requirements associated with differing system levels

	Manufacturing	Vehicles	Built environment
Device level	• Actuators • Real-time condition monitoring • Drives • Sensors	• Braking control • Driver controls & systems • Drive train Management • Engine management • Environmental systems • Sensors • Traction control	• Access control • Environmental control • Heating • Lighting • Monitoring • Security • Sensors • Ventilation
Local level	• Automated guided vehicles • Individual machine tools & robots • Islands of automation • Interaction & interoperation • Local inventory control • Materials handling • Task-based groupings • Work groups • Work transfer systems	• Autonomous operation • Driver support & assistance (Cruise Control, parking assist, 360° vision) • Environmental controls • In-vehicle Entertainment • Navigation & route planning	• Traffic flows • Smart homes • Building performance • Smart drainage • Facility management • Construction automation
Cloud level	• Analytics • Digital twin • Factory wide inventory control • Logistics • Planning & scheduling • Plant operation • Storage management • Supply chain management	• Advance information (i.e. Parking) • Alerts & alarms • Analytics • Digital twin • Interaction with other vehicles (internet of vehicles) • Interaction with local environment • Local routing • Maintenance scheduling • Remote operations (locator, lock/unlock, etc.) • Route optimisation	• Flood defence • Digital twins • Emissions monitoring • Building Information Management • Traffic flow management

4.3.1 Heilmeier's Catechism

The following list of questions about the design process is referred to as *Heilmeier's Catechism* were formulated by George H Heilmeier, the director of the Defense Advanced Research Projects Agency (DARPA) in the USA from 1975 to 1977, to

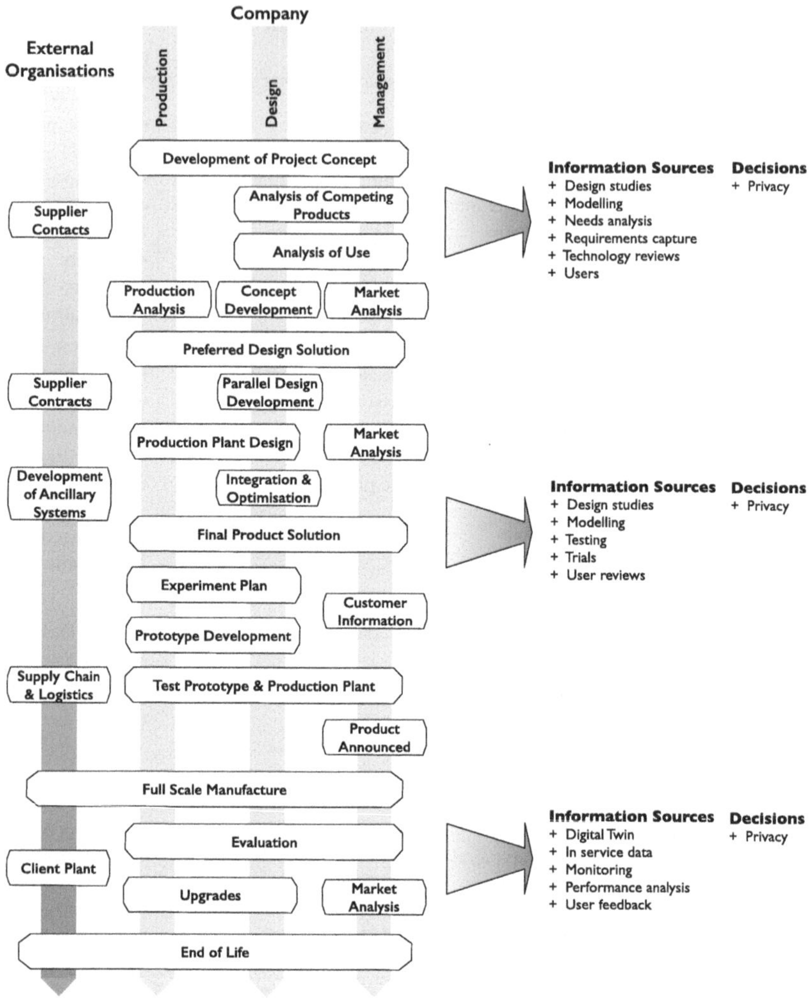

Fig. 4.3 Representation of the design process showing the relationship between Production, Design & Manufacturing within the company and with external organisations and incorporating decision-making structures

aid in the thinking through and evaluate of prospective, and highly complex, research programmes [20].

- What are you trying to do? Articulate your objectives using absolutely no jargon.
- How is it done today, and what are the limits of current practice?
- What's new in your approach and why do you think it will be successful?
- Who cares, and if you're successful, what difference will it make?
- What are the risks and the payoffs?
- How much will it cost?

- How long will it take?
- What are the midterm and final '*exams*' to check for success?

Considering each of the individual elements of Heilmeier's Catechism it can be seen that they force the design team to adopt a decision-making approach at each stage of the design process, based on a process of analysis, data collection and interpretation in order to reach a conclusion. Some of the analysis may be technical, as for instance relating to costs and risk, while other elements, as for example when concerned with novelty or '*who cares*' will be associated with user responses and reactions.

By forcing the design team to conform to these issues in a semi-formal manner, a degree of discipline in decision-making is imposed. This then means that the nature of these decisions can be readily explained to others to facilitate communication and reduce the potential for communication errors within and across the design team.

In illustration of this, consider the (oft-reported) incident of the *Mars Climate Orbiter*. Following the initiation of its orbital insertion manoeuvre in September 1999, radio communication was lost as it passed behind Mars some 49 s earlier than expected and was never regained. It was eventually established that the spacecraft was at a lower altitude than intended and broke up in the Martian atmosphere and that the main underlying cause lay in a communication failure between two systems groups with the result that part of the software, supplied by Lockheed Martin, produced results for the thrusters in pound-seconds while the trajectory calculation software which used these results was expecting them to be expressed in newton-seconds. Discrepancies between calculated and measured positions had in fact been detected and reported but were ignored [21].

While the intent of procedures such as the Heilmeier Catechism is to enable all members of the design team to understand the underlying decision-making processes in relation to project objectives, these processes can not necessarily encompass all issues, nor remove unintended bias resulting from the composition of the design team, its background and experience [22].

4.4 Design as a Decision-Making Process

In his 1962 work, '*Introduction to Design*' Asimow [23] suggested that there are fourteen fundamental principles of design that can be usefully grouped into two closely related categories. Eight of these are concerned with the nature of the design process. These categories identify the iterative nature of design resulting from the existence and emergence of sub-problems whilst the main problem is being considered. These principles are then related as follows:

Morphology	→ A design is a progression from the abstract to the concrete through a …
↓ Design Process	→ which is essentially iterative and contains …
↓ Subproblems	→ which are uncovered whilst addressing the main problems, and which must be solved in order to solve the main problem. This process leads to a …
↓ Reduction in uncertainty	→ about the success of the artefact. In solving sub-problems …
↓ Minimum commitment	→ should be made at any stage of the design since firm commitments at an early stage of the design may limit choice at a later stage. The extent of the design process can be established by reference to …
↓ The economic worth of the evidence	→ which measures the cost of evidence against the significance of that evidence to the success or failure of the artefact and to …
↓ The basis for the decision	→ which is the point that sufficient confidence has been gained in the design to allow the next stage to proceed. A design will have its existence to the extent to which it can be expressed through …
↓ Communication	

Then in 1972 Markus [24] proposed that '*horizontal*' iterations were where sub-problems are being solved within the general '*vertical morphology*' of the design solution as shown in Fig. 4.4.

The iterations during the phases in the Markus model are in fact simplistic models of the decision-making process. In 1988, Howard [25] proposed a similar three-stage model for general decision-making but uses the terms '*formulate*' and '*evaluate*' in place of analysis and synthesis, although the description of the model is virtually identical to that of Markus. It is evident from Fig. 4.4 that the design process is essentially viewed as a series of decisions, and therefore the nature of these decisions must be understood if an effective and efficient design is to be achieved.

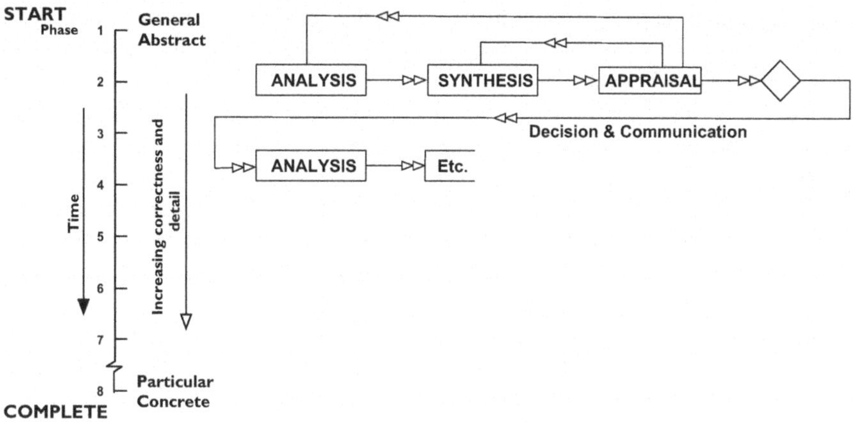

Fig. 4.4 Markus's model of the design process

4.4.1 *The Nature of the Decision-Making Process*

Rational decision-making is a multi-step process for making choices between options. The process of rational decision-making favours logic, objectivity and analysis over subjectivity and insight. There are many models of the decision-making process but most include the following stages:

- Problem identification.
- Data and information gathering.
- Consideration of the various outcomes and alternatives.
- Decision-making based on appropriate criteria.
- Evaluation of the decision and its consequences.

The applicability of this approach to system configuration and operation design decisions will be influenced by the types of decision problems that are required during the design process at the three systems levels defined previously in Fig. 4.2, namely **Device, Local** and **Cloud**.

Problems can be well-structured. These have a clear definition of the project outcomes, clear and measurable assessment criteria, and sufficient data to fully assess criteria which enables a systematic method of comparing options resulting in the selection of the most beneficial outcome. Well-structured problems can be solved using a well-defined step-by-step approach; however, this may be restricted by time and data constraints. Well-structured problems may be most likely to emerge during the system design process at the Device Level.

Ill-structured problems have no clear definition of the project outcomes, lack clear and measurable assessment criteria and have insufficient available information to allow a full assessment of different outcomes. Ill-structured problems are not amenable to well-defined decision-making approaches and may require designers to apply judgement, experience and heuristics in the selection of the most beneficial outcome. Ill-structured problems may emerge most often during the system design process at the Cloud Level.

McCall and Burge [26] provide a useful review of the concept of Wicked problems. Wicked problems are ill-structured and often involve multiple stakeholders with differing perspectives. They have no '*right*' or '*optimal*' solution and cannot be solved by the application of standard (or known) methods. They require creative multi-disciplinary solutions that involve a range of stakeholders. Figure 4.5 then shows 10 of the properties associated with a Wicked Problem [26, 27]. They may emerge most often during the design process at the Cloud Level.

The nature of design work and the time and resources required during the design process at the three systems levels are influenced by the '*degree of rationality*' [28] of the decision-making process. Rational decisions would lead to the optimal choice being made but would require a highly specified and clearly defined environment. Many recent publications on decision-making refer to the work of March and Simon [28], Lindblom [29] Simon [30] from 1958, 1959 and 1976, respectively, who all noted that decision-making in practice is seldom structured and that often '*satisfactory*' solutions are reached on an ad-hoc basis. Astley et al. [31], reviewing the above, concluded that '*decision-making can be seen as a process of muddling through*

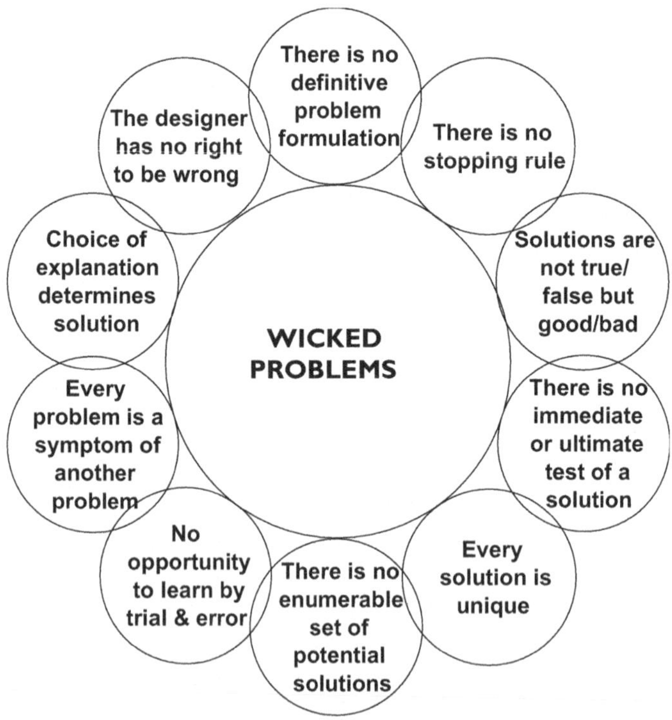

Fig. 4.5 Properties of a Wicked Problem (after McCall & Burge [26])

towards a satisfactory and sufficient outcome as opposed to necessarily obtaining the optimal solution'.

March and Simon [28] however question the ability of the '*rational*' decision-maker to make optimal choices and distinguish between optimal and satisfactory solutions to problems. They describe a solution as being optimal if:

- There exists a set of criteria that permits all known alternatives to be compared;

  And

- The alternative in question is preferred, by these criteria, to all other alternatives.

  An alternative is satisfactory if:

- There exists a set of criteria that describes minimally satisfactory alternatives;

  And

- The alternative in question meets or exceeds all these criteria.

They conclude that most human decision-making is concerned with the discovery and selection of satisfactory rather than optimal alternatives and describe this process as '*satisficing*' as represented by Fig. 4.6.

This concept of '*satisficing*' is comparable with Asimow's principle of the '*bases for decision*' within the design process [23]. These bases for decision must be related to evidence, and the '*economic worth*' of that evidence must be considered, that is the cost of obtaining that evidence against the value of the evidence to the correctness of the decision. In essence, the quality of the solution may vary depending upon the time and effort expended to produce a solution and although many '*satisfactory*' solutions may exist some will be closer to the optimal solution than others.

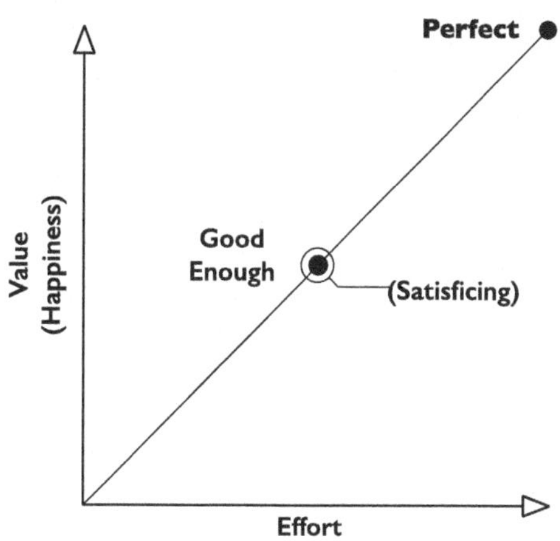

Fig. 4.6 Concept of satisficing

4.4.2 Information and Decision-Making

It was suggested above that structured problems may be more common during the design process at the Device Level and ill-structured problems may be more common at the Cloud Level. However, fully integrated systems will require systems at each level to communicate and exchange data across levels as shown in Fig. 4.7. Therefore, designers of systems that operate at each level will have to consider functional and data requirements at other levels and this will influence individual project outcomes, assessment criteria and data availability. This will tend to increase the complexity of decisions at the Local and Device levels.

It is therefore necessary to understand the nature of the information that flows between designers and across levels to ensure integration between Cloud, Device and Local level systems. Information flows should be mapped and key decision points identified across levels to ensure that the right information is available to the right person at the right time at each step of the decision-making process.

This information can have a few forms which will again influence the complexity of decisions at all levels. The '*Ashen*' model, discussed in depth in Snowden [32], can be used to classify sources of information to support decisions. This will allow differentiation between decisions that are principally driven by computation or by established and documented decision-making procedures and those that were based on the decision-maker's own experience and perceptions. A simplified definition of the Ashen classifications is:

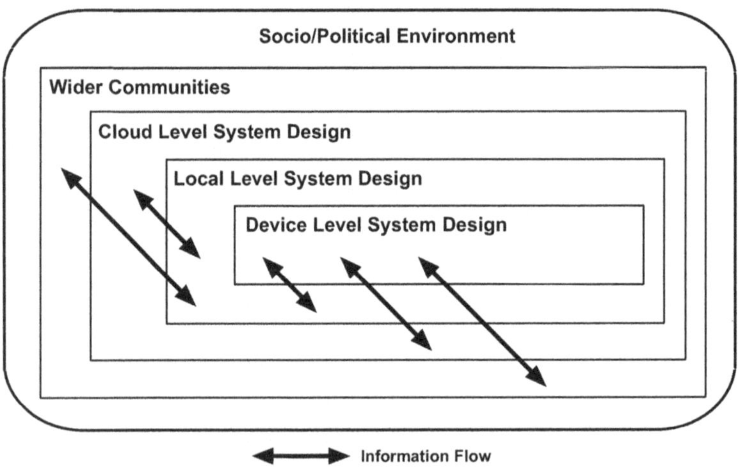

Fig. 4.7 Decision Levels and information flows

Artefact	All existing explicit knowledge and/or codified information within an organisation, e.g., documents, databases
Skills	Expertise, practised ability, dexterity, tact that can be identified, a tangible measure of their successful acquisition
Heuristics	Rules of thumb, often used to make decisions
Experience	Actual observation or practical acquaintance with facts or events and the knowledge resulting from this
Natural talent	A special aptitude, faculty or gift

Decision points can be identified by the way that designers use information to identify decision points as shown in Fig. 4.8. The concept behind this and Fig. 4.3 is that the designer, on receipt of the input information can do a number of things with it:

- Store the information as received without using it.
- Communicate the information as received to another stakeholder or stakeholder(s).
- Use the information as received in reaching a decision, the outcome of which is communicated to another stakeholder or stakeholders.
- Analyse and transform the information using a tool and store the information without using it.
- Analyse and transform the information using a tool and then communicate the processed information to another stakeholder or stakeholder(s).
- Analyse and transform the information using a tool and use this information in reaching a decision, the outcome of which is communicated to another stakeholder or stakeholders.

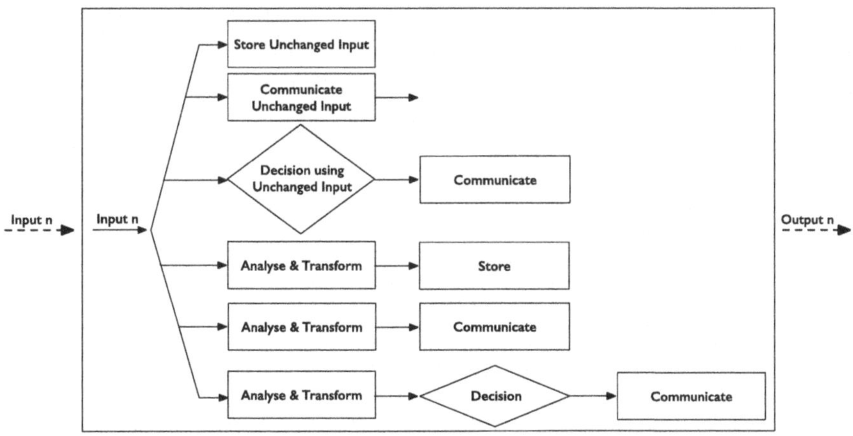

Fig. 4.8 Conceptualisation of decision process

In this way, the decision process can be described for each phase. The tools used in the information transformation within the process should be identified and catalogued, and key decision nodes within the sphere of influence of the interviewee will be identified.

4.4.3 Design Costs and the 80:20 Rule

A key issue in the engineering design process is the balance between the allocation of costs and actual expenditure with some 80% of the overall project cost being allocated by the end of the concept development and evaluation phases as suggested by Fig. 4.9. However, that same time period only accounts for 20% or less of the project spend, leading to misunderstandings of the contribution of these early phases to added value.

The other cost issue associated with this, apparent, imbalance between allocation and expenditure lies with the costs associated with remediation in that the later an error is detected in the design process, the greater the costs are to remedy this. Thus, a problem undetected until pre-production testing rather than during concept evaluation could have a remediation cost potentially of orders of magnitude greater than if it had been detected at the earlier stage.

The 80:20 rule, otherwise referred to as the Pareto Principle, thus argues that as suggested by Fig. 4.10 for many cases, a performance level of the order of 80% results from 20% of design elements, the '*vital few*'. Attempting to achieve the remaining 20%, the '*useful many*', then results in significant increases both in project time and costs, often many times greater than those associated with achieving the 80% level.

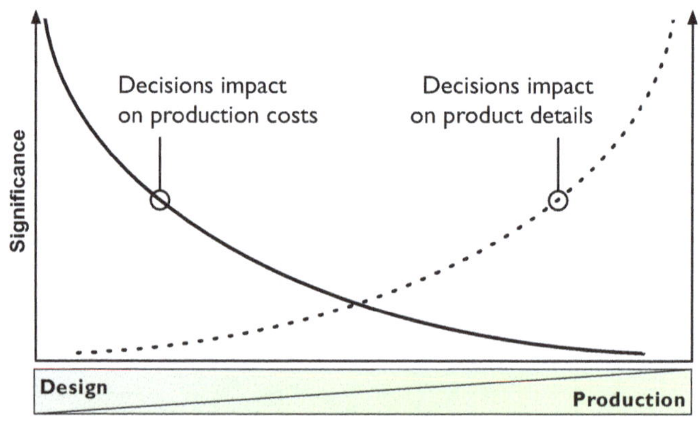

Fig. 4.9 Cost allocation associated with decisions in design

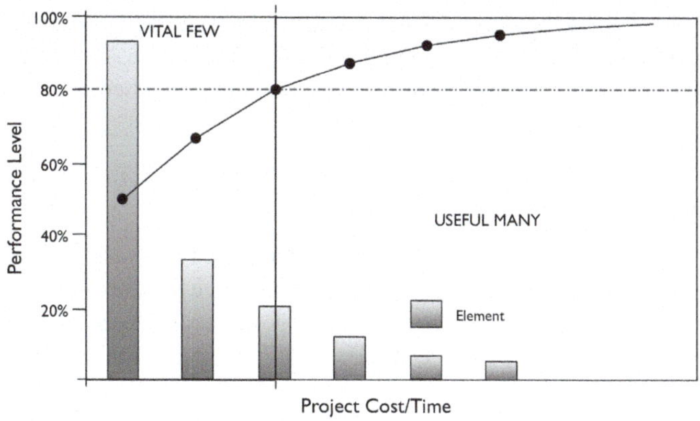

Fig. 4.10 80:20 rule

4.4.4 *The User*

The targeted end-user is an important influential element of the design process, and their relationship with the overall design process is therefore of critical importance to the success or otherwise of a design. Figure 4.11 suggests something of the nature of this relationship between the designer and the user. From this figure, it can be seen that the user has maximum impact on the design process in the early or conceptual stages with the designer then taking the lead as the product moves to production.

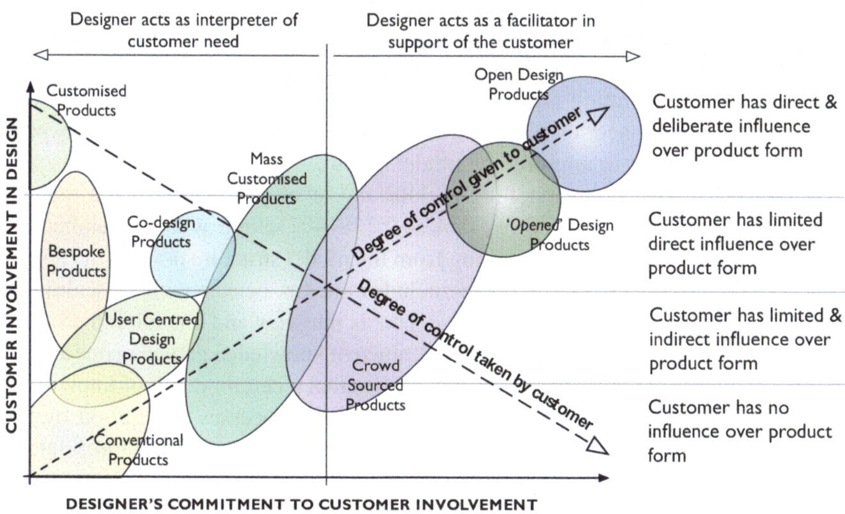

Fig. 4.11 Designer as interpreter or facilitator in relation to product type

An insight into the problems associated with this relationship between the user and designer may be seen in the comment in the 2022 joint report on Automated Vehicles [33] by the Law Commission & Scottish Law Commission where they commented that:

> *Systems are largely designed by young men who tend to overlook the diverse needs of those affected by their work.*

This form of associated bias has long been recognised and various attempts have been made over the years to overcome this, as for instance the use of the '*ageing suit*' by companies such as Nissan and Ford to allow designers to experience the use of a vehicle by older individuals and the studies at Volvo using all-female design teams [34, 35].

Further, as identified by Caroline Criado Perez in her book *Invisible Women* [36], many systems are designed using data derived largely from the male population, resulting in systems that are difficult for women to use.

4.5 Modelling

Modelling has a very important role to play in supporting the decision-making process by allowing designers to examine and evaluate different options as part of that process [37]. However, the design of the models themselves is not necessarily straightforward, and decisions made as to the structure of the model can significantly influence the outcomes, and hence the validity or otherwise of the associated conclusions. This is of increasing significance in relation to the increasing introduction of AI-based models and modelling where the availability of validated and robust training data sets is key to performance and acceptance, particularly where the AI-based systems can develop their own operating procedures and methods independently of their users. This is of particular concern when considering issues such as the ability of inexpert, or perhaps naïve, users to interpret the model output.

Factors associated with modelling include the level and nature of the type of model being used, as suggested by Table 4.2 [38–40], along with any embedded uncertainty in the model, either resulting from the model structure or from the nature of the available data. These uncertainties include issues associated with the scaling up of previous, successful, designs to move concepts into new and unexplored areas of applications, or areas where there is a transference of knowledge gained in one area to support developments in another. All this implies that a certain depth of knowledge and understanding of both the design and modelling processes is required by the modellers to ensure that the model outcomes are being interpreted appropriately. Issues for consideration then include:

- A feature of complex systems is emergent properties which cannot necessarily be predicted by an understanding of the simple components of the system. This

Table 4.2 Types of model

Model type	Description
Black box	A Black Box model is defined entirely by its inputs and outputs and associated transfer function with its inner workings hidden from and invisible to the user
Grey box	A Grey Box model uses a partially defined analytical structure known to the user with a data-driven element to determine the relationship between inputs and outputs
White box	In a White Box model operation is transparent to and known by the user

is a risk when implementing the outputs of such models, but it may also present unexpected opportunities.
- Complex models are no longer necessarily the result of significant research and design effort but are increasingly a starting point for finessing a design or product (or understanding).

This raises the question of whether complex outputs can reasonably be deployed by the naive user or are they too demanding and could only be developed by the original modellers?

- Black boxes are often central in the development of models but were definable with clearly specified inputs and expected outputs.

As models have become more complex it now looks as if the introduction of AI is adding a super black box in complex models—ones that can never be filled in by better understanding or more basic data.

- What is the value of an untestable model and how can successful or failed implementations from a model be used to develop the subsequent, better model?
- How are current models limited by the experience and knowledge of the modellers?
- Are models always going to be out of date, and if so, what is the cost or consequence of ageing models?
- Are AI-generated designs and outputs fundamentally trustworthy if the mechanics of AI decision-making are opaque and hidden from the user?
- How can measurements of fitness be provided for AI models, and do such models need to be compared with other models produced by separate AIs to obtain a consensus solution?
- Is there a way of using the potential of AI-driven '*inspiration*' to develop understanding?
- How to identify Black Swans?

The concept of **Black Swan events** as a metaphor for describing surprise outcomes was proposed by Nassim Nicholas Taleb [41–43] to explain:

- The role of high-profile, hard-to-predict and rare events that lie beyond the realm of normal expectations.

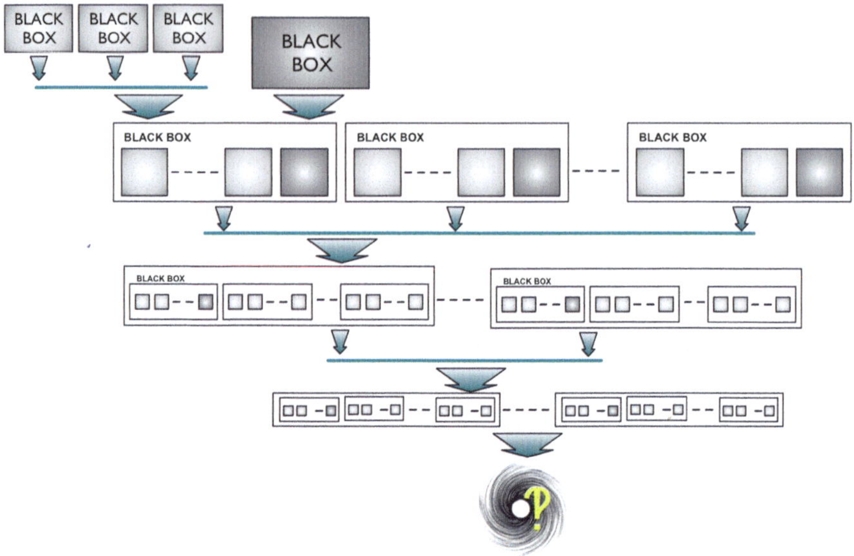

Fig. 4.12 Black boxes and chaos

- The non-computability of the probability of consequential rare events of small probability.
- The biases that blind people, both individually and collectively, to issues of uncertainty and the role of rare events.

Over time, models have developed from simple representations with a single black box (knowing the unknown) to more complex models with more black boxes, and then to AI models with potentially unknowable structures. The question then is as to whether this is a continuous trend or development to be associated with modelling, or is there something fundamentally different about the use of AI models with they themselves leading to the possibility of chaotic outcomes as suggested by Fig. 4.12?

Previously, models could do two things—make predictions—and allow the user to test the importance of specific relationships within the model. AI can only make predictions, and through these allow learning, but AI systems cannot be dissected and perhaps fundamental truths discovered by them can never be identified to or by people.

4.6 Artificial Intelligence

Frontier AI represents the forefront of artificial intelligence (AI) innovation and involves pushing the boundaries of existing AI technologies, exploring new methodologies and addressing complex challenges. At its core, Frontier AI signifies the dynamic pursuit of breakthroughs that redefine what is achievable, driving the evolution of intelligent systems and their practical applications across various domains. It encapsulates the relentless exploration and implementation of novel approaches, ensuring AI remains at the forefront of transformative progress.

In a limited illustration of the capabilities of current AI technology, the above introductory paragraph was produced by ChatGPT in response to the query '*What is Frontier AI in 100 words*?' and demonstrates the capability of AI software to respond to such unstructured requests, albeit in this case in a rather flowery, ambiguous manner.

In a wider context, developments in (AI) are beginning to suggest a potential for its use in support of various aspects of the design process ranging from the design of new materials to the potential shortening of the '*Design* → *Build* → *Test* → *Evaluate* → *Redesign*' process through the use of virtual and robotic laboratories [43–47].

The introduction of AI technologies also brings with it significant concerns, as were identified in the UK by the Science, Innovation and Technology Committee of the House of Commons [48] and set out here in Table 4.3. Similar issues were raised at the AI summit held at Bletchley Park in the UK in November 2023. Such challenges, issues and concerns have also been identified elsewhere, for instance in the United States and Europe [49–51].

4.7 Conclusions

As systems become increasingly layered and complex they put additional stress on the decision-making process associated with both the design and operational phases. This is particularly acute as the layered architecture means that designers at the upper layers must rely on what are essentially Black Box components defined by their inputs and outputs. Such components may have been in turn derived from specifications from and by designers operating at the upper levels of the system architecture, meaning that the vertical communication between the levels is crucial to effective decision-making.

This relationship between the system elements and components also extends to the way in which systems are modelled and the interpretation of those models. Here, the introduction of AI-based methods in attempts to enhance the modelling process with the aim of reducing the time absorbed by the '*Design* → *Build* → *Test* → *Evaluate* → *Redesign*' sequence also brings with it significant risks in terms of the ability of designers to interpret the results of such models.

The chapter has therefore attempted to set out and identify some of the key issues and questions associated with the establishment of an effective decision-making process in support of the design and implementation of complex systems.

Table 4.3 AI challenges (adapted from [48])

Challenge	Comments
Bias	AI can introduce or perpetuate biases that society finds unacceptable
Privacy	AI can allow individuals to be identified and personal information about them to be used in ways beyond what the public wants
Misrepresentation	AI can allow the generation of material that deliberately misrepresents someone's behaviour, opinions or character
Access to data	The most powerful AI needs very large data sets, which are held by a few organisations
Access to Compute	The development of powerful AI requires significant computing power, access to which is limited to a few organisations
Black box	Some AI models and tools cannot explain why they produce a particular result, which is a challenge to transparency requirements
Open source	Requiring code to be openly available may promote transparency and innovation; allowing it to be proprietary may concentrate on market power but allow more dependable regulation of harms
IPR & copyright	Some AI models and tools make use of other people's content: policy must establish the rights of the originators of this content, and these rights must be enforced
Liability	If AI models and tools are used by third parties to do harm, policy must establish where the developers and providers of the technology bear any liability for harm that is done
Employment	AI will disrupt the jobs that people do and that are available to be done. Policymakers must anticipate and manage the disruption
International coordination	The development of powerful AI requires significant computing power, access to which is limited to a few organisations
Existential	Some people think that AI is a major threat to human life: if that is a possibility, governance needs to provide protections for national security

Acknowledgements Sadly, Andrew Spiers died suddenly and unexpectedly just as the first drafts of this chapter were being completed and his co-authors would therefore like to dedicate the chapter to him. A good friend and colleague who will be greatly missed.

References

1. Corning PA (1998) Complexity is just a word! Technol Forecast Soc Chang 59(2):197–200
2. Mayfield M, Punzo G, Beasley R, Clarke G, Holt N, Jobbins S (2018) Challenges of complexity and resilience in complex engineering systems, ENCORE Network + White Paper
3. Morales GA, Nizamis K, Bonnema GM (2023) Engineering complexity beyond the surface: discerning the viewpoints, the drivers, and the challenges. Res Eng Des 34(3):367–400
4. Bradley D, Russell D, Hehenberger P, Azorin-Lopez J, Watt S, Milne C (2020) From mechatronics to the cloud, reinventing mechatronics, Xiu T Yan, Bradley D, Russell D & Moore P (eds), Springer, pp 17–33

5. Kopetz H, Steiner W (2022) Internet of things, real-time systems: design principles for distributed embedded applications, Springer, pp 325–341
6. Poutanen P, Soliman W, Ståhle P (2016) The complexity of innovation: an assessment and review of the complexity perspective. Eur J Innov Manag 19(2):189–213
7. Hassannezhad M, Cantamessa M, Montagna F, Clarkson PJ (2019) Managing sociotechnical complexity in engineering design projects. J Mech Des 141(8):081101
8. Grogan PT (2021) Perception of complexity in engineering design. Syst Eng 24(4):221–233
9. Vasconcelos DR, Andrade RMC, Severino V, Souza JD (2019) Cloud, fog, or mist in IoT? That is the question. ACM Trans Internet Technol 19(2):1–20
10. Javed W, Parveen G, Aabid, e Rubab SU, Ikram S, Rehman KUU, Danish M (2021) A review on fog computing for the internet of things, IEEE 2021 International conference on innovative computing, pp 1–7
11. Ketu S, Mishra PK (2022) Cloud, fog and mist computing in IoT: an indication of emerging opportunities. IETE Tech Rev 39(3):713–724
12. Yaacoub JPA, Salman O, Noura HN, Kaaniche N, Chehab A, Malli M (2020) Cyber-physical systems security: Limitations, issues and future trends. Microprocess Microsyst 77:103201
13. Marwedel P (2021) Embedded system design: embedded systems foundations of cyber-physical systems, and the internet of things, Springer Nature
14. Tekaat JL, Anacker H, Dumitrescu R (2021) The paradigm of design thinking and systems engineering in the design of cyber-physical systems: a systematic literature review, IEEE International symposium on systems engineering, pp 1–8
15. Moustafa N (2021) A systemic IoT–fog–cloud architecture for big-data analytics and cyber security systems: a review of fog computing, Secure Edge Comput, 41–50
16. Sandhu AK (2021) Big data with cloud computing: discussions and challenges. Big Data Mining Anal 5(1):32–40
17. Pahl G, Beitz W (2013) Engineering design: a systematic approach, Springer Science & Business Media
18. Buede DM, Miller WD (2016) The engineering design of systems: models and methods. John Wiley & Sons
19. Cross N (2021) Engineering design methods: strategies for product design. John Wiley & Sons
20. www.darpa.mil/work-with-us/heilmeier-catechism. Accessed December 2023
21. Mars Climate Orbiter Mishap Investigation Board Phase I Report (1999) @ www.ftp.hq.nasa.gov/pub/pao/reports/1999/MCOreport.pdf. Accessed November 2023
22. Simpson E, Bradley D, Hehenberger P (2018) Innovation and failure in mechatronics design education, 16th Mechatronics forum international conference: reinventing mechatronics, University of Strathclyde Publishing, pp 17–24
23. Asimow M (1962) Introduction to design. Prentice Hall, Englewood Cliffs
24. Markus TA (1972) Building performance. Applied Science Publishers, London
25. Howard RA (1988) Decision analysis; practices and promises. Manage Sci 34(6):679–695
26. McCall R, Burge J (2016) Untangling wicked problems. AI EDAM 30(2):200–210
27. Rittel HW, Webber MM (1973) Dilemmas in a general theory of planning. Policy Sci 4(2):155–169
28. Marsh JG, Simon HA (1958) Organisations. Wiley, New York
29. Lindblom CE (1959) The science of muddling through. Public Adm Rev 19(2):79–88
30. Simon HA (1976) Administrative behaviour. Free Press, New York
31. Astley WG, Axelson R, Butler RJ, Hickson DM, Wilson DC (1992) Complexity and cleavage; dual explanations of strategic decision making. J Manage Stud 19(4):357–375
32. Snowden D (2000) Organic knowledge management: Part I The ASHEN model: an enabler of action. Knowl Manag 3(7):14–17
33. Law Commission & Scottish Law Commission (2022) Automated vehicles: joint report https://www.lawcom.gov.uk/project/automated-vehicles/. Accessed November 2023
34. Groza HL, Sebesi SB, Mandru DS (2017) Age simulation suits for training, research and development, International conference advancements of medicine and health care through technology (MEDITECH 2016), Springer International Publishing, pp 77–80

35. Backman M, Börjesson S (2006) Vehicles for attention creation: the case of a concept car at Volvo Cars. Eur J Innov Manag 9(2):149–160
36. Perez CC (2019) Invisible women: data bias in a world designed for men, Abrams
37. Buede DM, Miller WD (2024) The engineering design of systems: models & methods. John Wiley & Sons
38. Guidotti R, Monreale A, Ruggieri S, Turini F, Giannotti F, Pedreschi D (2018) A survey of methods for explaining black box models. ACM Comput Surv 51(5):1–42
39. Reynders G, Diriken J, Saelens D (2014) Quality of grey-box models and identified parameters as function of the accuracy of input and observation signals. Energy Build 82:263–274
40. Loyola-Gonzalez O (2019) Black-box vs. white-box: understanding their advantages and weaknesses from a practical point of view. IEEE Access 7:154096–154113
41. Taleb NN (2010) The black swan: the impact of the highly improbable, 2nd edn. Random House
42. Ale BJ, Hartford DN, Slater DH (2020) Dragons, black swans and decisions. Environ Res 183:109127
43. Yüksel N, Börklü HR, Sezer HK, Canyurt OE (2023) Review of artificial intelligence applications in engineering design perspective. Eng Appl Artif Intell 118:105697
44. Wang H, Fu T, Du Y, Gao W, Huang K, Liu Z, Chandak P, Liu S, Van Katwyk P, Deac A, Anandkumar A (2023) Scientific discovery in the age of artificial intelligence. Nature 620(7972):47–60
45. Xu Y, Liu X, Cao X, Huang C, Liu E, Qian S, Liu X, Wu Y, Dong F, Qiu CW, Qiu J (2021) Artificial intelligence: a powerful paradigm for scientific research. Innov 2(4):21
46. Cath C (2018) Governing artificial intelligence: ethical, legal and technical opportunities and challenges. Philos Trans R Soc Math Phys Eng Sci 376(2133):20180080
47. Hagendorff T (2020) The ethics of AI ethics: an evaluation of guidelines. Mind Mach 30(1):99–120
48. House of Commons Science, Innovation and Technology Committee (2023) The governance of artificial intelligence: interim report, UK Gov www.committees.parliament.uk/publications/41130/documents/205611/default/. Accessed 29 November 2023
49. Government Office for Science (2023) Future risks of frontier AI (Annex A), UK Gov www.assets.publishing.service.gov.uk/media/653bc393d10f3500139a6ac5/future-risks-of-frontier-ai-annex-a.pdf. Accessed 29 November 2023
50. The White House (2022) Blueprint for an AI Bill of Rights: Making automated sytems work for the American People, US Gov @ https://www.whitehouse.gov/wp-content/uploads/2022/10/Blueprint-for-an-AI-Bill-of-Rights.pdf. Accessed November 2023
51. European Parliament (2023) Artificial intelligence act, European Parliamentary Research Service @ www.europarl.europa.eu/RegData/etudes/BRIE/2021/698792/EPRSBRI(2021)698792EN.pdf. Accessed November 2023

Chapter 5
Mechatronics, the Eternal Cyber Security System Lifecycle and the IoT, and the IIoT .... Oh and OT!

Natalie Coull, Ian Ferguson, and Lynne Coventry

Abstract Critical cybersecurity challenges associated with Mechatronic systems, the Internet of Things (IoT), the Industrial Internet of Things (IIoT), and Operational Technology (OT) are examined. The vulnerabilities in interconnected systems, as demonstrated by attacks such as Stuxnet, are shown to have potentially severe consequences. The expansion of IoT has been identified as a factor that significantly increases the attack surface, thereby raising both the likelihood and impact of cyberattacks. A cybersecurity lifecycle model is proposed, which emphasises continuous improvement through iterative phases of system building, attacking, and investigating. Secure software engineering practices, threat modelling for IoT, the integration of blockchain technology, and forensic techniques for IoT devices are discussed as potential solutions. The paper concludes that the inherent vulnerabilities in these systems necessitate a proactive focus on secure design and collaboration across disciplines to reduce cyber risks.

5.1 What's the Problem?

Sometime in 2010, someone (no one knows who for sure) threw a USB stick over the car-park fence of the Nantanz nuclear enrichment facility in Iran. Sometime later, someone else found it, picked it up, took it into the plant and plugged it into their computer—probably just to see what was on it. The malware that had been deliberately placed on the stick prior to throwing it over the fence was then activated and discreetly explored the plant's internal network. Moving from machine to machine it was searching for a very specific system: the PLCs controlling gas centrifuges used to enrich uranium hexafluoride into weapons-grade nuclear fissile material.

Upon finding the controller, the malware raised the speed limit of the centrifuges allowing them to spin slightly faster than usual—a change sufficiently stealthy that

N. Coull · I. Ferguson (✉) · L. Coventry
cyberQuarter, Department of Cybersecurity and Computing, Abertay University, Dundee, UK
e-mail: ian.ferguson@abertay.ac.uk

P. Hehenberger and D. Bradley (eds.), *Mechatronic Futures*,
https://doi.org/10.1007/978-3-031-83571-1_5

no-one would notice, but enough that the resulting spin speed increased wear-and-tear on the highly stressed mechanisms, meaning they began to fail in service more frequently. The incident was named the Stuxnet attack by those who investigated it and the resulting delay to the Iranian nuclear weapons program has been estimated to be of the order of six months.

There are several important lessons here. Cybersecurity is important in both a Mechatronics and a wider systems context. Why? Because:

- Computers control critical equipment, not just information systems.
- Everything is connected—whether the centrifuge controller should have been on the same network as the plant email system is debatable, but it was.
- There was no defence against malware at the controller level—because no-one had ever thought it necessary.
- There is always someone who wants to attack your systems.

In this chapter we shall look at some of the cybersecurity challenges facing Mechatronic systems and some of the potential solutions—both of which turn out to have human aspects as well as technological ones.

5.1.1 IIoT, IoT OT—Setting the Scope for the Chapter

Recognising that Mechatronic systems have a long history and have been referred to by many names over the years, for the purposes of this chapter we will treat the terms Mechatronics, IIoT (Industrial Internet of Things), IoT (Internet of Things), OT (Operational Technology), embedded systems, ubiquitous computers, disappearing computers, sensor nets and smart-cities as synonyms on the basis that they all refer to the class of systems dependant on small computers, often controlling sensors and actuators, connected mainly, but not exclusively, by the Internet. We will hence use the term 'IoT' as a shorthand to refer to all of these.

5.1.2 The Scale of the Problem

What is the real risk posed to the IoT by cyber-attacks? Stuxnet was an example of states attacking each other to slow nuclear proliferation but how likely is *your* system to be attacked? Risk has two parts: the likelihood of a bad event and the severity of the consequences of that event. The number of connected IoT devices is notoriously hard to estimate but figures from networking giant Cisco [1] estimate the number to be between 20 and 30 billion devices in 2023. This provides a massive '*attack surface*' which needs safeguarding. In a recent experiment, a fake IoT device connected to the Internet (a '*honeypot*') was attacked within 12 h of its being connected and received around 60 known attacks in two weeks of operation. The potential impact of IoT vulnerabilities has been demonstrated by the Mirai and Persia botnet attacks [2].

Due to poor security, miscreants were able to use IoT devices to instigate distributed denial of service (DDoS) attacks on Dyn (a DNS service provider), affecting GitHub, Twitter, Reddit, Netflix and Airbnb, rendering these services unavailable. Whilst the impact of these services being taken offline may have just been an inconvenience to their users, the cost implications of investigating and recovering from the attack were substantial.

5.1.2.1 Consequences

Whilst the days of pacemakers, baby monitors and robot surgical machines with unprotected interfaces exposed on the public Internet [3] are hopefully behind us, it is still relatively simple to use search engines such as Shodan to find examples of industrial plant and control systems which are exposed without access controls. The *Stuxnet* attack demonstrated the ability to damage a physical system via malevolent software. It can surely only be a lack of awareness and prioritisation of security on the part of system creators and operators that continues to allow such open access. Further, as shown with more recent botnet attacks, IoT networks can be used to attack systems with no actual device compromise required.

The likelihood and potential severity of consequences of a cyberattack on an IoT system mean that cyber-issues in IoT technology pose a real risk. Whilst errors such as leaving default credentials unaltered on a production system can be regarded as '*operator error*' and not the responsibility of the system developer we suggest that such a distinction is not useful (*someone* should have ensured that did not happen!). Further, we have formed a cyclical view of systems which is an evolution of the '*prototyping*' view of the software development lifecycle. In the model, security is seen as a process of constant improvement based upon prior experience and ever-evolving threat.

5.1.3 Secure Software Engineering—As a Concept

Before examining the cybersecurity problems specific to IoT systems, we must first look at systems development and software development in general with respect to cybersecurity.

5.1.3.1 The Problem

For many years, security simply was not a concern for software engineers who prioritised functionality and agility. Such an attitude may have its roots in the history of computing: When computers were mainframes and the property of large organisations, no one had one to attack from. The Internet was thus designed largely without security in mind. All subsequent cybersecurity is an attempt to put a patch over this.

Even in more modern times, coding (and designing) securely are not features of a typical developer's work or education—indeed a survey of current computer science offerings in the UK will still show that cybersecurity usually forms at most one module in the typical undergraduate course.

Given such a history it is perhaps easy to appreciate why legacy systems may have deep and inherent security vulnerabilities. However, we are still not universally adopting security as a first-class concern during software development. Partly this is because cybersecurity is hard (further complicated by the number of legacy languages still in use), partly because developers are human (and make mistakes) and partly due to the business-focus on the fast release of new functionality and patching it later.

The consequence is that we are left with software that has been built to work but where being intrinsically secure is not seen as being part of '*working*' but rather as an optional, '*bolt-on*' extra. An example here is the shipping of software with default credentials enabled which the operators of the system are then supposed to change to something more secure. Is there not a better solution? Could the developers not have come up with a solution to provide initial access to the system to allow configuration. Probably yes, but that would have taken longer, been more complex and thus more expensive. Could the system not be secure by design?

5.1.3.2 The Response

It is axiomatic within the software engineering discipline that the earlier in the life-cycle of a system that a problem can be identified, the easier and cheaper it is to fix. For this reason, considerable research effort is being put into creating secure systems and secure system development processes. Techniques termed Secure Software Engineering Practices (SSEPs) have emerged to support the concept of Secure by Design. Some (by no means exhaustive) examples of these practices are discussed below.

5.1.3.3 Continuous Integration

Secure Software Engineering is characterised by using a modern development process: Continuous Integration (CI) involves automatically integrating code changes from multiple contributors into a shared repository several times a day. Each integration is verified by an automated build and tests to detect integration errors as quickly as possible. This practice is supported by various software tools which manage changes to the codebase and keep track of known issues, respectively. These tools facilitate the rapid identification and resolution of conflicts and defects, ensuring that each integration maintains the software's quality and security. CI encourages developers to commit changes frequently, reducing complexity and allowing teams to address problems promptly.

5.1.3.4 Code Review

Code review is a critical quality assurance technique in software development where developers systematically examine each other's code to find mistakes, improving both the software's functionality and security. It often involves the use of guidelines and checklists to ensure thoroughness and consistency in the review process. For instance, industry standards like the *OWASP Code Review Guide* [4] offer structured directives for identifying security vulnerabilities, while checklists such as the *IEEE Standard for Software Reviews and Audits* [5] or the *SEI Checklist for Inspecting C + + Code* [6] provide broader criteria for assessing code quality and maintainability. Code review is frequently a team-based activity, fostering a collaborative environment where multiple developers can share insights, propose improvements, and collectively enhance the codebase. This collaborative aspect not only leverages diverse expertise to spot and resolve issues more effectively but also aids in maintaining a cohesive coding standard across the team.

5.1.3.5 Static Analysis Tools

Static analysis is a method used in software engineering to examine the code without executing it, providing an efficient way to detect errors, enforce coding standards, and identify security vulnerabilities early in the development cycle. In the context of secure software engineering, static analysis is particularly critical as it helps ensure that the software does not contain security flaws that could be exploited once deployed. Control-flow analysis, as an example of static analysis, specifically examines the paths along which a program executes, helping to uncover logical errors and insecure coding practices.

5.1.3.6 Dynamic Analysis Tools

Dynamic analysis is a testing and evaluation method that involves analysing a program by executing it with various inputs and observing its behaviour in real-time. Dynamic analysis allows developers to see how their software performs under different conditions, which can reveal issues that static analysis might miss, such as runtime errors, memory leaks, and synchronisation issues. Dynamic analysis is particularly effective at identifying security problems because it can simulate attacks or unusual conditions to see how the software behaves. It can uncover vulnerabilities like buffer overflows, SQL injections, and other security flaws that only become evident when the program is running.

5.1.3.7 Use of Cybersecurity Tools by Developers

The cybersecurity and penetration-testing community provides extensive knowledge on system vulnerabilities, catalogued in the NVD-CVE (National Vulnerability Database—Common Vulnerability Enumeration) and NVD-CWE (National Vulnerability Database—Common Weakness Enumeration) databases [7]. CVE lists publicly disclosed cybersecurity vulnerabilities, including an ID, description, and reference for each. CWE categorises common software security weaknesses and programming errors that lead to vulnerabilities. While CVE applies to specific software versions, CWE offers broader categorization. Although it is unclear how widely *developers* currently use this knowledge, it could serve as a valuable resource for learning from past mistakes.

5.1.3.8 (More)? Secure Languages—Rust Versus 'C'

Many commonly used programming languages, such as C, originated decades ago when security was not a primary concern. While it is often impractical to rewrite an organisation's entire codebase in a modern language, choosing a more secure language like Rust for new projects can significantly reduce the risk of certain security vulnerabilities.

C is powerful but inherently risky due to features like manual memory management, lack of bounds checking, and pointer arithmetic, all of which can lead to serious security flaws if mishandled. Rust, in contrast, is designed with safety in mind, using a unique ownership system and strict compile-time checks to prevent common vulnerabilities like buffer overflows and data races, making it a more secure choice for systems-level programming.

5.1.3.9 Coding Standards

Coding standards play a crucial role in ensuring safe and secure code by providing a systematic framework to avoid common security vulnerabilities and promoting consistent, maintainable code. They help prevent errors by prohibiting unsafe practices, such as using vulnerable functions or improper memory management, and enhance readability, making it easier for developers to identify potential issues during code reviews. Additionally, coding standards support more effective code reviews by offering a clear checklist for verifying secure practices and can be enforced using automated tools for early detection of security flaws.

These standards often incorporate industry best practices for security, guiding developers on key aspects like encryption, input validation, and session management, which are critical for building secure software. They also play a significant role in meeting legal and compliance requirements in regulated industries, helping organisations adhere to necessary security specifications. Some key coding standards include the *CERT Coding Standards* [8], *OWASP Secure Coding Practices* [9], *MISRA guidelines* [10], *ISO/IEC* 27,001 *recommendations* [11], and *Microsoft's Security Development Lifecycle (SDL)* [12], all of which provide valuable frameworks for developing secure and reliable software.

5.1.3.10 Formal Approaches

Formal approaches to software development, such as those used in the F* (F star) programming language [13], offer significant cybersecurity advantages by applying rigorous mathematical methods to ensure software security, safety, and correctness. F* combines functional programming with a powerful type system,[1] allowing developers to express detailed invariants that the compiler can automatically check for correctness. This approach increases reliability and security by enabling the verification of critical properties in code, such as preventing common vulnerabilities like buffer overflows and unauthorised data leaks. Additionally, F* supports automated verification against specifications, reducing the reliance on manual reviews and ensuring that software adheres strictly to security policies.

Despite the benefits, formal methods like F* come with challenges, including a steep learning curve, increased development time, and limited library support due to their relative immaturity. However, these methods provide a high level of assurance that software behaves correctly, making them especially valuable for applications where security is critical, such as in finance, healthcare, and critical infrastructure. By promoting modularity and enabling formal proofs of security properties, formal methods help developers build more secure, reliable, and maintainable software systems, even in the face of evolving cybersecurity threats.

5.1.4 Secure Software Engineering—As Applied to IoT

Despite the wide range of tools, techniques, and approaches discussed in the previous section from the realm of generic software engineering, IoT systems remain uniquely vulnerable. The vulnerabilities stem not only from technological limitations but also from human and organisational challenges that are distinctively pronounced in the IoT context.

[1] A type system refers to a logical set of rules that assign a property called a type, as for instance; integer, floating point or string, to every term such as a word, phrase or set of symbols.

5.1.4.1 Technological Challenges: The Limitations of Cryptography in Low-Capacity Devices

A fundamental aspect of securing any system involves restricting access to authorised users and systems only. In many traditional setups, this security is achieved through cryptographic techniques that safeguard the transmission and storage of information. In the context of IoT, the main challenge lies in the deployment of these cryptographic methods on devices with limited computational power. These devices, often end-nodes in sensor networks or IoT systems, struggle with the demands of complex encryption algorithms. The argument that '*encryption is too challenging to implement on low-capacity devices*' still holds merit. For instance, implementing RSA encryption with robust key lengths (such as 4096 bits) remains impractical on simple microcontrollers [13]. However, advancements in technology, propelled by Moore's law, predict that increasingly capable low-power devices will eventually be able to support sophisticated cryptographic operations, potentially overcoming current limitations.

5.1.4.2 Human and Organisational Problems: The Overlooked Priority of Security

In the realm of the IoT, there seems to be a persistent under-prioritisation of security, overshadowed by the emphasis on functionality. This issue is exacerbated by a legacy of operational technology (OT) professionals and environments that have not traditionally integrated cybersecurity into their frameworks. Moreover, there is a pronounced communication gap between various stakeholders in IoT projects—ranging from Mechatronic engineers to software developers and penetration testers—who often do not share the same technical language or security-centric perspectives. One suggestion to improve IoT cybersecurity is to more tightly integrate the '*clans*' that appear in Fig. 5.1. This disconnection is critical because many IoT professionals come from an electronic engineering background where software engineering is equated merely with programming. This perspective often limits their view to immediate software functionality rather than encompassing broader cybersecurity measures. Their approach generally assumes that without traditional computing layers, such as file systems, the systems are inherently secure against hacking, an assumption that overlooks the complexities of network-level vulnerabilities and software exploits.

Addressing these challenges requires a paradigm shift to elevate the priority of security in the design and development of IoT systems, coupled with efforts to enhance interdisciplinary communication and education on secure coding practices.

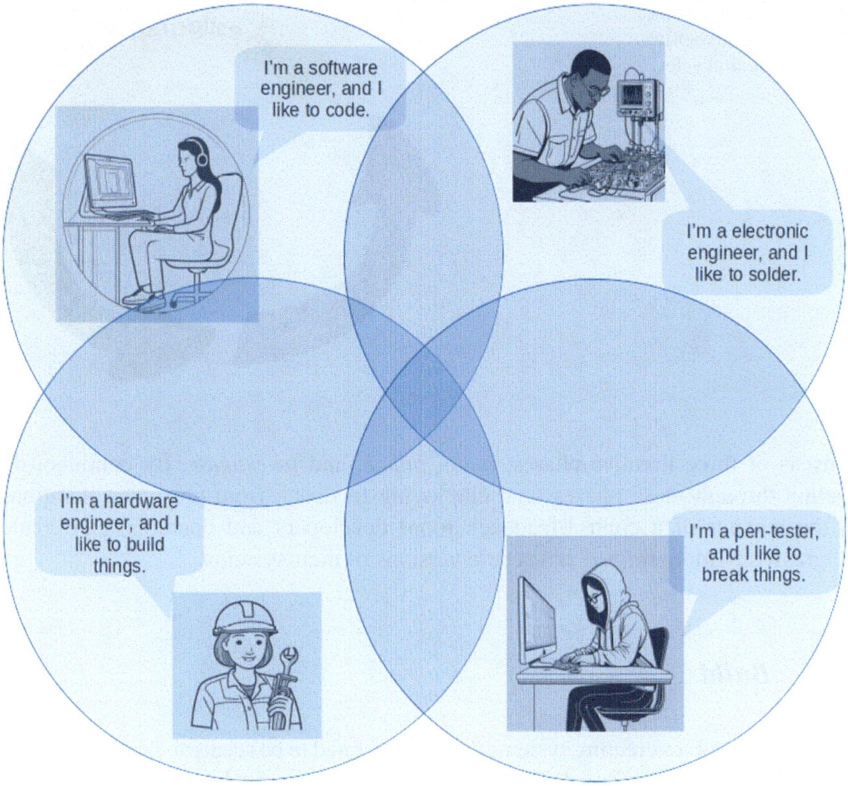

Fig. 5.1 The battle of the clans

5.2 The Polo-Mint Model of the Cybersecurity Lifecycle

The authors' approach to cybersecurity is rooted in the perspective of penetration testing and ethical hacking.[2] We operate under the premise that to ensure the security of a system, it must be rigorously tested through simulated attacks. Cybersecurity is a constant arms race; attackers are perpetually equipped with the initiative, time and resources to find and exploit new vulnerabilities. This dynamic results in a continuous cycle of exploitation and patching—a process initially perceived as a superficial fix but which we recognise as a vital part of a much deeper evolutionary cycle in both technology and developer expertise.

Recognising the limitations of one-time testing, we propose a (literally!) more rounded model: the '*Polo-mint Model*' of cybersecurity. This model envisions the lifecycle of technology systems—both software and hardware—as a never-ending loop of improvement, symbolised by the Polo-mint diagram (Fig. 5.2). The model

[2] Abertay University launched the world's first undergraduate degree programme in ethical hacking in 2006.

Fig. 5.2 The polo-mint model of a secure software development lifecycle

consists of three iterative phases: *build*, *attack*, and *investigate*. By continuously cycling through these phases (and employing feedback from preceding iterations in the manner of a control-feedback loop) developers and operators can create progressively more refined and secure versions of their systems.

5.2.1 Build

This phase involves creating systems that are designed to be secure using Secure Software Engineering Practices. By employing defensive security knowledge of previous cybersecurity flaws as well as developers' own expertise, systems can be created and redesigned to build on lessons learned.

5.2.2 Attack

During the *attack* phase, systems are rigorously tested to identify vulnerabilities. Penetration testers, operating under controlled and consensual conditions, simulate attacks to uncover weak points. Upon discovering vulnerabilities, these are documented in CVE entries, detailing the nature of the flaw and the successful attack vectors. This documentation is crucial for broader awareness but is typically withheld from public release until remedial patches are deployed, thus safeguarding the information from potential exploitation by malicious parties. On the other hand, when systems face actual unauthorised attacks, the insights gained will emerge through analysis during the later *investigate* phase.

5.2.3 *Investigate*

Following a real (non-test) security incident, systems and underlying processes are examined to understand their root causes and antecedent vulnerabilities. Countermeasures are identified, which may include recommending software updates, network changes, or updating security policies. Knowledge of this is fed forward into the *build* phase of the next generation of more secure systems by capturing the understanding of the attack as CVE entries and antipatterns.

5.3 Application of the Polo-Mint Model to IoT Systems

Whilst the Polo Mint model is typically applied through the lens of standard computer systems, networks and organisations, we now examine several potential solutions for the consideration of these phases within IoT. Our attempts to contribute to solutions for these problems amounts to trying to give developers of IoT systems three capabilities:

- Knowing how you are vulnerable and how to minimise impact of vulnerabilities—(information derived from *attack* and *investigate*—used in *build*).
- Being able to defend yourself (Having the right hardware/software/techniques to do it).
- Being able to work out what happened (and learn from it) when things go wrong (*investigate*).

Some of the example projects and techniques (from the authors and their research students) outlined in subsequent sections address problems clearly situated in one phase (e.g., forensic investigation in *investigate*) whilst others encompass activities which must take place in multiple phases (e.g., tested in *attack* feeding forward to *build*).

5.3.1 *Threat Modelling Methodology for IIoT*

The Polo-mint model is necessarily reactive in nature. In order to narrow the gap between a built system and its attackers either the loop should be traversed rapidly and often, or possible attacks should be anticipated and proactively defended against. This latter approach is the concern of threat modelling. Threat modelling is a creation-phase technique that addresses the problem of identifying and assessing potential security weaknesses in a system before they can be exploited by attackers. It provides a systematic approach to anticipate, prioritise, and mitigate security risks in the design and development phases, enhancing the overall security and resilience of the system. Figure 5.3 below [14] captures some of the possible attack vectors for IoT systems.

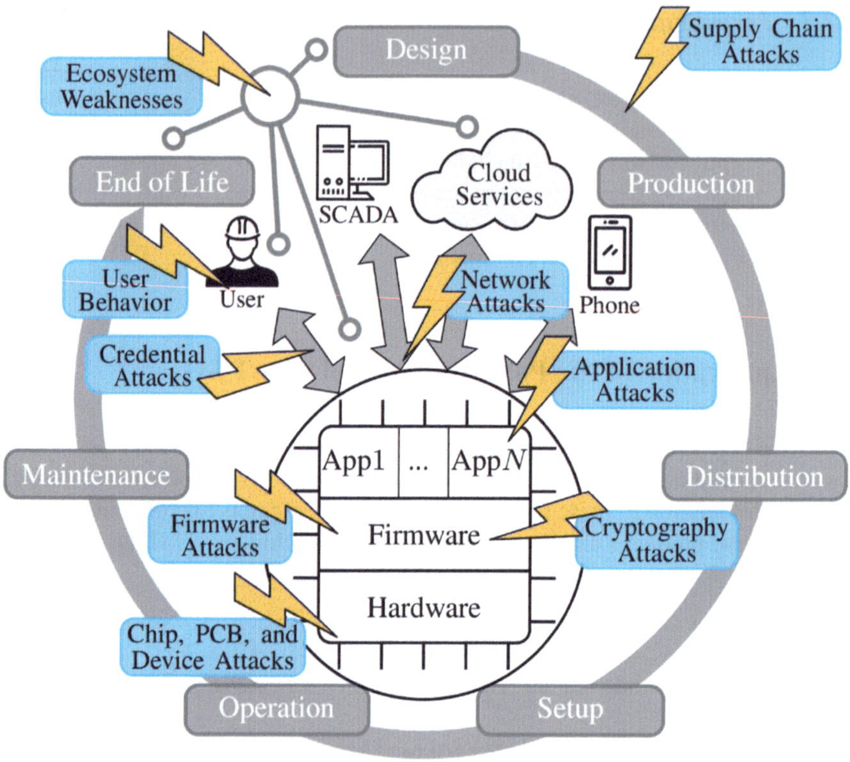

Fig. 5.3 Attacks on IoT systems

Threat modelling is particularly appropriate and beneficial when building Internet of Things (IoT) systems for several key reasons: IoT ecosystems consist of a wide range of devices, often with differing capabilities, operating systems, and levels of security. Vulnerability modelling helps in understanding and managing the security risks associated with such diverse components. IoT systems typically feature numerous interconnected devices, creating a large and complex attack surface. Each node and communication link can potentially serve as an entry point for attackers. Vulnerability modelling allows developers to identify and secure these potential entry points systematically. Many IoT devices operate with limited computational power and memory, which can restrict the use of robust security measures such as complex encryption or continuous security updates. Vulnerability modelling can help identify the most critical areas that need protection, optimising the allocation of scarce security resources. IoT devices are often deployed for long durations without regular maintenance or updates, making them vulnerable to evolving threats. Through vulnerability modelling, potential long-term security issues can be anticipated and mitigated during the design phase, reducing the need for frequent updates. With increasing regulatory focus on the security of IoT devices, vulnerability modelling provides a framework to ensure that systems comply with relevant standards and laws from the outset, avoiding legal and financial repercussions. IoT systems often integrate newer technologies with existing legacy systems. Vulnerability modelling helps to assess how new interactions between these systems could introduce security weaknesses.

Existing threat modelling techniques are not explicitly focussed on IoT and concentrate on the software of a system rather than having a '*holistic*' approach and considering security problems with their origins in the hardware design and implementation [15]. One response to this is TTM—Thing Threat Modelling—a modelling technique that employs enhanced data-flow diagrams (cyber-physical DFDs—CPDFD) effectively capturing critical security information concerning potential weak points within a system's architecture. This technique is versatile, offering both a big-picture overview and the capability to zoom in for a fine-grained analysis. It focuses particularly on the interfaces between components—whether they be hardware-to-hardware, hardware-to-software, or software-to-software—where data is most vulnerable, both during transmission and storage. This approach extends down to the device and chip level, ensuring comprehensive coverage from high-level operations to the details of hardware implementation. It also integrates insights gained from detailed investigations of previous security incidents or compromises, incorporating these lessons directly into the modelling process.

Information captured through this method is visualised in diagrams within an online tool, facilitating easy access and manipulation. These diagrams (an example is shown in Fig. 5.4) serve as a dynamic repository that not only documents current system configurations but also allows for direct comparison with other designs or future iterations. This capability enhances ongoing security assessments and aids in the continuous improvement of system architecture, making it an invaluable resource for developing robust, secure systems.

By employing vulnerability modelling, developers of IoT systems can better understand potential security flaws and implement effective mitigation strategies during the design and development stages, rather than reacting to breaches after deployment. This proactive approach not only enhances the security of IoT systems but also improves their reliability and trustworthiness, which are crucial for their widespread adoption and success.

5.3.2 Vulnerability Anti-Patterns

One of the characteristics of a technological discipline that makes it an engineering discipline is the ability of its practitioners to learn not only from their own mistakes but also from the mistakes of others. Cybersecurity anti-patterns address the problem of recurring security flaws that arise from common mistakes in software design and implementation. In essence, they are a way of describing how not to do something. Given that IoT environments often involve a diverse array of devices with varying levels of security capabilities, recognising these anti-patterns helps developers not

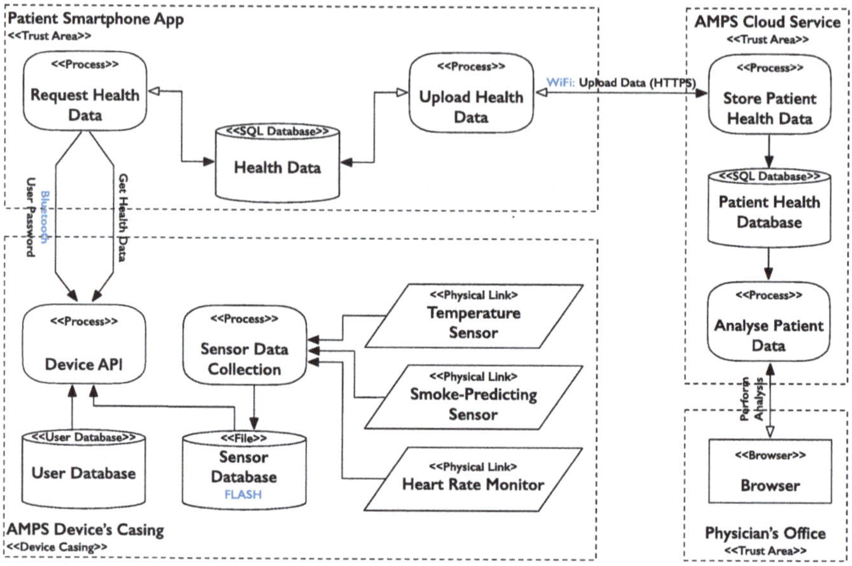

Fig. 5.4 An example TTM diagram showing the modelling of hard and software system components and their communication

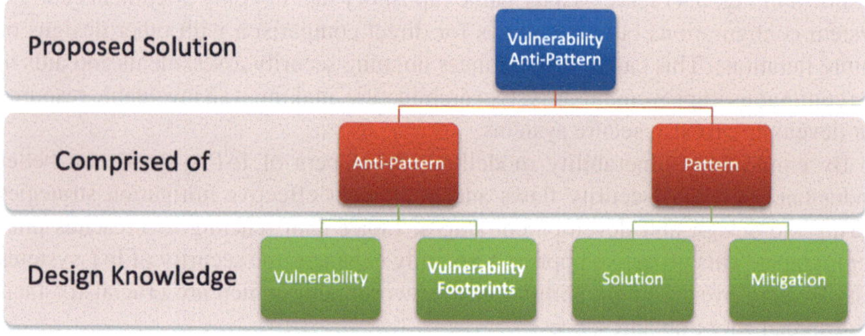

Fig. 5.5 Vulnerability anti-pattern conceptual model

to repeat common mistakes as they refine and update their systems in the context of the polo-mint model.

Vulnerability Anti-Patterns (VAPs) [16] are a technique developed to collect the information emerging from the Investigation phase and make it available in a manner easily understood by developers in the *build* phase. The approach is two-pronged: (a) A pattern language has been defined which allows the semi-formal documentation of an identified, exploited vulnerabilities by creating an abstract representation of what went wrong and (b) a means of collecting and cataloguing the patterns (via a potentially publicly available repository) has been developed Fig. 5.5.

5.3.2.1 Vulnerability Anti-Patterns

A Vulnerability Anti-Pattern (VAP) is composed of:

Anti-pattern: ineffective security practices or poor solutions that aid the process of identifying and communicating unsuccessful design intents.

Pattern: introduces refactored solutions, which suggest safe alternative procedures. Against each anti-pattern there is a set of solutions in the form of patterns. The pattern aids developers to solve the identifiable anti-pattern. This work has important implications for developing secure training methods and providing an effective understanding about vulnerabilities.

5.3.2.2 An example VAP—'Use of Potentially Dangerous Function'

In this VAP, the **Anti-Pattern** describes the use of dangerous functions during the development phase that are likely to lead to exploitable behaviour while safer alternatives are available. An instance of this anti-pattern is the use of *C's* `strcpy()` function, which provides no inherent safeguards against incorrect source or target buffer sizes, frequently resulting in faults such as buffer-overflow vulnerabilities. Vulnerabilities resulting from improper use of `strcpy()` are so common—as evidenced in the CVE database—that some standards prohibit its use entirely [17].

In this case a **Pattern** would present a solution to the '*Use of Potentially Dangerous Function*' anti-pattern that proposes a corresponding solution and mitigation. For example, the use of the safe function `strcpy_s()` using bounds-checking.

5.3.3 Blockchain and the IoT

We turn now to an example from the *build* phase of work aimed at providing a technique to help defend systems—the use of blockchain technology to support IoT systems. Blockchain technology has been proposed as a potential solution to some of the problems facing IoT systems such as security, scalability, and interoperability due their matching decentralised architectures which offer several properties desirable in distributed systems operating over numerous and heterogeneous networks such as decentralisation, scalability, anonymity secure networking over untrusted parties.

The architectures of IoT systems built to take advantage of the blockchain are based upon off-loading the high-intensity operations from end-nodes to local proxies which then handle onward secure communication although some consensus algorithms upon which blockchain algorithms depend have their own issues with computational intensity.

Security and Trust: IoT systems are potentially vulnerable to security threats due to their large '*attack surfaces*'—i.e. the number of connected devices and their varied security protocols. Blockchain technology enhances security through its decentralised nature, which removes single points of failure and the immutability properties of the blockchain ledger makes it exceedingly difficult for attackers to manipulate data across the network. The use of cryptography in blockchain also ensures that data transactions are secure and tamper-proof, building trust in device interactions.

Data Integrity: With IoT devices continuously generating large volumes of data, maintaining the integrity of this data is crucial. Blockchain's immutable ledger, ensures that once data is recorded, it cannot be altered. This is particularly important for critical applications in fields like healthcare, supply chain management, and financial services where data integrity is paramount.

Decentralisation: Traditional IoT systems often rely on centralised models for data processing and storage, which can lead to bottlenecks and points of vulnerability. Blockchain decentralises data management, allowing data to be processed and stored across multiple nodes in the network. This not only enhances security but also improves fault tolerance and reduces the impact of a single node's failure on the entire network.

Automated Transactions and Smart Contracts: Blockchain enables the use of smart contracts, which are self-executing contracts with the terms of the agreement directly written into lines of code. In an IoT context, smart contracts can automate processes and transactions between devices without human intervention, streamlining operations and reducing the potential for errors.

Improved Scalability: Although blockchain technology itself faces challenges with scalability, newer blockchain architectures are being developed to support large-scale IoT networks. These include solutions like sharding, state channels, and side chains, which can handle large volumes of transactions generated by IoT devices, thus enhancing the overall scalability of IoT systems.

Interoperability and Standardisation: IoT devices often operate in silos with proprietary protocols, making it difficult for devices from different manufacturers or systems to work together. Blockchain can provide a standardised protocol for all IoT devices, facilitating interoperability and seamless data exchange across different platforms and systems.

5.3.3.1 Problems in Combining IoT and Blockchain Approaches

Integrating blockchain technology with IoT sensor nodes presents significant challenges due to the limited processing power and energy constraints of these devices, which struggle to handle the computationally intensive tasks required for blockchain operations. Indeed, some of the requirements of blockchain are in direct opposition to limited-capability devices:

- The blockchain '*proof-of-work*' consensus protocol is fundamentally computationally intensive

- Mining of blocks is time consuming whilst low latency is often desirable in IoT devices
- BC scales poorly as the number of nodes in the network increases. IoT networks are expected to contain many nodes.
- The underlying BC protocols create significant overhead traffic, which may be undesirable for certain bandwidth-limited IoT devices.

Additionally, the typically low storage capacity of IoT sensors complicates the maintenance of the blockchain's decentralised ledger, posing further difficulties in implementing this technology effectively across IoT networks.

5.3.3.2 An Alternative to Proof-Of-Work

One possible solution the computational intensity of the proof-of-work protocol is the *Balance Authentication Mechanism* (BAM) derived by McKenzie et al. [18, 19] which employs an alternative approach based on based on classic fourteenth century double-entry book-keeping of Pacioli [20] and provides all of the desirable properties of security, scalability, safety and liveness. Designed to operate in a distributed asynchronous network, this approach increases scalability, while maintaining a high transactional throughput, even when faced with Byzantine failure [21].

5.3.3.3 Conclusion

The prototype BAM system ensures the security of data at rest and in transit by utilising *smart-contracts* as encrypted containers. Whilst BAM offers a solution to the blockchain problems of Byzantine failure and computationally intensive proof-of-work consensus mechanisms and has some desirable properties that other alternative consensus protocols do not, it still places a certain amount of the load of crypto-processing on limited-capability leaf node devices. Thus, the ability to create systems in which limited-capability end devices securely communicate high-volumes of data remains an open problem.

5.3.4 Forensic Investigation of IoT Devices

The severest cybersecurity incidents may constitute criminal offences, making system '*investigability*' crucial (as well as beneficial to the *build* phase). Digital forensics on IoT devices is challenging due to diverse hardware and software configurations, complicating standardisation. Additionally, limited processing power and storage in many IoT devices make it difficult to retrieve and preserve volatile evidence. Using hardware hacking techniques, methods (based on the following outline) for acquiring stored data from various IoT architectures are being developed.

Footprinting: The first step in gaining access is understanding the device's architecture, identifying its data-storage capabilities (e.g., NAND Flash memory), and finding interfaces (e.g., JTAG, serial, i2c) that can provide access to the stored data.

Acquisition: In IoT devices, especially low-end ones, there may be no OS to aid in reconstruction, requiring reverse engineering of the device firmware to interpret data. Physical connections to the device (e.g., serial or JTAG ports) are usually needed for data acquisition, though network hacking may be an alternative.

Integrity: To ensure data integrity, traditional forensic techniques such as hashing should be applied to ensure any acquired data remains unchanged, with repeatable processes.

Analysis/Reconstruction: Initial analysis can be done by viewing data in a hex dump or extracting printable characters, potentially revealing credentials. Further analysis may include executing code in a simulator or reverse engineering to approximate the original source code.

Documentation and Sharing: Document findings and methods from each investigation and share them with the community to improve forensic procedures for similar devices.

5.3.4.1 Collecting, Cataloguing and Characterising Attacks on IoT

Understanding how malware successfully compromises an IoT system is key in knowing how to engineer a version of a system or device without that vulnerability. In a project addressing the *attack* and *investigate* phases, attacks and malware targeted specifically at IoT systems collected, catalogued and characterised. This resulted in the production of a taxonomy of IoT Malware strains by analysing the common approaches taken by the creators of malware. Analysing incidents in this manner provides useful material for further penetration testing as well as feeding back to the *build* phase. This work directly examined how IoT systems are attacked and from the understanding gained developed a toolset and methodology for reversing engineering and analysing malware samples.

A literature review showed that IoT malware often exploits common weaknesses such as improper input validation and authentication, with a trend towards leveraging less complex CVEs for exploitation without user interaction. Additionally, the geolocation of IoT malware incidents showed a global distribution. Following this, samples of IoT malware and reports of analyses of IoT malware were collected with the aim of identifying the common attack vectors. The samples collected were catalogued and formed the basis of a repository.

An analysis methodology and toolset for reverse engineering the catalogued samples, based upon the techniques of 'standard' malware analysis (use of code debuggers to step through the executing malware, decompilers to recover an approximation of the source code etc.) but in an IoT context, was developed. By analysing samples of malware and performing meta-review of analyses of incidents available online it proved possible to categorise them, based on several properties including type of threat, exploitation method, type of vulnerability and severity.

Once the common IoT malware strains, the techniques (attack vectors) are known and understood, the indicators of compromise are easier to recognise, thus accelerating incident response.

5.3.4.2 A Build Phase Example: Hardware Security for Key Enclaves

Hardware security, the security of the processors and their instruction sets is something of poor-relation (although the IoT-appropriate CHERI architecture [22] has recently been gaining traction) and yet the layers of operating systems and software between the processor and the user are profoundly shaped by the fundamental way in which computation takes place. The work of Sutherland [23] showed the feasibility of using the (legacy) facilities of commodity-level hardware to provide a secure key-enclave for keys and crypto primitive operations. By effectively disconnecting a section of memory in which they are stored from the remainder of address space a guarantee of tamper-proofness can be provided.

The crucial features are also available on families of microprocessors typically used in IoT applications such as the ESP32 range. The project involved the creation of a secure key enclave using memory isolation in Systems Management Mode (SMM) to protect cryptographic operations and keys from tampering in cloud computing environments. Traditionally, securing sensitive data like encryption keys required dedicated features such as Intel's SGX, but vulnerabilities like side-channel attacks have raised concerns. The research leveraged SMM, specifically the Systems Management Mode RAM (SMRAM), to create a secure memory enclave for storing cryptographic keys and related code. By locking and rendering the SMRAM inaccessible to userland code, the approach aims to provide better security isolation than existing techniques like process separation. The study demonstrates how this method can be used to establish a secure enclave, with potential generalizability to other processor architectures like ARM that offer similar facilities.

The approach was validated by creating a secure webserver with an encryption key protected using the proposed method. Even with full Administrator access, intruders were unable to extract the encryption key, showcasing the effectiveness of the secure key enclave. By focussing on securing network-connected systems against remote or physical attacks, the research emphasises the importance of safeguarding cryptographic keys to prevent unauthorised information disclosure. The approach not only protects server authentication materials but can also be extended to enhance security for client-side cryptographic keys, stored passwords, and payment mechanisms, highlighting its versatility and potential for broader application in securing sensitive data.

5.4 Conclusions and Future Work

Security is an ongoing challenge and an ever-evolving arms race between those who create systems and those who seek to exploit them. The key lies in narrowing the gap between the design and development of systems, and identifying any fundamental security issues with those systems, ensuring that security is treated as a first-class concern throughout the entire lifecycle of a system. By prioritising swift responses and proactive measures, we can better defend against the inevitable threats that emerge.

Achieving this level of security requires a concerted effort across the developer and cybersecurity communities, emphasising the importance of sharing best practices and integrating cybersecurity knowledge into software and systems engineering education. Organisations must overcome the challenges of adopting secure development practices, recognising that while it may be expensive and time-consuming, it is essential for safeguarding against threats. The development of standard, open operating systems and libraries with rigorous quality control and security measures will also be crucial in building a more secure technological future.

Acknowledgements The authors would like to thank the following colleagues for the many valuable discussions which have contributed to this chapter: Dr Adam Sampson, Dr Simon Leibl, Dr Tayabba Nafees, Dr Beverley McKenzie, Dr James Sutherland, Mr Jamie O'Hare, Mr David McLuskie and Mr Ross Heenan.

References

1. CISCO (2024) Annual internet report (2018–2023) www.cisco.com/c/en/us/solutions/collateral/executive-perspectives/annual-internet-report/white-paper-c11-741490.pdf. Accessed September 2024
2. Kolias C, Kambourakis G, Stavrou A, Voas J (2017) DDoS in the IoT: mirai and other botnets. Computer 50(7):80–84
3. Erven S, Collao M (2015) Medical devices: pwnage and honeypots, DEF CON 23, DEF CON https://doi.org/10.5446/36435. Accessed September 2024
4. Open Web Application Security Project (2017) OWASP Code review guide release 2 www.owasp.org/www-project-code-review-guide/assets/OWASP_Code_Review_Guide_v2.pdf. Accessed September 2024
5. IEEE (2008) IEEE standard for software reviews and audits, IEEE Std 1028–2008, 1–53, https://doi.org/10.1109/IEEESTD.2008.4601584. Accessed September 2024
6. Ballman A (2016) The SEI CERT C++ Coding standard, rules for developing safe, reliable, and secure systems in C++, 2016 Edition www.resources.sei.cmu.edu/downloads/secure-coding/assets/sei-cert-cpp-coding-standard-2016-v01.pdf. Accessed September 2024
7. Byers R, Turner C, Brewer T (2022) National vulnerability database, National Institute of Standards and Technology, https://doi.org/10.18434/M3436. Accessed September 2024
8. SEI (2024) CERT Coding standards, Carnegie Mellon University, Software Engineering Institute, www.wiki.sei.cmu.edu/confluence/display/seccode. Accessed September 2024
9. Open Web Application Security Project (2024) OWASP Secure coding practices-quick reference guide, www.owasp.org/www-project-secure-coding-practices-quick-reference-guide/. Accessed September 2024

10. Motor Industry Software Reliability Association Guidelines, www.misra.org.uk/. Accessed September 2024
11. International Organisation for Standarization (2022) ISO/IEC IS 27001:2022—information security, cybersecurity and privacy protection—Information security management systems—Requirements, ISO www.iso.org/standard/27001. Accessed September 2024
12. Microsoft (2024) Microsoft security development lifecycle (SDL) @ www.microsoft.com/en-us/securityengineering/sdl. Accessed September 2024
13. Swamy N, Hriţcu C, Keller C, Rastogi A, Delignat-Lavaud A, Forest S, Bhargavan K, Fournet C, Strub PY, Kohlweiss M, Zinzindohoue JK, Zanella-Béguelin S (2016) Dependent types and multi-monadic effects in F*, Proceedings 43rd Annual ACM SIGPLAN-SIGACT Symposium Principles of Programming Languages, POPL '16), Association for Computing Machinery, pp 256–270, https://doi.org/10.1145/2837614.2837655. Accessed September 2024
14. Liebl S (2024) A threat modelling approach to enhance the security of internet of things devices, Abertay University Doctoral Thesis, www.rke.abertay.ac.uk/en/studentTheses/a-threat-modelling-approach-to-enhance-the-security-of-internet-o, www.iariajournals.org/security/tocv14n12.html
15. Liebl S, Lathrop L, Raithel U, Aßmuth A, Ferguson I, Söllner M (2021) Analyzing the attack surface and threats of industrial Internet of Things devices. Intl J Adv Secur 14(1&2):54019
16. Nafees T, Coull N, Ferguson RI, Sampson A (2017) Engineering secure software and systems, Proceedings 9th international symposium ESSoS 2017, lecture notes in computer science; 10379, Bodden E, Paye M & Athanasopoulos E. (eds.), Springer, 133–142 10
17. Howard M, Lipner S (2006) The security development lifecycle: a process for developing demonstrably more secure software, Microsoft Press https://scholar.google.co.uk/scholar?as_q=The+security+development+lifecycle&as_epq=&as_oq=&as_eq=&as_occt=any&as_sauthors=Lipner&as_publication=&as_ylo=2005&as_yhi=2007&hl=en&as_sdt=0%2C5. Accessed September 2024
18. Mackenzie BA, Ferguson I, Razaq A (2022) A secure and scalable IoT consensus protocol, 5th International conference on safety and security with IoT, SaSeIoT 2021, EAI/Springer innovations in communication and computing, pp 65–80, https://doi.org/10.1007/978-3-030-94285-4_5. Accessed September 2024
19. Mackenzie B (2022) To ascertain if blockchain can facilitate the internet of things security and scalability requirements, Abertay University Doctoral Thesis, www.rke.abertay.ac.uk/en/studentTheses/to-ascertain-if-blockchain-can-facilitate-the-internet-of-things-. Accessed September 2024
20. Pacioli L, Brown RG, Johnston KS (1963) Paciolo on accounting, Facsimiles-Garl
21. Shostak R, Pease M, Lamport L (1982) The byzantine generals problem. ACM Trans Prog Lang Syst 4(3):382–401
22. Woodruff J, Watson RNM, Chisnall D, Moore SW, Anderson J, Davis B, Laurie B, Neumann PG, Norton R, Roe M (2014) The CHERI capability model: revisiting RISC in an age of risk, Proceedings 41st international symposium computer architecture, ISCA 2014, www.class.ece.iastate.edu/tyagi/cpre581/papers/ISCA14CHERI.pdf. Accessed September 2024
23. Sutherland JA, Coull N, Ferguson RI (2023) On the creation of a secure key enclave via the use of memory isolation in systems management mode, 14th International conference on cloud computing, grids, and virtualisation, pp 30–40, www.thinkmind.org/index.php?view=article&articleid=cloud_computing_2023_2_20_28005. Accessed September 2024

Chapter 6
Privacy

Christopher Milne and Steven A. Watt

Abstract The International Association of Privacy Professionals opening remark when addressing the question "*What does privacy mean*?" perhaps quizzically state "*Well, it depends who you ask*." While privacy is multifaceted we advocate utilisation of definitions from UK and European data protection law, the accompanying data protection principles, data protection impact assessment/privacy by design and default, and practical learnings from case law. There is a common language and body of knowledge which can be applied to inform and embed privacy by design and default into the thinking and work of engineers, organisations, where privacy (practical application and responses to each of the data protection principles) is suitably hardcoded into the fabric of product/service design, and operation. The language, jurisprudence and practical application of data protection law should not remain siloed and the sole preserve of data protection officers. To provide effective privacy to individuals as technologies, data, big data, the Internet of Things, and Artificial Intelligence become ubiquitous across all aspects of our lives, we advocate that a multidisciplinary approach is fundamental. Privacy by design and by default is a realistic methodology to meet current and future privacy challenges. However, essentially the language, thinking and practical application of data protection law is largely isolated and evident from the work of data protection practitioners. Where the language and knowledge gap is bridged across both disciplines, essentially developing and applying a multidisciplinary approach to data protection by design and default, we may create pathways where privacy solutions can be usefully identified, developed, applied and operate at component, product, data, system and service levels. The argument that privacy law needs to catch-up with technology is often played out. Arguably not; perhaps multidisciplinary thinking and working is needed?

C. Milne (✉) · S. A. Watt
University of St Andrews, St Andrews, UK
e-mail: cm230@st-andrews.ac.uk

P. Hehenberger and D. Bradley (eds.), *Mechatronic Futures*,
https://doi.org/10.1007/978-3-031-83571-1_6

6.1 Introduction

In this chapter, for reasons of practicality the assessment of privacy (and concerns for the privacy of the individual)[1] is explored in the context of personal data as defined and legislated in United Kingdom (UK) and European data protection laws, as these have the potential to provide a common language and baseline to define the scope of privacy and the requirements to be met in protecting privacy.

The UK General Data Protection Regulation (UK GDPR) incorporates the General Data Protection Regulation (the GDPR) into UK domestic law. Both are essentially the same; the differences that exist have little if any practical relevancy when considering questions of privacy in respect of Mechatronic and related systems and their designers.

The legislative and regulatory frameworks that accompany the UK GDPR and GDPR provide a toolkit which can be utilised to provide a roadmap/pathway focussed on facets crucial to the understanding of privacy and how that can be addressed in systems and product design. Privacy can be defined and responses to that understood through use of the following toolkit.

6.2 Introducing the Toolkit

The primary legislation, in this instance the UK GDPR/GDPR provides:

Definitions—Fundamentally personal data, special categories of personal data and biometric data, and personal data breach. Privacy can mean many things to different people. Working with a precise definition of what must be protected, and what protection means, provides focus and a degree of assurance; tasks can be quantified so as to inform thinking and decision-making; and objective analysis of success or otherwise can be measured.

The data protection principles—The foundations for establishing and protecting privacy are codified in 6 principles. These were originally derived from the *Council of Europe, (1981), Convention of the protection of individuals with regard to the automatic processing of personal data, European Treaty Series—108, Article 5.* This set out 5 duties, which did not at that time include data security. Updated in 2018, *Council of Europe, (2018), Convention 108 Convention of the protection of individuals with regard to the processing of personal data,* provides a clear and concise narrative of the fundamentals of data protection, which although separate from UK GDPR/GDPR closely align. Returning to the data protection principles in UK GDPR/GDPR, these provide that personal data shall be:

1. Processed lawfully, fairly, and transparently in relation to the individual who is the focus of the personal data.

[1] www.iapp.org/about/what-is-privacy/ (accessed October 2024).

- Lawfully means relying in advance on a specific legal basis/condition set out in Article 6 and Article 9 of the said Regulations where special categories of personal data are in use, as for instance medical information about individuals.
- Fairly means that use cannot be beyond the reasonable expectations of individuals, and individuals were not deceived when their personal data was collected.
- Transparency means that the entity that uses the personal data is open and honest, and individuals' have received sufficiently clear information on the use of their data, the identity of the entity, who that data may be shared with, and specific rights available to them.

There are of course exceptions, e.g. it would undermine the purposes of the prevention and detection of crime where a Police force was required to give notice of a criminal investigation. For the purposes of this discussion on privacy, it is not necessary to explore exemptions in UK GDPR/GDPR and from domestic law, e.g. the Data Protection Act 2018, but to have some awareness of their existence.

2. Collected for specific stated and legitimate purpose(s) and not used in a manner incompatible with those purposes. (There are exceptions for scientific and social sciences research.) Therefore, if individuals used software to help draft curriculum vitae, which used Artificial Intelligence, it may be unlawful for that information to be used for the secondary purpose of the service provider using those materials to improve machine learning, unless that was a legitimate purpose. Or where a smartwatch provider extracted data from user interaction to understand activity and used that to make inferences about health or otherwise.
3. Adequate, relevant and limited to what is necessary for the purpose, e.g. an online job application form need not collect national insurance information or bank details—as such is only necessary post recruitment where an applicant is successful and has accepted a job offer.
4. Accurate, necessary and up to date. How do systems/services verify data accuracy; how many individuals view their personal data and if appropriate have the means to update and/or question that?
5. Kept for longer that is necessary (in an identifiable form) to support the purpose(s) for which it was collected. Where personal data are destroyed when there is no further legitimate use that will of course eliminate any privacy risk to individuals, as will anonymising the data so that cannot be linked to individuals.

This also minimises risk for service providers, who are liable for any infringement of the UK GDPR/GDPR, which may include failure to protect against a personal data breach. On 16 October 2020, the Information Commissioner fined British Airways (BA) £20 million (Sterling) as it failed to ensure that appropriate [cyber] security measures were in place to protect personal data from third-party attack.[2]

---

[2] www.ico.org.uk/media/action-weve-taken/mpns/2618421/ba-penalty-20201016.pdf (accessed October 2024).

Data Protection Impact Assessment, Article 35—This requires that before the collection and use of personal data that may involve the use of personal data and/or where a new use of personal data may high risk to the rights and freedoms of individuals, that an assessment of the proposed processing is undertaken [against each of the data protection principles] to identify any risk (that processing against a principal cannot be met) and if so, what mitigation is available.

Case law from UK and European Courts (notably the Court of Justice of the European Union (CJEU) and commentary of case law from legal professionals, provides insight into significant matters of law which can have significant practical effect/input in product/service design and/or in the operation and delivery of the same.

County Court at Oxford, (October 12, 2021) Fairhurst v Woodard (Case No: G00MK161)[3]— This found that a neighbour's use of their doorbell camera (which captured images and audio) breached UK GDPR. The defendant claimed use of the camera was legitimate to protect their property from crime. However, as the Court found that the imagery captured from the camera extended into the property of the claimant (some 60 feet beyond the defendant's property boundary) that was disproportionate to the needs of protecting their (the claimant's) property.

- This suggests that product design perhaps should have been informed by blurring imagery beyond calibrated distances, thus protecting legitimate use and the privacy of others where imagery from other properties is not captured.

CJEU—C-446/21—Maximilian Schrems v Meta Platforms Ireland Limited (04 October 2024)[4]— The claimant claimed that the social media platform Facebook misused details of his sexual orientation by using that to push him targeted advertising, despite him not providing Facebook with those details. The claimant asserted that Facebook gathered details of his sexual orientation from third-parties, or plug-ins, and kept that data for an indefinite period. And, from that used those details to push adverts/events aimed at gay people. The CJEU responded to two questions of law, although for the purposes of this discussion only the first question is considered namely:

> *Does the data minimisation principle (Article 5(1)(c) GDPR) mean that all personal data held by a platform may be aggregated analysed and processed for the purposes of targeted advertising without any restriction?*[5]

The Court found for the claimant on both questions. For the first question, it was found that Meta had built a data pool going back 20 years and had used that

---

[3] www.judiciary.uk/wp-content/uploads/2022/07/Fairhurst-V-Woodard-Judgment-1.pdf (accessed October 2024).

[4] curia.europa.eu/juris/document/document.jsf;jsessionid = 5CE53D5E3FCC1ABA77F2ACD5AAC2F038?t = &docid = 290,674&pageIndex = 0&doclang = en&mode = req&dir = &occ = first&part = 1&cid = 1,306,139 (accessed October 2024).

[5] www.gdprhub.eu/index.php?title = CJEU_-_C-446/21_-_Maximilian_Schrems_v_Meta_Platforms_Ireland_Limited (accessed October 2024).

to identify interests hence to push adverts. This breached the third data protection principle (minimisation of personal data), it was not necessary for Meta to use all the data collected to provide targeted advertising.

- In terms of service design, Meta now needs to consider how they capture and retain data, ensuring that they only extract and use data from a data pool for specific purposes. It is likely that the CJEU's decision will impact on Meta's ability to monetize data, not least as personal data that is excessive will now need to be deleted so to comply with the fifth data protection principal.

Guidance from regulators (for the UK the Information Commissioner's Office; members of the European Union each have their own national supervisory authority. The European Data Protection Board (EDPB) (established in EU law via GDPR, Part 3, Articles 68 through 76) is a collective of data protection leads from each member state supervisory authority. The main input from the EDPB into this proposed toolkit are the opinions, binding decisions, guidance, recommendations and best practices which the EDPB may examine and produce under its own initiative on any aspect of/question covering the application of the said Regulation.

- In terms of the establishment and protection of personal data (aka privacy) the EDBP have published a range of materials on data protection by design and by default, including *Guidelines 4/2019 on Article 25: Data protection by design and by default. Version 2.0, Adopted 20 October 2020.*[6]

  These guidelines:

  > … *Contain guidance on how to effectively implement the data protection principles in Article 5, listing key design and default elements as well as practical cases for illustration. The* ***controller***[7] *should consider the appropriateness of the suggested measures in the context of the particular processing in question.*

Here, the emphasizing of "***controller***" is to draw attention to a constraint. While it is offered that the multifaceted nature of privacy can be usefully codified for the purposes of Mechatronic System Design and the designers of these by focussing on the protection of personal data and drawing upon the tools introduced, above, the intellectual development of privacy by design and default, and commentary on its application is naturally derived from, and largely remains allied to data protection regulators, practitioners and controllers.

A controller, is defined by UK GDPR/GDPR, Article 4, point (7) as follows:

> 'Controller' *means the natural or legal person, public authority, agency or other body which, alone or jointly with others, determines the purposes and means of the processing of personal data.*

Data protection legislation and its application naturally focuses on the responsibilities of controllers and processors. Crucially, a controller is several steps removed

---

[6] www.edpb.europa.eu/sites/default/files/files/file1/edpb_guidelines_201904_dataprotection_by_design_and_by_default_v2.0_en.pdf (accessed October 2024).

[7] Author emphasis.

from product/service designers, and they (a controller) must remain independent and accountable. Controllers, normally through an appointed data protection officer, are mainly well versed and aware of their duties as prescribed by UK GDPR/GDPR. This perhaps begs the questions if a multidisciplinary approach is to be taken, who then provides advice and comment to design teams on data protection elements of relevance (emanating from application of the toolkit introduced herein); should design teams have access to the specialist knowledge of a data protection officer or an equivalent? UK GDPR/GDPR, Article 25, provide that a controller is responsible for data protection impact assessment, and taking a proactive approach to the protection of personal data. Ideally, the principles of data protection by design and default should also be codified into product and service design—the development of a multidisciplinary thinking and approach, where the lexicon of data protection and its development is not the preserve of legislators, regulators and practitioners.

A review of articles on "*data protection impact assessments*" indexed in Web of Science from 2020 to October 2024, shows a clustering of publications within titles allied to law (Table 6.1). This is illustrative of a knowledge divide, and perhaps of a lack of a multidisciplinary approach, which dilutes the ability to protect privacy, there is an artificial separation of roles and responsibilities, derived from the philosophy that data protection is compliance with the law.

The IEEE Digital Privacy resource, *Understanding Privacy in the Digital Age*[8] presents a multifaceted view of privacy.

> … *Privacy in the digital age comprises a multifaceted spectrum, with each facet presenting its unique challenges and considerations. The conversation around digital privacy goes beyond the surface-level discourse of controlling one's personal information—it also involves aspects such as consent, transparency, and trust.*

The role of data protection law features in that discussion.

> … *On the legislative front, privacy laws might evolve as technology advances. As evidenced by the introduction of the GDPR in the EU and the CCPA in California, there is a growing push for clear guidelines and strong protections for data privacy. However, the challenge lies in creating laws that are both robust enough to protect privacy and flexible enough to adapt to technological advancements continuously.*

However, the discussion is in its own way is a continuation of a siloed approach, as, there is no recognition that many of the privacy challenges introduced therein, notably the perceived lack of control over personal data (the observation that user consent does not work) and invasive targeted advertising within social media platforms arguably now have some measure of resolution. As noted herein, above, the courts and notably the CJEU by assessing and enforcing application of the data protection principles from the challenges brought by Schrems[9] are steadily addressing fundamental privacy concerns.

---

[8] www.digitalprivacy.ieee.org/publications/topics/understanding-privacy-in-the-digital-age (accessed October 2024).

[9] www.Noyb.eu/en (accessed October 2024).

Table 6.1 Data protection impact assessment articles by publication—articles allied to law in bold

Publication	Count
Computer Law Security Review	7
British Medical Journal Open	2
Future Internet	2
AI and Ethics	1
Air Space Law	1
Applied Sciences	1
Applied Sciences Basel	1
Asia Pacific Journal of Health Law Ethics	1
Boletin de la Asociacion Internacional de Derecho Cooperativo (International Association of Cooperative Law Journal)	1
Cities	1
Clinical Linguistics Phonetics	1
Computer Law and Security Review	1
Computer Science and Information Systems	1
Computers Security	1
Connection Science	1
Data Policy	1
Em Questao	1
Enterprise Modelling and Information Systems Architectures (An Internationakl Journal)	1
European Journal of Risk Regulation	1
European Labour Law Journal	1
Exercer la Revue Francophone de Medicine Generale	1
Forensic Science International Digital Investigation	1
Frontiers in Genetics	1
Geoforum	1
IEEE Access	1
Informatica (An International Journal of Computing and Informatics)	1
Information communications Technology Law	1
Information Society	1
International Data Privacy Law	1
International Journal for Equity in Health	1
International Journal of Law and Information Technology	1
Journal of Korean Information Law	1
Journal of Law and the Biosciences	1
Journal of Medical Ethics	1
Journal of Risk Research	1

(continued)

Table 6.1 (continued)

Publication	Count
Language Resources and Evaluation	1
Modern Law Review	1
Oxford Journal of Legal Studies	1
Review of Central and East European Law	1
Security and Privacy	1
Social Sciences Basel	1
Soft Computing	1
Software and Systems Modelling	1
Software Systems Modelling	1
The American Psychologist	1
Young Consumers	1

Significant changes have been made and others will come, as social media providers respond to rulings, and the courts and regulators. Where a multidisciplinary approach can be taken this may increase the likelihood of getting fundamentals right first time, thereby placing privacy at the centre where the application of the data protection principles and the thinking around those and its development are a shared endeavour. The thinking behind data protection by design and default is sound, except, the origins and presentation as part of legislative compliance is siloed. The recognition of and understanding of the data protection principles perhaps ought to feature equally as part of the lexicon of mechatronic systems and their designers.

We introduce a toolkit, above, to explore the concept of how key elements of data protection law and governance, may be utilised to support understanding and application in a multidisciplinary approach, and to prompt further discussion around that. To provide further context, we next explore where we see key themes arising in and from the practitioner's perspective offered in the earlier edition of this work, where those related to privacy where we view key changes and challenges that now arise.

6.3 Changes—The Emergence of an International Standard for Privacy by Design

On 31 January 2023 the International Organization for Standardization ("ISO") published ISO 31700 Consumer protection—Privacy by design for consumer goods and services. This standard has 2 parts: Part 1[10] sets out the 30 high level requirements to meet the standard; Part 2[11] provides illustrative use cases for online retail, a fitness company, and smart locks for home front door installation.

Commentaries on the standard,[12,13,14] note:

> "*Products and services are the front line of consumer protection, and it is important to remember that organisations have control over their operations. However, they do not have control of the way consumers use the products and services. As such, the only protection is the functionality that is embodied within the hardware and software. The new standard follows the data protection by design and by default principle of the General Data Protection Regulation and brings value by bringing uniformity to privacy by design, albeit not a conformance standard.*
>
> [The intended audience is] *consumers, engineers and organisations at large.*
>
> *ISO 31700 lays down 30 requirements for embedding data privacy into consumer products and services. Like the EDPB's approach, it does not specify thresholds or steps but keeps the ruleset high level and provides examples for better understanding.*
>
> *ISO 31700 is not directly linked to the EU data protection framework, but some overlaps do exist. For example, it adopted a* "GDPR-ish" *definition of personal information, and many of its requirements overlap with those from the GDPR.*
>
> [Part 1 of the standard provides generic guidance on and advice]

- *To develop features that enable consumers to enforce their data protection rights,*
- *To assign relevant roles and powers,*
- *To provide privacy information to consumers,*
- *To conduct data protection risk assessments,*
- *To define and document requirements for data protection controls,*
- *On the design of data protection controls,*
- *On data management over the entire life cycle and*
- *To prepare for and deal with data protection breaches.*"

It remains unclear what impact ISO 31700 may have, as a non-binding, international standard. This is a welcome development in that the core elements of UK GDPR/GDPR, Article 25 have been transformed into a useable form for engineers and organisations. There is scope, perhaps, for ISO 31700 to become a catalyst to

[10] www.iso.org/standard/84977.html (accessed October 2024).

[11] www.iso.org/standard/84978.html (accessed October 2024).

[12] www.pwc.ch/en/insights/regulation/welcome-to-the-new-iso-31700-standard-for-privacy-by-design.html (accessed October 2024).

[13] www.privacyworld.blog/2023/02/iso-31700-the-latest-tool-to-operationalize-gdpr-privacy-by-design-compliance/ (accessed October 2024).

[14] www.activemind.legal/guides/iso-31700/ (accessed October 2024).

assist in the development and establishment of the multidisciplinary approach that we advocate for whereby a common language and toolkit can emerge which bridges the lexicons of data protection and mechatronics systems and their design.

6.4 Challenges—Developing a Multidisciplinary Approach to Privacy by Design and Default

Much thinking and work has been undertaken by legislators and supervisory authorities to develop the concept of privacy by design and default, to embed that into data protection law (UK GDPR/GDPR, Article 25) so to protect the fundamental rights and freedoms of individuals. And, to regulate where controllers have failed to protect and maintain the confidentiality of personal data by designing, building and operating systems/services that failed to provide basic protections, which ought not to have been missed.

As an example of regulating, to enforce a controller's obligations to undertake data protection by design and default, the Data Protection Commissioner (the "DPC") (the Irish supervisory authority) following investigation (28 November 2022), fined Meta Platforms Ireland Limited (the controller of the social media platform, Facebook) €265 million for a breach of GDPR, Article 25(1) and Article 25(2).[15] Meta was also required to make a number of remedial actions within an undisclosed period. The personal data of 533 million individuals was compromised via cyber-attack and published. This included phone numbers, locations, dates of birth, Facebook IDs, names and email addresses of Facebook users (2018–2019). The DPC found that the controller (Meta) had not foreseen particular risks and mitigated for those. It remains to be seen whether regulatory enforcement in this area will meaningfully change operating models, where a proactive approach to the protection of privacy is taken, or if controllers may accept the risk of regulatory action as a consequence and cost of business?

While arguably, the enforcement actions by the courts and supervisory authorities when regulating UK GDPR and GDPR has some effect, this is of course after the event, and prevention is better than cure. While in recent times privacy by design and default has been conceived and developed to enhance the protection of individuals via data protection law, the levels of protection that were envisioned and intended to come from a proactive response to data protection also need to be "*hard coded*" into the design and testing phase. We see that working to develop a multidisciplinary approach to privacy by design and default is required. Otherwise, emphasis and action is likely to remain the preserve of data protection officers, which as demonstrated in some instances will be a missed opportunity.

[15] www.dataprotection.ie/en/news-media/press-releases/data-protection-commission-announces-decision-in-facebook-data-scraping-inquiry (accessed October 2024).

6.5 Generative Artificial Intelligence and Large Language Models (Generative AI/LLM)

This is an emerging area with some knowns and many unknowns. From a privacy perspective as practitioners our present approach is focussed on protecting personal data by controlling the supply chain. In-line with UK Government guidance to civil-servants on the use of generative artificial intelligence[16] and the Generative AI Framework for HMG[17] there are prohibitions on the use of personal data and/or commercially confidential information in public generative AI applications or APIs, so to uphold the first and second data protection principles. Essentially, providing personal data to a third-party for their own use to support their machine learning, is likely to be unlawful secondary use of personal data, and potentially exposes individuals and organisations to data leakage, where data/information are recycled, potentially allowing for individuals to be identified and/or commercially sensitive content is recycled and presented to others.

From our understanding of the risks and our duties to protect personal data and to maintain confidences, such is a proportionate response. Of course, we do not seek to stifle innovation nor to deny opportunities to enhance the quality of the services we provide, which can of course include generating efficiencies where Generative AI/LLM may free individuals from some tasks, safely. Personal data and/or commercially confidential data/information can be used with approved services. Due diligence must be completed to ensure that any supplier is correctly contracted, with all of the protections for third-party contractor (i.e. a processor) use of personal data, as intended by UK GDPR, Article 28 in place. Such, includes the instruction that a processor cannot make use of personal data provided to them for any purposes not authorised by the controller—which includes secondary use of data to support machine learning, unless that is to provide services directly to the controller and the controller alone.

As noted, this is the current position at the time of writing, and as Generative AI/LLM develops, it is fundamental that there is a full understanding of the operation of that technology and identification of the full spectrum of risks and potential mitigations.

Therein lies the fundamental challenges, which is to have knowledge and understanding so to effectively conduct risk assessment (essentially privacy by design and default) and to be able to do so in a timely manner.

This chapter opened by highlighting barriers to comprehensive and effective privacy by design and by default, by noting the emergence of that approach from a data protection perspective, where language and primary application remained within the field of data protection law and practitioners and regulators thereof, and the emergence of a silo approach, which could be addressed by multidisciplinary

[16] www.gov.uk/government/publications/guidance-to-civil-servants-on-use-of-generative-ai/guidance-to-civil-servants-on-use-of-generative-ai (accessed October 2024).

[17] www.gov.uk/government/publications/generative-ai-framework-for-hmg (accessed October 2024).

working. On 21 May 2024, the Council of the European Union approved the Artificial Intelligence (AI) Act.[18] This will fully become part of EU law in May 2026. The said Act and the legislative framework that surrounds that will provide details of how to protect personal data and by extension privacy. Domestic legislation from other major economies is likely to follow. And, given that personal data will be a significant input into Generative AI/LLM, it is expected that that legislation will be harmonised with UK GDPR/GDPR, without affecting the application of existing laws.

Should history repeat itself, and governance and regulation of Generative AI/LLM emerge from legislation, and largely remain regulator and practitioner led, there risks a continuation of the silo approach that is outlined. Perhaps now is the time for a wider consideration of how a multidisciplinary approach can be developed to help provide a solid foundation for embedding risk analysis and protection i.e., privacy by design and by default of the use of personal data in Generative AI/LLM not only to protect privacy, but also to provide the other fundamental rights and freedoms.

UK GDPR/GDPR, Article 22 provides that data subjects (individuals) –

> *… Shall have the right not to be subject to a decision based solely on automated processing, including profiling, which produces legal effects concerning him or her or similarly significantly affects him or her.*

How is that right to be recognised and effectively provided where personal data are processed by AI applications?

6.6 Conclusions

Privacy perhaps need not be so complex, as there is a potential to coalesce by working with relatively universal definitions of personal data, special categories of personal data and the data protection principles of UK GDPR/GDPR, which derive from broader international convention. It remains to be seen how ISO 31700 may be applied by engineers and organisations in product/service design and delivery. However, given that the international standard for privacy by design overlaps with the language and philosophy of GDPR, relying on the definitions, the data protection principles, and case law that derive from the said Regulation would provide a relevant common baseline.

It is not sufficient for privacy by design to be delivered by controllers with the subject expertise of data protection practitioners. That is a limited partial response; given that mechatronic components and devices are often primary data sources, the integration of privacy by design, meaningfully into design processes and philosophies has the potential to meaningfully solve many of the privacy problems, both known and yet to emerge. The publication of ISO 31700 signals that there is fundamental

[18] www.consilium.europa.eu/en/press/press-releases/2024/05/21/artificial-intelligence-ai-act-council-gives-final-green-light-to-the-first-worldwide-rules-on-ai/ (accessed October 2024).

agreement (at least within the standards community) that the philosophy of privacy by design is sound, and that it has a fundamental role in protecting the privacy of individuals, by engineering privacy into product design.

As engineers, organisations and other key stakeholders such as educators and professional bodies begin to think about how they will respond to ISO 31700, and perhaps also Generative AI/LLM, that may also provide an opportunity to consider how multidisciplinary may be fostered, developed, enhanced and sustained? Project management is also likely to have a fundamental role when identifying and bringing together key skill sets and thinking and ensuring that those are available when needed. Success is likely to depend on bringing the relevant skills sets and thinking out of what were perhaps once natural silos and working collaboratively so that issues and responses can be identified and suitably addressed throughout the product and service lifecycle. Privacy then may become easy or at least easier.

Chapter 7
Mechatronics in an Era of Sustainability and Climate Change

David Bradley, Peter Hehenberger, and Edward Simpson

Abstract Sustainability and the Circular Economy will determine much of the structure of future life. As an interdisciplinary subject, Mechatronics can contribute a great deal to the achievement of sustainability goals. On the one hand, Mechatronic systems influence a wide variety of processes and contribute to longevity, energy efficiency and waste avoidance. On the other, mechatronic products should fulfil sustainable criteria. In this chapter, the various factors influencing sustainability within. A Mechatronics context are discussed and analysed and explicitly applied in a case study.

Keywords EcoMechatronics · Sustainability · Circular economy · Sustainable production processes

7.1 Introduction

The concept of sustainability in achieving an effective, long-term balance between the appropriate use of Earth's resources with respect to the well-being of people and society can perhaps be expressed by the concept of '***do no harm.***' This simple expression builds on the oft-quoted statement from the 1987 *Our Common Future: Report of the 1987 World Commission on Environment and Development* [1] that:

> '*Sustainable development is development that meets the needs of the present without compromising the ability of future generations to meet their own needs.*'

However, this same report also goes on to say that:

> '*The concept of sustainable development does imply limits—not absolute limits but limitations imposed by the present state of technology and social organization on environmental resources and by the ability of the biosphere to absorb the effects of human activities*'

D. Bradley · E. Simpson
Abertay University, Dundee, UK

P. Hehenberger (✉)
University of Applied Sciences Upper Austria, Wels, Austria
e-mail: peter.hehenberger@fh-wels.at

P. Hehenberger and D. Bradley (eds.), *Mechatronic Futures*,
https://doi.org/10.1007/978-3-031-83571-1_7

This then establishes:

- The concept of *'needs,'* and in particular the essential needs of the world's poor, to which priority should be given.

And

- The idea of limitations imposed by the state of technology and social organisation on the environment's ability to meet present and future needs.

These basic concepts were further developed through the introduction of the concept of a Circular Economy (CE) as a way of expressing the increasingly perceived need to more effectively manage the:

$$\text{Resource Extraction} \rightarrow \text{Production} \rightarrow \text{Use} \rightarrow \text{Disposal}$$

cycle. In particular, the aim is to emphasise that in a CE, resource use is improved by minimising the extraction of natural resources, maximising waste prevention and optimising environmental, social, material and economic values throughout the lifecycles of materials, components and products [2, 3].

This implies that problem identification and understanding is key to achieving the desired outcomes. For example, rather than treat transport in terms of the individual modes of transport; cars, trains, aircraft, ships and so forth, it needs to be considered as an integrated whole, with an understanding as to how decisions taken in respect of one area of technology impacts on the functioning of the system as a whole.

A further factor that needs to be considered is the political dimension in which the nature and availability of technology is often secondary to local and national political decisions and requirements, as illustrated by the debates over the wording of the closing document for COP28 (and indeed for earlier COPs) [4] or the routing of transmission lines in Australia [5].

All of this has implications for the ways in which products and systems are designed, manufactured, used and disposed of at end-of-life. In the context of Mechatronics, the idea of an *EcoMechatronics* approach to the development of future Mechatronic (and other) systems has been postulated to take account of the potential impact that Mechatronic components, devices, sub-systems and systems have in achieving sustainability goals not only through their operation but also with regard to their potential to provide information on system behaviour and performance [6].

The chapter therefore considers various aspects of sustainability and places them in the appropriate context for the achievement of long-term sustainability goals. Key to this is the recognition of the role of Mechatronics, and in particular *EcoMechatronics*, as a driver for change.

7.2 Sustainability—Issues and Challenges

7.2.1 Circular Economy

Perhaps the defining definition of a Circular Economy (CE) is that put forward in 2013 by the Ellen MacArthur Foundation which reads:

'An industrial economy that is restorative or regenerative by intention and design.'

And is generally represented diagrammatically as in Fig. 7.1 [2, 3].

The MacArthur Foundation then goes on to add that this model:

'... replaces the end-of-life concept with restoration, shifts towards the use of renewable energy, eliminates the use of toxic chemicals, which impair reuse and return to the biosphere, and aims for the elimination of waste through the superior design of materials, products, systems, and business models.'

Since the publication of the MacArthur Foundation reports, a wide range of other definitions for a CE have been put forward and in their 2017 paper Kirchherr et al. [7] identified 114 separate and distinct definitions!

When considering the concept of a CE it is important to recognise that it does not represent a closed loop, indeed the actions of 2nd Law of Thermodynamics, expressly mitigate against it being such. This means that in practice there will always be a need for an input of new materials and energy to make up for the system losses, and

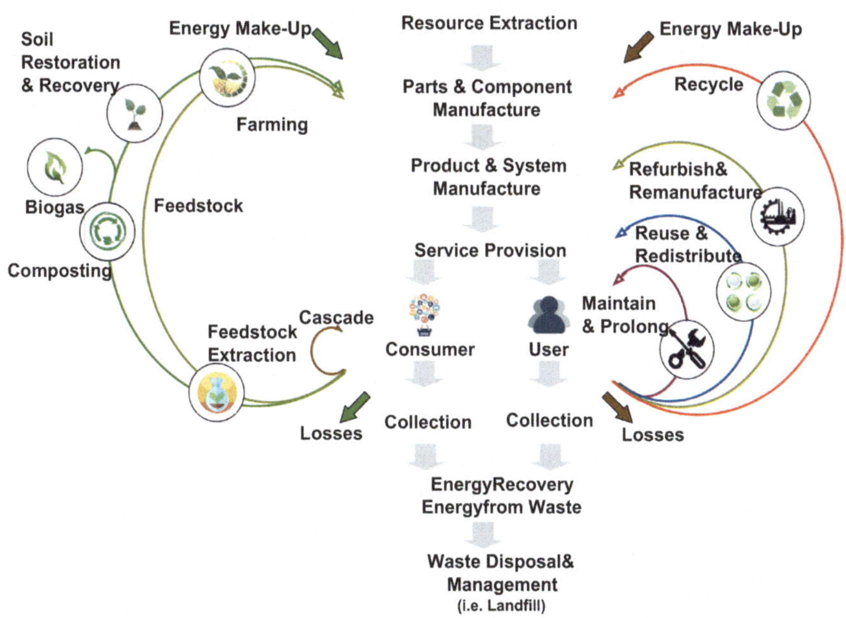

Fig. 7.1 A circular economy (after [2, 3])

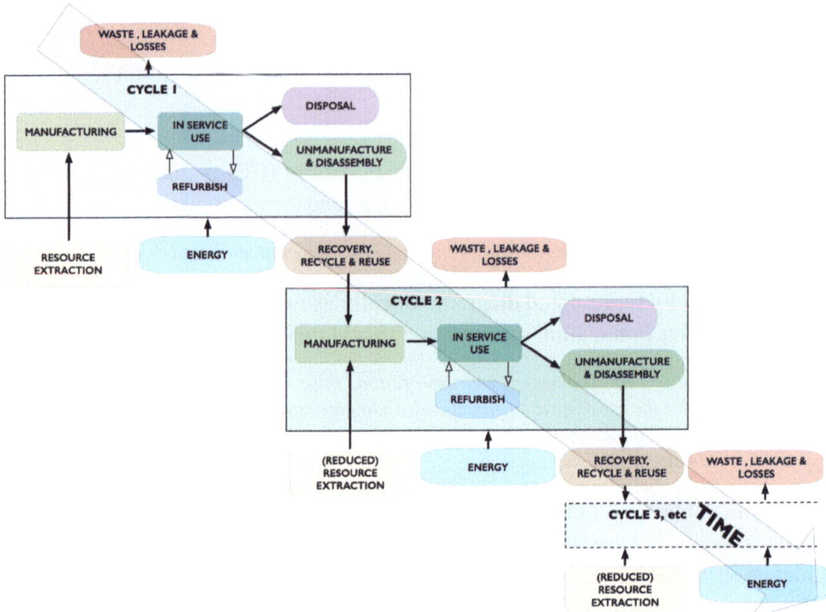

Fig. 7.2 A circular economy with multiple iterations over time

materials and energy flow analyses therefore need to take this into consideration. This situation is then perhaps better represented by a relationship such as that suggested by Fig. 7.2.

It is also important to note that resource recovery and reuse may not always be the optimum solution at end-of-life and in some instances the energy, environmental and other costs associated with recovery may well outweigh those same issues in relation to the use of new source material. In these cases, the emphasis will be on utilising the most effective; energetic, environmental and economic, means of disposal at end-of-life.

7.2.2 Decarbonization and Global Net Zero

Decarbonisation and Global Net Zero references the situation where CO2 and other greenhouse gas emissions resulting from human activities and their removal from the atmosphere are in balance over a given period. As of 2023, more than 140 countries have a Net Zero target covering 92% of global GDP, 88% of emissions and 89% of the world population. However, many current claims may be considered to be of low credibility, reflecting a lack of binding regulation.

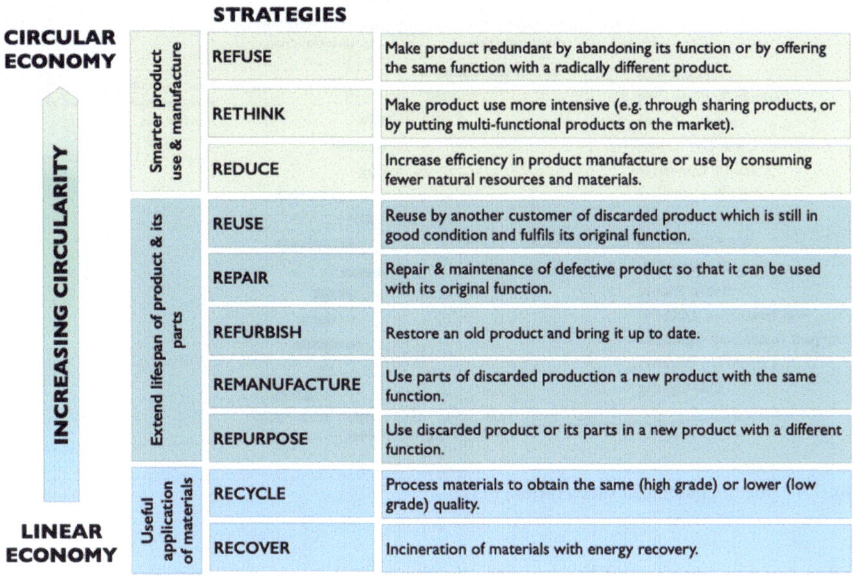

Fig. 7.3 From a linear economy to a circular economy (after [8])

The achievement of the Net Zero goal is associated with many issues, not least of which is developing the necessary infrastructure to support the change. For instance, consider the potential replacement of petrol and diesel engine cars in the UK. In 2022 according to the Digest of United Kingdom Energy Statistics,[1] car travel in the UK consumed around 20 million tonnes of oil, approximately equivalent to 233 TWh of energy. Assuming 30% efficiency at the vehicle, this represents a useful output of 69.9 TWh. For an electric vehicles assuming 90% efficiency at the vehicle and transmission losses of 10% this would represent a requirement of around 86 TWh or some 27% of the UK electrical demand for 2022 of 320.7 TWh.

7.2.3 Design

The concept of sustainability is forcing designers to address a range of questions and issues that have not conventionally been considered other than peripherally [9, 10]. For example, the balance between the requirements of Design for Assembly and those for Design for Disassembly, which may often be in conflict. It is also clear that the introduction on a mass scale of technologies that are ecologically and environmentally responsible within the context of sustainability generates a requirement for design simplification to facilitate both the update and repair of products

[1] Digest of United Kingdom Energy Statistics (DUKES) 2023 @ www.gov.uk/government/statistics/digest-of-uk-energy-statistics-dukes-2023 (accessed February 2024).

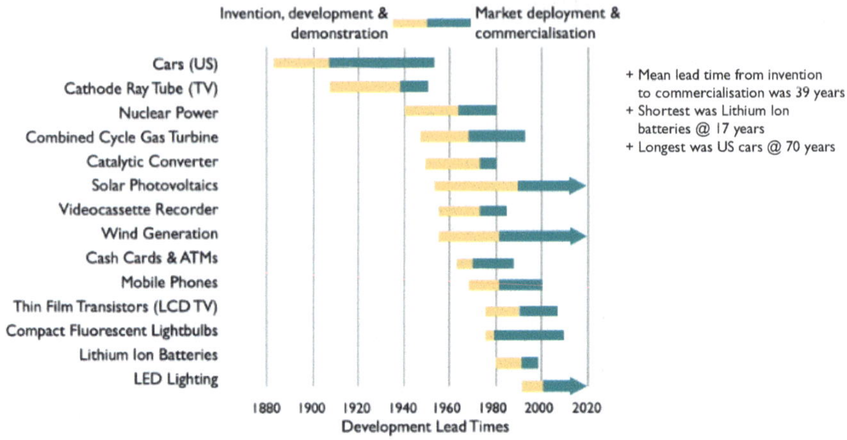

Fig. 7.4 Development lead times (after [11])

and processes, supported by the introduction of relevant and appropriate standards for their management. This will require the collaborative effort of governments and regulators along with that of industry, along with its supply chains, and education.

Development Lead Times

It also needs to be remembered that, as indicated by Fig. 7.4, the time taken from concept to commercial implementation of a product or system is typically of the order of 30–40 years [11]. In terms of sustainability, this may well impact on the end-of-life situation and the need to develop new technologies to support disassembly and resource recovery, as for instance for batteries, and implies that the design process must implement some form of future-proofing to consider implications for end-of-life.

Obsolescence, Update, Repair or Replace

Over recent years technology has been marketed on a process of replacement rather than repair in which products such as mobile phones are subject to continuous '*improvements*' such as better cameras and more storage, with older models then being scrapped, with obvious issues for materials recovery and waste management. While there are increasing moves to supply technologies that can be upgraded, this remains an issue for the designer, along with managing user expectations. In terms of sustainability, this relates to the differing concepts of obsolescence as suggested by Table 7.1, and the management of a transition from a situation where the emphasis is on new product development to upgrade and preservation with obvious implications for resource utilisation.

Table 7.1 Types of obsolescence

Type	Description
Technological	This results when newer, often better performing, products enter the market and support for the existing products is removed
Functional	This results when a device can no longer adequately perform to its intended functions. For instance, when hardware is not able to support new software, or software is not supported by a newer hardware
Economic	This results when it is no longer economically viable to continue using the product or system. This could occur in situations where would be more expensive to repair a product than to buy a new one
Regulatory	This occurs results when a product, or part of a product, no longer conforms with or to standards or regulations. An instance of this would be seen in the banning of older, higher emission, vehicles from city centres
Inventory	This occurs when the means of producing a product themselves cease to be available, potentially as a consequence of their own obsolescence!
Style/Aesthetic	This occurs when the product is no longer considered desirable because of changes in trends and fashions

Cradle-2-Cradle

Cradle-to-cradle design as represented by Fig. 7.5 maps industrial processes onto natural processes, with materials viewed as nutrients. The approach emphasises that industry must protect and enrich ecosystems and biology while maintaining a safe and productive technical conditions. The model can be applied to a wide range of situations and circumstances such as urban environments, buildings, economic models and social systems [12].

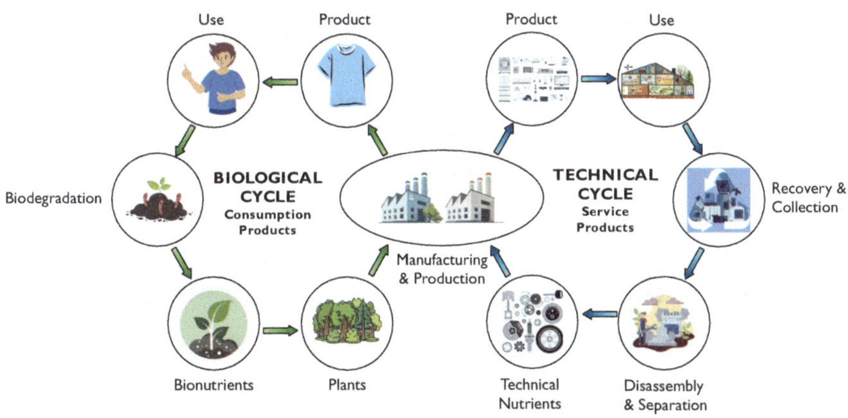

Fig. 7.5 Cradle-2-cradle concept (after [12])

7.2.4 Manufacturing

The manufacturing process is important to both sustainability and the achievement of a CE, in respect of the production of the initial product as well as to disposal, including unmanufacture and disassembly, at end-of-life. This requires that manufacturing processes, their energy consumption and waste production, must be included within the environmental assessment of a product or system. From a Mechatronics perspective, the various elements comprising an individual Mechatronic device, component, system or sub-system all need to be manufactured and subject to a resource recovery and reuse process at end-of-life. This implies that the Mechatronics designer must take these into consideration.

7.2.5 Lifecycle Management

The ability to monitor and record behaviour over the lifecycle of a product or system is increasingly important in terms of an ability to respond to changes over time in a way that minimises the potential environmental impacts and avoids unnecessary activities. In illustration consider how the use of a Digital Twin can be used to support the life cycle monitoring of a vehicle. In the representation of Fig. 7.6, an individual vehicle can communicate with other vehicles, infrastructure elements, personal devices and The Cloud.

Thus, the vehicle can be in continuous communication with its Digital Twin, which is also receiving information from other vehicles in the class. Via the Digital Twin, the need for a maintenance appointment, defined by need rather than by a timed schedule, is identified. The garage is then contacted and a booking made, and entered into the user's diary. Components are ordered, or produced in-house by the

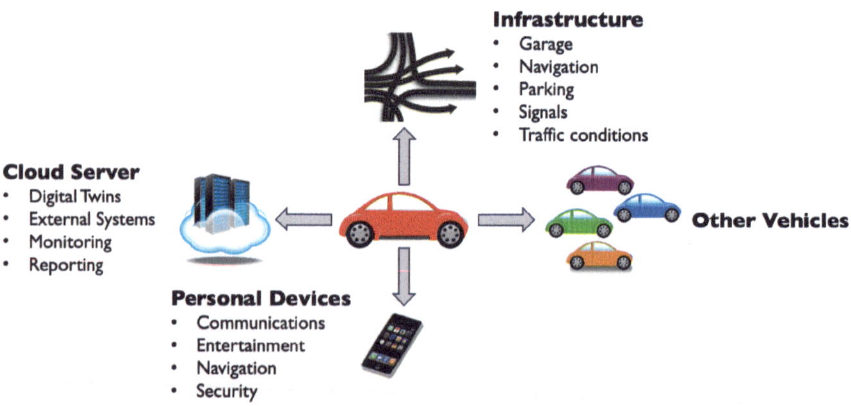

Fig. 7.6 Internet of vehicles

garage using 3D printer technology so that they are ready when the vehicle enters the garage. In this way, a complete record of the vehicle's life is produced and available such that at end-of-life component and materials recovery may be facilitated. Similar relationships can be found for buildings, power generation, energy management and a range of domestic systems.

7.2.6 Waste

In his 1971 book *The Closing Circle: Nature, Man, and Technology* [13] Barry Commoner proposed the *Four Laws of Ecology* as being:

- *Everything is connected to everything else*—there is one ecosphere for all living organisms and what affects one, affects all.
- *Everything must go somewhere*—there is no '*waste*' in nature and there is no '*away*' to which things can be thrown.
- *Nature knows best*—humankind has fashioned technology to improve upon nature, but such change in a natural system is likely to be detrimental to that system.
- *There is no such thing as a free lunch*—exploitation of nature will inevitably involve the conversion of resources from useful to useless forms.

Further, as has already been seen, the Second Law of Thermodynamics means that all systems will require some make-up either of energy, which in the natural world is in the form of solar radiation, or physical resources. It is also the case that there will be instances where the energy, environmental and other costs of recovery outweigh those same issues with respect to the use of new source material. In those cases, the emphasis will be on the most effective, energetically, environmentally and economically, means of disposing of the artefact containing this material at end-of-life.

A further element of waste management is the concept of downcycling where Helbig et al. [14] suggested the following set of conditions:

- Reduction of material quality.
- Suitability only for less-demanding applications.
- Use in less-valuable products.
- Loss of alloying elements during recycling.
- Different material system with no functional use.
- Need for additional virgin raw material.

Which again identifies the need for careful consideration of life cycle properties at the design stage.

7.3 Role of Mechatronics

What is the role of Mechatronics in the global striving for sustainability? While Mechatronics is a broad discipline the more specific area of *EcoMechatronics*, as a specialism within Mechatronics, embeds sustainability as a core objective where the decision-making is influenced by the cutting-edge of sustainability innovation in problem-solving. The challenge is to identify how Mechatronics applications can, as a whole, contribute towards achieving more sustainable solutions.

Where problem-solving is constrained by the need to meet a range of competing criteria where sustainability is not the main focus, the process outcomes require compromises such as cost being prioritised over sustainability. Limiting solutions to decision-making models enables the point at which the decision to stop paying for the benefits of better sustainability that could be achieved may be identifiable, but not necessarily result in an optimum sustainable solution being adopted.

Mechatronics has the potential to contribute to more sustainable solutions for society, but is society prepared to accept the challenges such solutions may require? At what point do other priorities such as cost and lifestyle change result in the loss of a viable solution that may be more sustainable if fully developed and exploited?

Mechatronics is an interdisciplinary engineering subject involving a broad range of skills and knowledge that includes aspects of mechanical engineering, industrial engineering, production engineering, engineering electronics and computer systems together with any further specialism relevant to a particular system problem and incorporating advanced methods for the design, construction, and operation of '*intelligent*' (but not necessarily '*smart*') products and processes. Indeed, there are at least 11 identifiable technical areas of competence within the remit of Mechatronics [15].

The multi-disciplinary profession of Mechatronics is not just concerned with higher system performance in and of itself, or economic factors, but in terms of how the system and its parts impact upon sustainability. However, technical roles do not tend to cover the diversity of aspects of Mechatronics to influence the decision-making process around what gets solved, with the focus generally more towards how to solve a problem. The ability to influence the decision-making process through the knowledge that Mechatronics professionals can provide is a critical factor in making progress towards understanding the consequences of compromising on sustainable issues.

The functions of Mechatronics, in particular control and monitoring, serve to inform decision-making and therefore subsequent actions. What is measured and what is controlled (to the extent that anything can be measured accurately for monitoring and also controlled for the purposes of an intervention) determine what is measured (or not measured) to evaluate the success or limitations of any solution to a problem. Assuming suitable criteria are identified and measurable then how are the '*improvements*' expected to be benchmarked as having been achieved?

Following this logic it would also be necessary to ensure that, as far as can be understood and determined, any negative impacts on sustainability that could and should be measured are identified. But it is not the sole responsibility of the Mechatronics discipline to disaggregate a complex problem to a range of '*mini-problems,*' or to understand the systemic problem and guide the overall solution instead of each part of the solution. Mechatronics can be the driver of sustainability solutions dealing with a complex system problem as a whole and with the problems within the disaggregated sub-systems.

The future of sustainability is a global problem shared by all humanity, crossing geo-political lines, cultural divisions and different generations. As such, the global problem is large and complex, for which Mechatronics can contribute solutions in relation to the needs of all humans in respect of quality of life, including how we address sustainability priorities. Whilst we can adopt Mechatronics to help in fields varying from agriculture to zoos [16], what and how we measure is limited to very specific contextualised problems that do not necessarily have sustainability as the focus for a solution. Any resulting improvement in sustainability may then be incidental to the application of Mechatronics. In making improvements for efficiency, the wider sustainability issues around the environment, social impact and economy require compromises to be fully understood. Efficiency gains made in areas such as energy recovery systems reduces energy losses in the system while adopting more natural-based and recyclable materials may be less harmful to the environment as one part of sustainability, but these solutions still have inherent losses within a Circular Economy model of sustainability. Mechatronics applications can be implemented to close the gap from what is lost to the Circular Economy model, and more ambitiously, possibly even reverse certain losses.

This raises the matter of the extent to which Mechatronics can influence the decision-making processes concerning sustainability. Whilst Mechatronics can contribute to advances and innovations around, for example, energy recovery in transport systems, or making flight more fuel and energy-efficient, where are the bigger questions around:

What forms of transport are needed?

And

What changes in human lifestyles are required?

Being both asked and answered?

Thus as Mechatronics would be an integrated aspect of any SMART city, where is Mechatronics influencing and future-proofing the design of a genuinely sustainable modern city that is currently being built? Is Mechatronics being involved in the decision-making process too late? There are also further challenges to identifying the point in the decision process where Mechatronics can have the greatest impact and influence. What is required to motivate the kind of thinking that focuses innovation towards a genuinely sustainable future, of which Mechatronics is an integrated aspect?

Theoretical claims, assertions and proposals are no good for the commercial world, proof of concept evidence attracts greater funding potential and influence on what is addressed. How can Mechatronics identify and monitor what is required to influence the decision-making process around increasing sustainability that is, on balance against all other criteria, sufficiently attractive to become the driver for positive change?

Sustainability requires an understanding of the global ecosystem and the interdependencies among organisms whereby the needs of one organism are met in some form through the reuse of resources previously used by another. This is the basis of Cradle to Cradle thinking and goes beyond challenging the mindset of '*doing less bad*' [17] to recalibrate the problem in terms of '*doing more good.*'

Mechatronics professionals can therefore have within them the aspiration of improving through design innovation the sustainability of a wide range of systems with solutions that contribute more to sustainability during the system lifecycle than they remove.

7.4 Mechatronics as a Driver for a Circular Economy

Various properties of Mechatronic systems as enablers of the Circular Economy are presented using plastics recycling as an example.

7.4.1 Background

As is seen, the concept of a Circular Economy references a cyclical process beginning with extraction of raw materials, followed by manufacturing and production, distribution and transport, product usage and finally the recycling of materials (incorporating collection and processing). Figure 7.7 shows a Mechatronics oriented representation of the Cradle-to-Cradle approach associated with the development of circular products. In contrast to a linear economy, the life of the raw materials is extended over several product life cycles.

In all the elements of the product creation phase of Fig. 7.7, Mechatronic systems are involved, for example in production systems as well as in sensor systems for materials recognition and sorting within the recycling process. Often systems, as for instance a plastics extrusion machine, are used for both primary materials and recycled materials. With the help of process modelling techniques (a Process Twin), process parameters are determined to obtain the same product quality regardless of the starting material. Chouinard et al. [18] identify the potential of Circular Economy implementation in Mechatronics industry and vice versa.

For the optimised operation of a Mechatronics oriented production system, the modelling of both material and energy flows is necessary. A simplified way to map material flows is to use material and energy balances, which in principle are nothing more than the application of the First Law of Thermodynamics. In practice, they are drawn up in the form of operational, process and product balances.

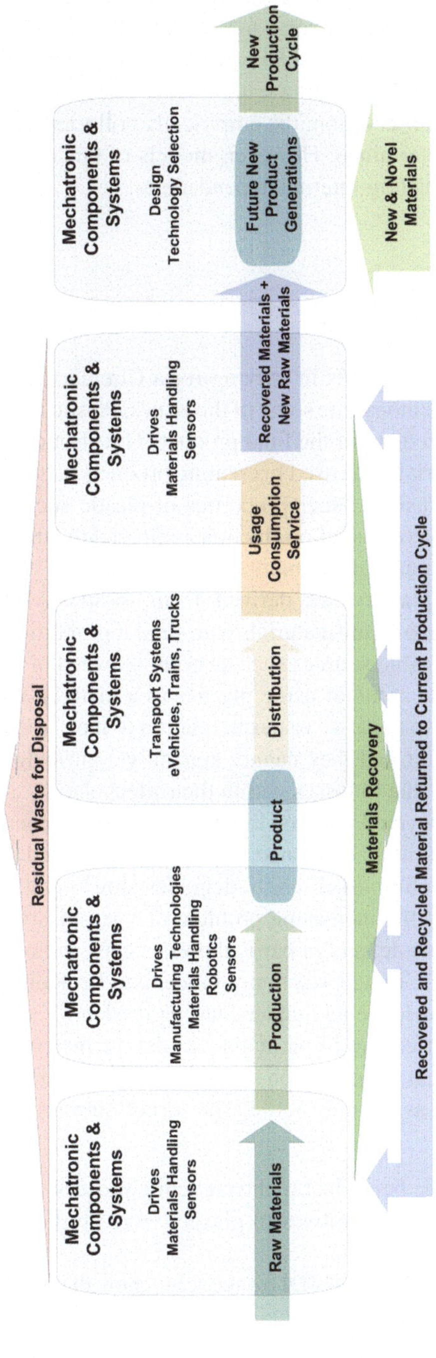

Fig. 7.7 Mechatronics as part of a cradle-to-cradle approach

Operational balances depict the material flows of a company while process balances are the result of a material and energy balance across individual or several process steps. Product balances then serve to visualise the material flows of a product over its entire life cycle. Balancing can be carried out in accounting terms by empirically recording all material, energy flows and stocks. Under the assumption of mass and energy balance and conservation, the empirically collected data can be checked and missing quantities determined. However, models of this type are only suitable to a limited extent for depicting internal dependencies.

7.4.2 Plastics Recycling

In analysing the inherent problems in striving for a Circular Economy model it is possible to use plastic to demonstrate some of the issues. Plastic is a material that uses polymers as the main ingredient of the final product. The plasticity properties of the polymers enable the material to be used in creating products using differing methods such as moulding or extrusion. Other properties of plastic such as its lightweight nature, durability and relative cost of production compared to other materials has led to its increasing use as a material for many everyday roles.

While most current plastics are derived from fossil fuel-based chemicals, researchers are attempting to create materials with similar properties to certain current plastics from more sustainable sources such as corn, seaweed or cotton derivatives. While there are positive benefits of using plastics as a material when compared to other options, there are also negative issues in relation to their use and impact on the environment with the result that they cannot generally contribute successfully to a Circular Economy model due to losses within their lifecycle.

While plastic can be recycled, the first loss to the Circular Economy model arises from plastic in the form of litter abandoned by people or collected and sent to landfill and therefore lost to recycling. Such waste degrades slowly and is giving rise to a contributory awareness of the increasing problem of microplastic pollution and its impact on the earth as a whole, and in particular on oceans and organisms.

Before this loss to the Circular Economy model can be addressed a number of issues require further knowledge and understanding to identify appropriate policies and actions that will be effective. In particular, whilst the material may be durable, that is a somewhat pointless benefit when a thermoplastic material is used as part of a single-use product such as a drinks bottle. The sort of questions to be answered at this stage include:

- To what extent is plastic being lost to the recycling process?
- What effect do the various additives to plastics have on the environment when plastic degrades?
- How do we capture all of the material for recycling now that there is an awareness of the problem at microplastic level?

The plastic retrieved after use for recycling and not lost as waste to litter or landfill is recovered and separated with the intention of recycling or incineration as the next typical lifecycle stage depending upon the properties of the plastic. Plastic for recycling can be sorted into the main streams of polypropylene (PP), high-density polyethylene (HDPE), low-density polyethylene (LDPE), PET and polystyrene (PS) using various rotary screen drums, near-infrared sorting and washing steps. The material from this process is granular and is reused in non-food-grade plastic products.

Where the plastic is considered suitable for incineration as part of a waste-to-energy option for managing the material the process gives rise to different problems in a Circular Economy model in that environmental issues such as the production of toxic gases and heavy metal-contaminated ash. How are these losses to be considered in a Circular Economy model? At each iteration of plastic retrieval and recycling energy is used along with other materials as a part of the material production process which will ultimately be other losses to the Circular Economy model [19]. Degradation of lorries used in transportation and the emissions from the fuel used (and the process for creating/providing the fuel), carbon output from energy used to re-process the material and subsequent further transportation as new product for next iteration in the life cycle of the material. These processes are illustrated by Fig. 7.8.

Plastics will often end up in household waste leading to recycled plastic with reduced quality, limiting the potential for closed-loop recycling. PE, PP, PS and PET are classified as thermoplastic polymers and are major contributors to single-use

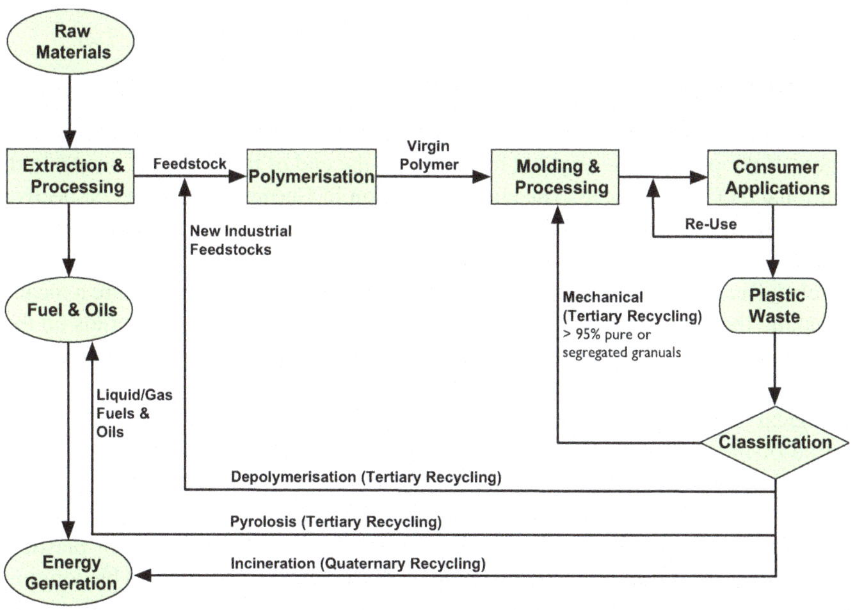

Fig. 7.8 Plastics and waste management

plastic waste. PET, PE and PP recycling methods require different solutions to address the limitations to closing the loop in a Circular Economy model. Potential degradation of the PET polymer can be reversed in a decontamination process, making PET waste well-suited for closed loop, multiple times recycling, even when the degree of heterogeneity in the waste is high [20].

Current recycling of mixed PP waste and separate recycling of individual PP waste will not facilitate recycling into new packaging products [20]. Therefore, the retrieval and reprocessing of PE and PP waste requires more effective sorting methods and regulatory harmonisation of product design, accounting for polymer properties and recyclability. Degradation of PP during recycling has been shown to be substantial, representing another important limitation for PP recycling, something necessary to address for the future [20].

There are different methods that can be used for plastic recycling that, as with any commercial process, give different results based on priorities and compromises between cost, feasibility and effectiveness. There is evidence in research that indicate that chemical recycling methods that can be upscaled for industry use may not be as effective as the potential of biological recycling either via microorganisms or biological-derived enzymes [21].

7.4.3 Selected Issues

According to Hehenberger [22], material and energy flows play an important role in analysing the sustainability properties of production processes (see Fig. 7.2).

Material flow analysis in plastics recycling

The European plastics industry is in transition to meet its 2050 net zero and circularity targets. If the plastics are recycled for energy at the end of their life, i.e. incinerated, additional CO2 is produced. The current situation in Germany from the perspective of the Circular Economy is that after one year, almost half of the newly produced plastics are still being used, 30% are being recovered for energy and 22% are being recycled. Recovered secondary raw materials from the domestic market are a means of making the economy less dependent on imports of low-cost raw materials on the world market. This is associated with greater planning security for the raw material refining industry. An expansion of the Circular Economy is thus inevitable from an ecological and economic point of view.

End-of-life products are recycled to reduce the burden on the environment or to save energy. Thus, information about the environmental impact of the recycling process itself, or its efficiency, is also important. The aim is to design a methodology for planning, describing and modelling material flows in the recycling of plastics. This includes the delimitation of the complexity of the recycling process by defining the system boundaries, the simulation of the material and energy flows as well as a life cycle assessment [23]. This should enable an efficiency assessment of the recycling process and knowledge of the environmental impacts.

Energy Flexibility in Production and Recycling

In a recycling process, there are usually a large number of energy consumers, e.g. electric motors. These are usually components in a hierarchical arrangement of components, machines or production cells. There are two possible approaches to determine the current energy consumption. On the one hand, the recording can take place top-down, e.g. the energy demand of the entire production/recycling site is measured and assigned to the consumers across several levels. Alternatively, the recording can also be done bottom-up, e.g. the production processes are modelled and the energy demand of the entire process [24].

The modelling of the energy flow within the Mechatronics systems plays an important role. When working with energy models, the design engineer will not consider all relevant influences on all definable properties parallel. In most cases, the focus is on the specific, as for instance on the overall energy consumption of a machine. Subsidiary models are then used to describe the behaviour of the internal components. These models differ in their complexity, which depends on the considered definable properties, simplifications (considered effects) and methods to describe the interrelation (numeric, analytic, graphical).

7.5 Conclusion

This chapter aims to consider specific aspects of the contribution of Mechatronics to Sustainability and Circular Economy. As an aspiration, Mechatronics professionals can proceed with the goal of improving design sustainability innovations with solutions that contribute more to sustainability during their lifecycle than they take out.

Acknowledgements This research was partially carried out in the course of the project NaKuRe and was financed by research funds granted by the Government of Upper Austria over the course of the programme FTI-Struktur, projectWi-2021-305611/21-Au/NaKuRe.

References

1. Brundtland GH (1987) Our common future: report of the 1987 world commission on environment and development @ sustainabledevelopment.un.org/content/documents/5987our-common-future.pdf. Accessed May 2024
2. Ellen MacArthur Foundation (2013) Towards a circular economy, Vol 1: an economic and business rationale for an accelerated transition @ www.ellenmacarthurfoundation.org/towards-the-circular-economy-vol-1-an-economic-and-business-rationale-for-an. Accessed May 2024

3. Ellen MacArthur Foundation (2013) Towards the circular economy, Vol. 2: opportunities for the consumer goods sector @ www.ellenmacarthurfoundation.org/towards-the-circular-economy-vol-2-opportunities-for-the-consumer-goods#:~:text=circular%20economy%20Vol.-,2%3A%20opportunities%20for%20the%20consumer%20goods%20sector,and%2075%25%20of%20municipal%20waste. Accessed May 2024
4. www.news.un.org/en/story/2023/12/1144742. Accessed May 2024
5. www.theguardian.com/australia-news/2023/aug/23/renewable-energy-new-power-lines-for-transmission-risks-concerns. Accessed May 2024
6. Hehenberger P, Habib M, Bradley D (2022) EcoMechatronics: challenges for evolution. Springer, Development and Sustainability
7. Kirchherr J, Reike D, Hekkert M (2017) Conceptualizing the circular economy: an analysis of 114 definitions. Resour Conserv Recycl 127:221–232
8. Potting J, Hekkert MP, Worrell E, Hanemaaijer A (2017) Circular economy: measuring innovation in the product chain, Planbureau voor de Leefomgeving, Pub. 2544 @ www.pbl.nl/sites/default/files/downloads/pbl-2016-circular-economy-measuring-innovation-in-product-chains-2544.pdf. Accessed May 2024
9. Bhamra T, Hernandez RJ (2021) Thirty years of design for sustainability: an evolution of research, policy and practice. Des Sci 7:2
10. Silberer J, Astfalk S, Planing P, Müller P (2023) User needs over time: the market and technology maturity model (MTMM). J Innov Entrepreneurship 12(1):39. https://doi.org/10.1186/s13731-023-00302-2.pdf
11. Hanna R, Gross R, Speirs J, Heptonstall P, Gambhir A (2015) Innovation timelines from invention to maturity, UK Energy Research Centre @ www.researchgate.net/profile/Richard-Hanna-2/publication/286025108_Innovation_timelines_from_invention_to_maturity_A_rapid_review_of_the_evidence_on_the_time_taken_for_new_technologies_to_reach_widespread_commercialisation/links/566567fd08ae192bbf922477/Innovation-timelines-from-invention-to-maturity-A-rapid-review-of-the-evidence-on-the-time-taken-for-new-technologies-to-reach-widespread-commercialisation.pdf. Accessed May 2024
12. Braungart M, McDonough W (2009) Cradle to cradle. Random House
13. Commoner B (1971) The closing circle: nature, man, and technology. Random House
14. Helbig C, Huether J, Joachimsthaler C, Lehmann C, Raatz S, Thorenz A, Faulstich M, Tuma A (2022) A terminology for downcycling. J Ind Ecol 26(4):1164–1174
15. Harashima F, Tomizuka M, Fukuda T (1996) Mechatronics—what is it, why, and how?, Editorial. IEEE/ASME Trans Mechatron 1(1):1–4
16. Takasaki T, Nagato R, Morishita E, Higuchi H, Ikutarou K, Hosaka H, Itao K (2001) Wild animal tracking by PHS (Personal Handy phone System). Elsevier Science, Human Friendly Mechatronics, pp 43–48
17. Braungart M, McDonough W, Bollinger A (2007) Cradle-to-cradle design: creating healthy emissions–a strategy for eco-effective product and system design. J Clean Prod 15(13–14):1337–1348
18. Chouinard U, Pigosso D, McAloone T, Baron L, Achiche S (2019) Potential of circular economy implementation in the mechatronics industry: an exploratory research. J Clean Prod 239:118014
19. Chen J, Wu J, Sherrell PC, Chen J, Wang H, Zhang WX, Yang J (2022) How to build a microplastics-free environment: strategies for microplastics degradation and plastics recycling. Adv Sci 9(6):2103764. https://doi.org/10.1002/advs.202103764
20. Eriksen MK, Christiansen JD, Daugaard AE, Astrup TF (2019) Closing the loop for PET, PE and PP waste from households: influence of material properties and product design for plastic recycling. Waste Manag 96:75–85
21. Lee A, Liew MS (2021) Tertiary recycling of plastics waste: an analysis of feedstock, chemical and biological degradation methods. J Mater Cycles Waste Manag 23(1):32–43
22. Hehenberger P (2022) A holistic and sustainable view on the product creation process for mechatronic systems, EcoMechatronics: challenges for evolution, development and sustainability (Eds: P. Hehenberger, M. Habib, D. Bradley), Springer, pp 65–84

23. Merschak S, Kneidinger C, Katzmayr D, Casata J, Hehenberger P (2024) Application of life cycle assessment for more sustainable plastic packaging—challenges and opportunities, IFIP 20th International conference on product lifecycle management. Leveraging digital twins, circular economy, and knowledge management for sustainable innovation (Eds C Danjou, R Harik, F Nyffenegger & Louis Rivest & A Bouras), Springer, pp 61–71
24. Hehenberger P, Leherbauer D, Penas O, Delabeye R, Patalano S, Vitolo F, Rega A, Alefragis P, Birbas M, Birbas A, Katrakazas P (2023) Holistic system modelling and analysis for energy-aware production: an integrated framework. Systems 11:100. https://doi.org/10.3390/systems11020100

Part II
Mechatronics Design

Chapter 8
The REMAC Framework for Requirements Engineering Success in Agile Mechatronic Development

Olivia Penas, Ferdinando Vitolo, Peter Hehenberger, Régis Plateaux, and Stanislao Patalano

Abstract In mechatronic design, agile methods address many of the emerging challenges of the dynamic, volatile, uncertain, complex, and ambiguous environment of rapid technological and market change. In the context of the introduction of the agile development process for hardware systems such as mechatronic systems, this paper proposes to tackle the following challenge: how to ensure the consistency and traceability of the requirements management all along an agile development of mechatronic systems. In this respect, we propose the analysis of the state of the art, before providing the REMAC (Requirement Engineering for Mechatronics in an Agile Context) framework including some key tools to manage the consistency of dynamic management of requirements in an agile context. Therefore, this framework supports engineers in the requirements engineering activities of their development process.

Keywords Requirements selection · Requirements engineering · Mechatronic development · Agility · Change management

8.1 Introduction

In an agile approach, the requirements engineering presents the main following features: (i) the system requirements are defined all along the development process, (ii) the stakeholders prioritize them as the product development progresses and (iii)

O. Penas (✉) · R. Plateaux
Laboratoire QUARTZ, EA7393, ISAE-Supmeca, Saint Ouen, France
e-mail: olivia.penas@isae-supmeca.fr

F. Vitolo · S. Patalano
Fraunhofer J-Lab IDEAS, Department of Industrial Engineering, University of Naples Federico II, Naples, Italy

P. Hehenberger
Institute for Agile Transformation, University of Applied Sciences Upper Austria, Wels, Austria

P. Hehenberger and D. Bradley (eds.), *Mechatronic Futures*,
https://doi.org/10.1007/978-3-031-83571-1_8

they are treated progressively in order of priority, with regular steps of verification and exchange with the stakeholders. Such properties allow us to take into account the evolution of the customers' needs during the product development process. Regarding mechatronic design, the agile context makes it difficult to consider the system with all its different views in each development cycle (sprint) [36], since a prototype has to be delivered in a limited period, considering only a part of the whole set of requirements. This raises the question of how to manage and ensure the consistency and completeness of the specification of the final system developed based on those requirements assigned to each sprint.

To tackle this research question, after an introduction to the agile requirements engineering for mechatronic design, Sect. 8.2 will describe the analysis of existing approaches to ensure requirements consistency and traceability in an agile context. Section 8.3 will then propose appropriate key Requirements Engineering (RE) management tools for successful mechatronic design in an agile context, through the REMAC (Requirement Engineering for Mechatronics in an Agile Context) framework. Section 8.4 provides the conclusions and considers future work.

8.1.1 Overview of Mechatronic Design

Mechatronic products offer many advantages such as precision, automation and extended functionalities. In most cases, these come at the cost of the increased complexity of the product and the related product development and design process. A mechatronic design process typically involves several disciplines (e.g., mechanical engineering, electrical engineering and, information technology including software engineering) [17, 26, 48]. A typical mechatronic design process model comprises six design phases: Requirements Design (1), Functional Design (2), Principle Design (3), Architectural Design (4), Preliminary Design (5) and Detailed Design (6). The core elements of such a model are the continuous elicitation, adaptation, fulfillment and documentation of requirements. Requirements of stakeholders (including customers, owners and users) regarding the product are clarified and specified at the beginning of the process, but are usually incomplete and may change throughout the process. Thus, assumptions are necessary to reach a solution.

Since results must be evaluated and validated against the requirements, external decisions are integrated into the mechatronic design process model. Iterations may become necessary if requirements appear out of reach. Furthermore, several design tasks in the design process model themselves require iterations.

Functions in the mechatronic system represent the expected behavior of components. Previously, mechatronic products were being seen as a dominant mechanical discipline and less software oriented. Due to technological advances, new trends have emerged over the past decades, such as the Internet of Things (IoT), which are more inclined toward software components. Functional allocation on software components has increased while maintaining the number of electronic devices on the same level throughout these 50 years of evolution in the mechatronics world [18]. Whatever

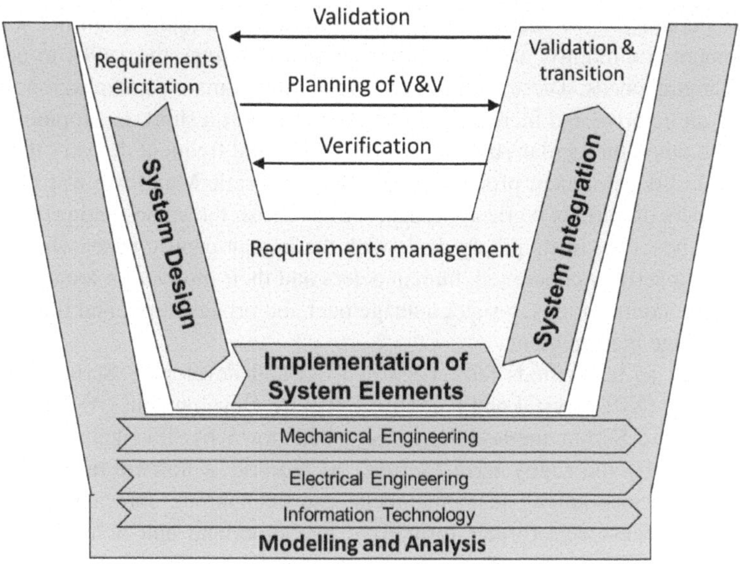

Fig. 8.1 V-Model based on [46, 47] with a focus on requirements engineering

these evolutions, the requirements engineering and management process still plays a crucial role in the success of the mechatronic design process, as represented in the V-model according to the VDI-guideline 2206 (Fig. 8.1).

In recent years, agile approaches have also taken place in mechatronics engineering [30, 51] and may be studied in the specific context of mechatronic design and notably for requirements engineering (RE) activities.

8.1.2 Agile Methods

The design process based on the V-model is a straightforward sequential process, with hints at iterations [15]. The V-model underlines the needs (i) for the definition of verification plans while the requirements development, (ii) for the continuous validation with the stakeholders and finally (iii) the criticality of a continuous evaluation of risks and opportunities [49]. Indeed, external (customer needs, technologies ...) and internal (complexity, competencies, ...) changes in the product development process lead to the limited effectiveness of traditional product development methods [5]. Due to the need for shorter development cycles and the constant change in the technology and business environment, organizations have to deal with business sectors where requirements are quite variable, and in parallel, find an alternative to the documentation-driven requirements management in order to be more adaptable to change.

In this context, agile methods have been developed, initially from the software development community, and they have then evolved since the 1990s to adapt to these changing needs. These methods are based on two main principles: on the one hand, on an iterative and incremental process to support a short development cycle and on the other, on regular customer collaboration and frequent delivery through a light and fast development process [7]. In 2001, an Agile Manifesto was produced by the leaders of various Agile approaches, to establish the values and principles of Agile methods. One of its pillars deals with the human-centric base: whatever the development activity concerned, human actors and their motivation are at the heart of decisions, contributions, project management and organization, and finally skills and knowledge management.

Numerous agile methods have been evolved and developed: Scrum, eXtreme Programming (XP), Crystal and Adaptive Software Development (ASD). The first concepts of the Scrum methodology, were laid down by Takeuchi and Nonaka [44], inspired by the rugby term '*Scrum*', to promote a holistic team approach, constant interaction among team members, and unchanging core team members. The Scrum process was further formalized by Sutherland and Schwaber [43] to provide a complete framework (value, actors, meetings, artifacts…). As a result, these agile approaches led to shorter development cycles, higher customer satisfaction and on-time delivery [38], lower bug rates and quicker adaptation to changing business requirements.

Focusing on mechatronic systems design, which mixes software and hardware development, Johnson underlines that numerous changes frequently happen also in hardware development, be it due to changing specifications, employee turnover, team dynamics or new emergent technologies [21]. Indeed, successful products evolve over time and require real collaboration within a team that has to be constantly synchronized with the needs of the customer, as the development team has to understand the context of its efforts and the business in which it operates. Therefore, Johnson argues that when hardware development is viewed as a creative process, the concepts of agility and related software practices can be exploited: rotating teamwork without disciplinary silos and physically brought together in one place with iterative hardware and test-driven developments, to, similarly, regularly demonstrate products, quantify progress and gather feedback.

Other researchers have also studied agile approaches for mechatronic product developments: Böhmer et al. propose a holistic methodology that integrates the perspectives of agile development for mechatronic engineering to cope with a high level of uncertainty and complexity, in accordance with the VUCA context described by Bennet et al. [4]. They underline that the numerous mechatronic integrations require processes to adapt to the continuous change in technology and user needs [5]. On their side, Kasauli et al. worked on large-scale agile system development with a multiple case study based on 7 large-scale systems companies, highlighting the corresponding requirements engineering challenges [22].

8.1.3 Agile Requirements Engineering

Regarding Requirements Engineering (RE) in the implementation of traditional methods, work commonly begins with the elicitation and documentation of a '*complete*' set of requirements, followed by architectural and high-level design, development, and inspection [8]. Yet, even if these traditional methods are still widely used in industry because of their straightforward, methodical and structured nature, as well as their predictability, stability, and high assurance, they provide slow adaptation to constantly changing business requirements and a tendency to be over budget and behind schedule with fewer features and functions than specified. Finally, they are unable to respond to changes that often determine the success or failure of a product [7].

In an agile context, RE is performed iteratively throughout the development process, not just in a defined phase at the beginning of the design phase. As a result, the main RE activities (elicitation, documentation, validation, negotiation and management) are not distinctively organized in an agile context. In this context, Inayat et al. define Agile Requirements Engineering (ARE) as an '*agile way*' of planning, executing and reasoning about requirements engineering activities [19]. Schön et al. propose a systematic literature review on ARE classifying research studies regarding the following specific aspects: stakeholder and user involvement, data gathering, user perspective, integrated methodologies, shared understanding, artifacts, documentation and non-functional requirements [38]. Later, Schön et al. detail specificities of agile requirements engineering activity, namely the requirements discovery (eliciting new requirements at each iteration), refinement (analyzing the impact of these new requirements), prioritization (in accordance with the corresponding benefit), review (comparing the implemented requirements with their expected related added value) and documentation (capturing collaborative discussions related to the requirements) [39]. Globally, most agile approaches share the same practices of RE: client participation and iterative interactions (observations, interviews, workshops and strong team collaboration) for elicitation, writing user stories, iterative requirements (incremental), continuous requirement prioritization, requirements change management and related tracking, '*just-in-time requirements*' (JIT requirements) and continuous planning, cross-functional teams to deal with requirements, prototyping and test before coding/developing and shared review and acceptance tests [12, 16, 19]. Even if most of the current ARE-related literature focuses on agile software development, most of these features remain relevant to deal with mechatronic development in an agile context.

8.2 Analysis of Existing Approaches to Ensure Requirements Consistency and Traceability in an Agile Context

Inayat et al. have reported that some challenges of traditional RE are covered by agile RE practices about communications issues, overscoping, requirements validation, requirements documentation, and rare customer involvement [19]. Still, in parallel, the practice of agile approaches for mechatronic development generates other challenges regarding ARE activities, notably for requirements consistency and traceability. Indeed, in previous work, we have presented the challenges related to the development of mechatronic systems in an agile context [28, 29, 36]. In this context, as this chapter focuses on the two critical issues related to the practical guidelines to efficiently ensure requirements consistency and traceability, in this section the proposed existing approaches that could meet this objective are analyzed.

In Software Engineering, aspect-oriented programming (AOP) [24] is a programming paradigm that aims to increase modularity by allowing the separation of cross-cutting concerns. It does so by adding additional behavior to existing code without modifying the code itself, instead separately specifying which code is modified via a '*pointcut*' specification, such as '*log all function calls when the function's name begins with* 'set'. This allows behaviors that are not central to the business logic (such as logging) to be added to a program without cluttering the code core to the functionality. AOP thus forms a basis for aspect-oriented software development. One challenge addressed in this chapter is to adapt this principle to the requirement engineering phase in mechatronic system design. Quality Function Deployment is an approach supporting this in an agile context.

Other researchers have focused on incremental changes during the development process providing approaches to support automatic response to change. Agile development is characterized by a continuous delivery of requirements that continually increments design contents, imposes priority changes, and adjusts targets. Decomposition techniques of complex problems combined with graph theory principles and tools have proven to be valuable in capturing requirement changes and supporting their consistency and traceability within the dynamic changes of the agile development process. Spijkman et al. [40] present a Twin Peaks model to assist the agile development team in: (i) achieving uniform granularity within and between the two Peaks (functional requirements and architecture specifications), (ii) establishing and maintaining alignment between the two Peaks. Furthermore, graph-based methods are widely used in literature to ensure traceability and consistency. Kuhn and Franke [25] present a graph-based data model for complex manufacturing traceability, while Rempel and Mäder [37] present three graph-based traceability metrics (number of related requirements, average distance to related requirements and requirement information flow) to assess the complexity of the traceability graphs for estimating the implementation risk of requirements. As a result, such graph-based approaches seem interesting to support ARE.

Other research work focused on the ontologies to support requirements consistency and traceability in an agile context. For example, Wibowo et al. propose two effective ontologies for requirements management (RTOnto for Requirement Traceability and RCMOnto for Requirements Change Management) to support the capturing of all the artifacts at every stage of the development life cycle along with a flexible granularity that fits the chosen development framework and project characteristics [50]. Murtazina et al. proposed an ontology-based approach to support requirements traceability as required by frequent requirements changes in Agile development [31]. They underline current challenges of requirements traceability in an agile context: its cost of time, required effort for manual data update, changes tracking, value identification of the different stakeholders involved, related organizational considerations and lack of traceability dedicated tools. Actually, the requirements traceability matrix is the most common traceability tool used to manage the traceability, either manually or automatically. Finally, even if some existing research work for software development shows that ontologies can be helpful to support RE activities in an agile context, they do not as yet address the impacts of multi-actors and time-shifts between short incremental cycles existing in such a context.

In parallel, to support requirements consistency and traceability, methods based on Model-Based Systems Engineering (MBSE) propose to ensure the consistency and the traceability process of the modeling data considering the whole lifecycle of the system [10, 13]. Such methods are often based on the SysML language and are implemented in software tools, in order to support capitalization and change management. Still, even if these tools allow to define requirements traceability links and provide users with some visualization means to check them (identifying missing or redundant links) and their consistency, they don't include agility artifacts to allow Systems Engineers to model and trace requirements change throughout the whole development time through the different short iterative incremental cycles. SysML/UML profiles can be used to extend these languages for a specific application domain or use: for example, for the mechatronic domain, we can cite ModelicaML (for the description of time-continuous and time-discrete/event-based system dynamics and the generation of executable Modelica code) [32], Therese (for thermal modeling) [2], Emile (for electromagnetic modeling) [23], Gertrude (for geometry modeling) [35], MARTE (for model-driven development of Real-Time and Embedded Systems) [14]. As a result, such a profile could provide additional semantic artifacts to the System model to enhance RE activities in an agile context related to the requirements consistency and traceability.

8.3 Key Instruments for RE Success in an Agile Context

Having analyzed existing literature approaches to support requirements consistency and traceability in an agile context, this section proposes a framework to provide key instruments to cope with their management in an agile context and notably within requirements changes management, be they from emergent changes (from the product itself) or from external changes (from SH /users, technological constraints, economic constraints, legislation...) [45].

8.3.1 Framework Overview

In this section, an overview of the framework REMAC, or Requirement Engineering for Mechatronics in an Agile Context, is presented. This framework consists of some main requirements related activities and the corresponding methods and tools proposed to tackle them in an agile context. Requirements engineering is a collaborative, iterative, incremental process of eliciting, analyzing, understanding and defining requirements to support verification and validation processes. Requirements can be found on several levels: these can be customer requirements, system requirements, specific requirements from the disciplines and tests. Only explicitly formulated requirements can be considered in the development process and we can differ them depending on their nature:

- **Functional requirements** describe the functional aspects of the considered system. To determine the functional requirements, it is helpful to ask the question: '*What will the future system do*?'. This means that functions, inputs and outputs of the future system are described. Inputs or outputs can be, for example, data, error messages, events and so forth. Functional requirements describe the important system states and the expected behavior of the future system with its environment. Often, only the functional requirements are in the foreground during product development. They directly support the customers in their activities and facilitate everyday work.

However, not only the functional but also the non-functional requirements are crucial for a successful development initiative.

- **Non-functional requirements** can be viewed and described from different perspectives. Non-functional requirements are in most cases underspecified. This is because they are often described by ambiguous adjectives such as, maintainable, portable or reliable.

Finally, whatever their nature, all these requirements that may dynamically change all along the agile project, have to be managed in order to keep them consistent and traceable from the beginning to the final delivery of the fully designed system. To tackle this challenge, Fig. 8.2 shows the main working activities addressed within the REMAC framework, with the proposed methods and/or tools, which are described in the following sections in more detail.

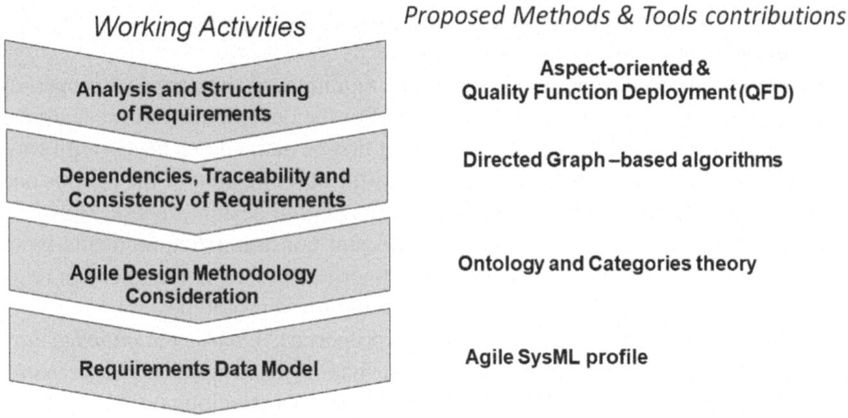

Fig. 8.2 REMAC (Requirement engineering for mechatronics in an agile context) framework

8.3.2 Aspect-Oriented Consideration with Quality Function Deployment (QFD)

In mechatronics, an aspect-oriented consideration results from the fact that function/component entities demarcation is often no longer to be found (e.g., decentralized control). Functions and tasks are often distributed among different systems, subsystems, components and features. Although these distributed functions cannot be directly assigned to any element, they still have to be designed, modeled, analyzed, tested and evaluated.

In order to be able to systematically record such cross-sectional functional requirements, the introduction of '*aspects*' is proposed. An aspect represents a '*way of considering*' a system—independently of all its module boundaries, and should be complementary of other aspects, in order that the summary of all aspects gives the overall view of the system. The prism analogy shows how this is done: light is split into its different colors (aspects) through a prism and transformed back into a white beam of light through another prism. In the same way, the overall view of a mechatronic system can be '*broken down*' into different aspects, which can be processed individually and finally merged back into the overall view of the system. These aspects can be used in an agile context, as an efficient way to analyze and structure the whole set of requirements (product backlogs in Scrum).

Indeed, unlike in an agile approach where product requirements are not entirely fixed, in a traditional approach, detailed requirements are given by customers in advance before planning the project. This approach is linear and process-oriented with exceptional long iterations. Organizational structure is also hierarchical; centralized authority steers the project by keeping control of most of the activities. Communications are mostly formal and need to go through the proper channel while keeping authorities in the loop. Formal statements are crucial while transferring knowledge, which makes the knowledge transfer process too bureaucratic. Also, reporting of work progress to the only manager and not the whole team creates misunderstanding within the team about work.

As a result, aspects can be used to translate requirements into product properties via suitable decompositions or classifications. For the development of mechatronic systems, this means that the set of aspects must first be defined in a meaningful way. The creation of the aspect hierarchy, as well as the classification of the aspects and properties, initially represents an additional effort in the design process. However, these aspects can be used to classify the different constraints/requirements types regarding the requirements selection process described in the previous section (e.g., organization aspects, specification for physical vs virtual prototypes ...).

For the transition of requirements to product properties, '*Quality Function Deployment*' (QFD) can be used, by supporting the distribution, dissemination and spreading of customer wishes into the different phases of product development [1, 6]. QFD performs the translation of customer requirements into technical product characteristics. This requires the establishment and distribution of quality functions within the product and a corresponding organization of the company departments. By systematically breaking down the technical properties derived from the requirements to ever-lower hierarchical levels of all involved product-relevant areas of the company, such as development, production or logistics, an orientation of the product to the customer (or to the customer's benefit) can be ensured. QFD is a matrix-based approach that, in a more or less extensive form, mainly aims at aligning the product development process with the customer's wishes in such a way that the finished product brings the highest possible ratio between benefit and effort as perceived by the customer. Accordingly, the method aims at a high level of integration of the individual specialist departments during the development process (see Fig. 8.3). In particular, in a SCRUM sprint the first two steps can be used to efficiently fix the requirements and their importance. Another QFD use case is the definition/formalization of the features optimization direction, required when reviewing the sprint deliverable during the Sprint Review meeting.

Finally, the complexity of managing requirements consistency within the agile design process can be more than mitigated by a better-structured procedure and a more complete definition, classification and prioritization of the requirements from the beginning. Especially in the further course of the design process, the number of changes can be reduced. The changes themselves can be traced and dealt with more quickly because the structuring through the aspects on the one hand allows a better insight into the system interrelationships and on the other enables a simple aspect-based filtering of the properties.

8.3.3 Requirements Dependencies Identification and Analysis

In the agile context, requirements engineering is performed throughout all the system lifecycle, i.e., it is not limited to a dedicated phase. Indeed, stakeholders' needs and requirements definition are iterative and recursive processes [20] performed without

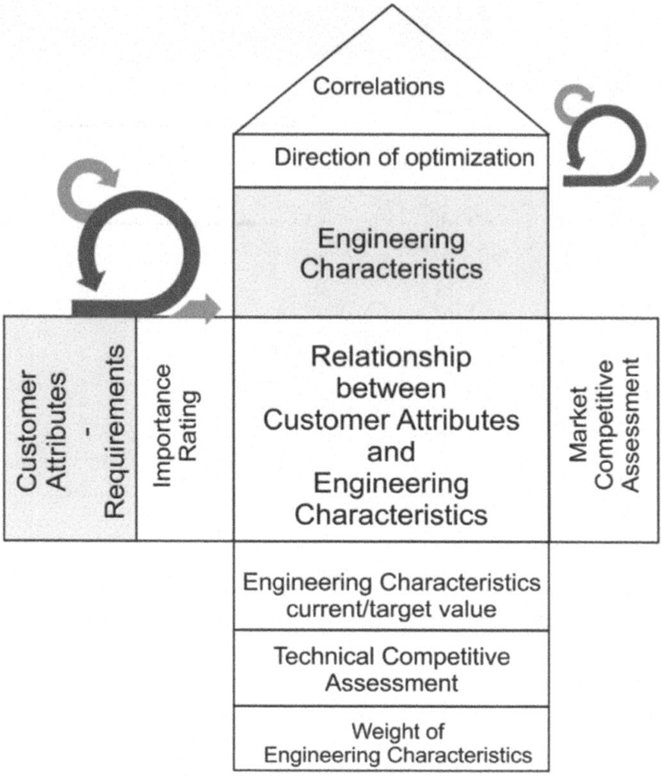

Fig. 8.3 Quality function deployment

a systematic approach during the system development process. As requirements can be continuously updated (refined, recorded, and managed), requirements definition has to be traced according to the developed product and its maturity. In such a context, a graph-based approach is the best way to manage and trace functional and non-functional requirements relationships.

According to the principles of the graph theory [9], it is possible to represent all dependencies/relationships among the involved elements by means of a directed graph (digraph). A digraph representation is based on nodes and arcs. In the following, according to Patalano et al. [33], nodes represent elements (like requirements, functions, and logical components) at different levels, while arcs represent the relationships among elements.

Although in the agile context, requirements can be defined at any time, it is important to provide a structured framework for their management. With reference to the requirements and related dependencies definition, traceability and consistency needs, we propose a multilevel V-model based on digraphs (Fig. 8.4). Descending and ascending branches are split up into four levels by increasing details from high to low level:

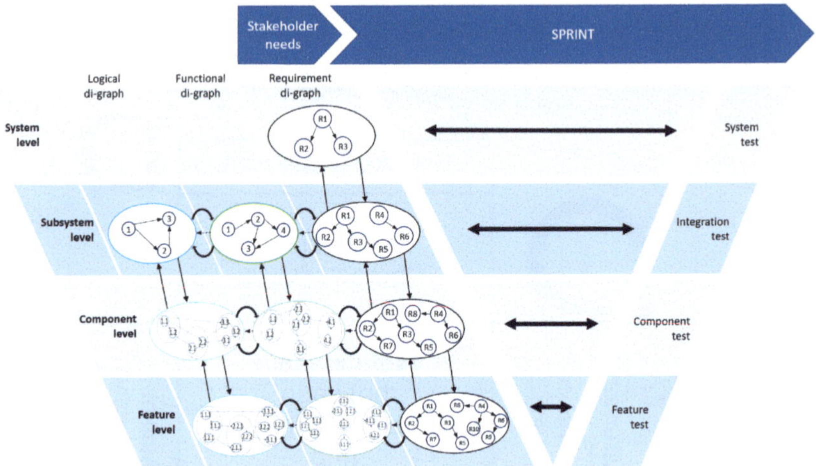

Fig. 8.4 Proposed multilevel V-model approach based on digraphs

- System level: higher level where there are only global mission and initial requirements. They represent the translation of stakeholders' needs into requirements;
- Subsystem level: the whole system is decomposed into subsystems according to the requirements at the system level. Therefore, functional, logical and requirements digraphs are generated;
- Component level: each subsystem is characterized by a set of components that interact to achieve the goal for which the subsystem was intended. At this stage, new induced requirements can appear, according to some component choice constraints;
- Feature level: each component can be decomposed into simplest elements (features) which describe a single and specific characteristic of the component as, for example, behavior, geometry and material properties. At this level, new requirements and dependencies appear regarding a specific 'aspect'.

For the descending branch, a triple concurrent definition of the system is carried out at each level in terms of requirements, functions that satisfy the defined requirements and logical components that are allocated to the defined functions and must satisfy the related requirements. Going down from system to feature level, requirements, functions, and logical components are deepened by an iterative and recursive agile process that operates a zigzagging (according to the axiomatic design principle [42] among the three system definitions, up to identify simplest elements i.e., the constituent elements which cannot be further detailed.

The approach is based on a two-step approach: (i) a draft structure—the first iteration within a sprint/cycle should be performed in a top-down manner in order to generate and allocate all available elements and their relations according to the increasing details; (ii) a detailed structure—requirements, functions, components can be added at any time and any level, all along the next cycles keeping the traceability and the consistency of the new elements against the whole system.

Regarding traceability, the framework presents horizontal (at the same level) and vertical (among levels) relationships. At a defined level (horizontal), each digraph has two types of relationships: internal—relationships among nodes belonging to the same graph; external—relationships among nodes belonging to different graphs. From high level to low level (vertical), increasing details of three digraph branches exist: requirements, functional and logical.

Focusing on consistency, the system definition is assumed consistent when it is coherent, complete, and non-redundant [34]. Therefore, a consistent definition is provided when the three following conditions exist:

- Coherency: all requirements, functional and logical elements have horizontal and vertical connections as well as internal and external connections. All relationships are mapped using an adjacency matrix and dedicated algorithms check the consistency of the matrix;
- Completeness: at each level, all requirements are satisfied at least by one functional element and its relative logical element. A dedicated algorithm named the Isolate algorithm detects that no isolated nodes exist at each level along the two directions (horizontal and vertical directions);
- Non-desired redundancy: more requirements can be satisfied by the same function, but a requirement cannot be satisfied by different functions, else it has to be split or more detailed. Therefore, at each level, requirement-function external relations can have only one arc with each requirement. The redundancy detection algorithm evaluates multiple-arc connections for each requirement at each level detecting the existing redundancies. Then, the designer will decide what the desired redundancies are and remove them from the data set to analyze.

When considering partial and progressive consistency, evolving during an agile process, we propose to assess it through consistency metrics by checking it with rule-based routines, to identify the missing elements required to meet the declared dependency links. This part can be illustrated through an ontology approach, as described in the next section.

Finally, the proposed model is composed of four levels and three types of digraphs that generate a four-by-three grid where new data can be added at any cell, according to the agile context. Then, due to the zigzagging approach and ad-hoc algorithms, the designer is driven through the definition of a coherence, complete and non-redundancies set of relationships. This model is very powerful for a mechatronic design where, due to the multiple domains involved, the system can be developed with a domain abstraction (or aspect) by adding new data for any domain at any level and then defining the needed relationships among all domain elements.

8.3.4 Design Methodology Formalization in an Agile Context

Many mechatronic design methodologies are based on multidisciplinary methods relying on a strong collaboration between actors throughout the design process. Yet, little research has been done on the adaptation of these methodologies in an agile context. Focusing on the requirements elicitation process, we propose to adapt the existing SE-READ (System Engineering based Requirements Elicitation & Architecture Design) methodology [27] to a double agile context.

The first consideration naturally arises when designing the system model that will be used to specify the system design, according to the previous section: how to ensure the consistency and traceability of the set of model artifacts (including requirements) generated by different actors in different formats (languages and tools) in an unplanned temporal sequence. The second aspect of agility that we propose to address is the middle-out/meet-in-the-middle approach describing a combination of top-down and bottom-up design strategies [11]. Indeed, few products are completely innovative, they are usually designed from existing parts, either to reduce time and cost or because the standardization of manufacturing equipment requires the reuse of some modular components. Thus, many current designs rely on the requirements related to the reuse of components (particularly in the automotive sector) or even existing structures (e.g., in aeronautics), while at the same time having to integrate new requirements (linked to regulatory changes or the integration of new innovative technologies). Similarly, with the significant development of agile design approaches, rapid and iterative prototyping of the system generates some elements of the architecture before the full set of requirements has been addressed.

Considering this context and all the constraints associated, the objective of the Methodology Analysis and Formalization bAsed on oLogs to Design Architecture framework (MAFALDA), based on CatSE (Category-based System Engineering) is to ensure the knowledge consistency of the views/models generated, with respect to both the described SE-READ design methodology and its constraints related to the agile and middle-out context. In other words, the order in which the different steps of the SE-READ methodology are addressed does not have to be predefined, to be compliant with an agile development process (Fig. 8.5).

The middle-out approach consists of coupling a bottom-up and a top-down approach, by integrating architectural elements in the initial requirements (case of reuse of standard components). This context is considered agile, as we assume that each step of the considered methodology can be performed by independent design actors with their own modeling tool/language, and at any time.

Focusing on the use of MAFALDA to support the requirements elicitation objective of the SE-READ black-box analysis, we propose a generic way to formally describe the links associated with the requirements (between requirements or for traceability of model artifacts) that come in an agile context, by using ontologies. Indeed, ontologies allow users to talk about common matters, even if they are expressed in different formalisms. They offer many benefits regarding knowledge formalization and representation: (i) they have formal semantics that can be

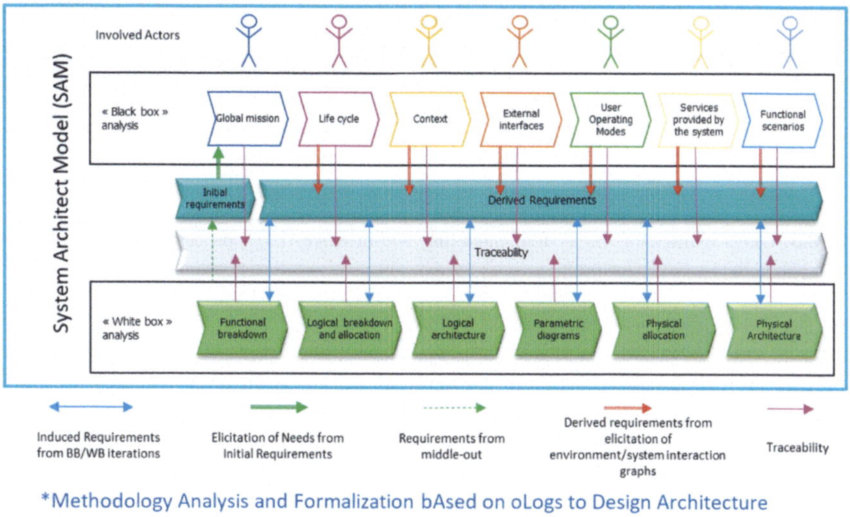

Fig. 8.5 MAFALDA* to support the SE-READ formalization to ensure data consistency within middle-out and agile context

exploited when part of the information is missing, (ii) they provide a mechanism for checking information inconsistencies, since they are encoded in formats that support automatic reasoning, unlike data models (which are developed for databases), (iii) they can be modified and extended even during their usage and be used together with the so-called logic reasoners or inference engines to deduce logic implications, and finally, (iv) they enable engineers and stakeholders to independently develop some partial descriptions of the same artefact and check consistency when the descriptions are combined [3].

As the allocation and traceability links between modeling artifacts and the requirements are only supported by the methodology SE-READ, and until now the manual resolution of all the constraints induced by these links makes the consistency assessment of the System Model iterative. For example, a requirement generated at the logical architecture step (white box) may produce new functional scenarios (black box) that have to be taken into account at the next iteration. Therefore, the first challenge of MAFALDA is to automatically support the consistency of all the links (defined between different points of view or multidisciplinary models), and more specifically here of the dependency links related to each viewpoint-derived requirements, that are defined by the methodology (excepting those which must remain under the designer control).

Technically, the system of interest in MAFALDA is the ontology of the SE-READ methodology, using the formalism of categories theory to generate an Olog, i.e. a model which describes the functional relationships between the different concepts of the ontology (artifacts of the System Model generated by SE-READ). In [41], Ologs are proposed as a formalization of ontology as a category. Ologs

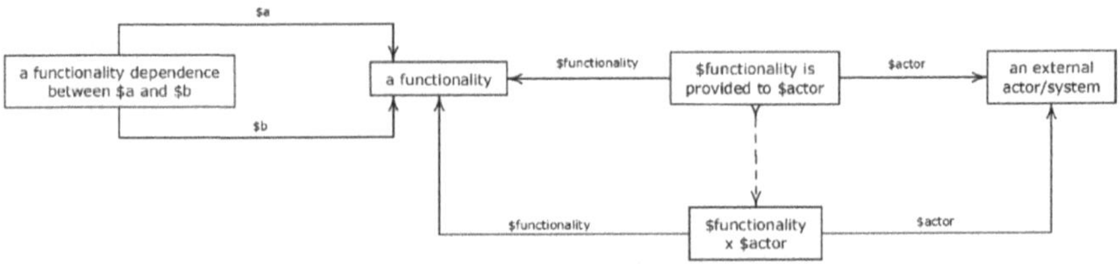

Fig. 8.6 Ologs extract for the system modeling within MAFALDA process (for the 'Services provided by the system' viewpoint)

could be considered as a graph structure. Their objects represent concepts and are nominal sentences (called types in the original work). Their arrows represent functional relations between concepts and are verbs (respectively called aspects in [41]. Figure 8.6 presents an extract of the whole Olog of the MAFALDA (agile SE-READ methodology) process.

Regarding the requirements elicitation activity within an agile context, MAFALDA can automatically generate the whole set of consistent derived requirements by using the Olog of the SE-READ black box analysis. Indeed, we have formalized each step of the methodology (providing a model as an object of an arbitrary formalism F) and how it contributes to the generation of the set of derived requirements (defining a functor R: F → 'Set'). For each cycle, the first step consists in populating a part of the MAFALDA ontology with the content of the models, and then the second step will unfold the ontology, to enumerate all the facts induced by the content of the ontology (updated requirements). The comparison with the previous version of the results (previous sprint/cycle) allows to trace the requirements changes and their impact or rationale on the system model artifacts to check consistency preservation or, if not, the missing elements.

8.3.5 *Specification of an Agile SysML Profile to Support RE Datamodel*

To complement former considerations with a means to enrich their traceability with the agile process elements, we propose to specify an agile SysML extension that supports the RE activities presented in the previous sections, including the Scrum agile process artifacts. To this respect, we propose the following data model (Fig. 8.7). This profile will allow users to enrich the system model with all the modeling artifacts needed by the REMAC framework, while supplementing them with the agile process elements, in order to know for example the rationale and origin of a *Requirement*, when resulting from a *Sprint review* decision, which requirement has been selected as *Sprint backlog* in a given *Sprint* to meet a *Goal*. To this respect, we could keep in the

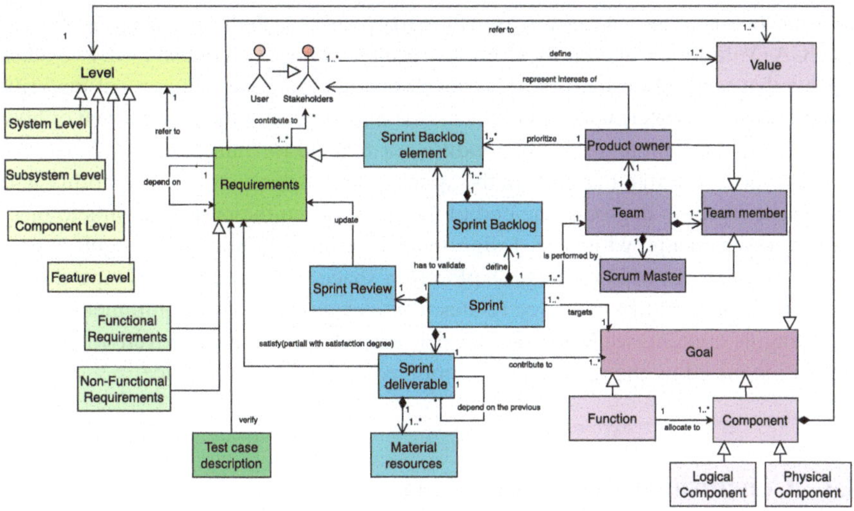

Fig. 8.7 Agile SysML profile specification

SysML system model the whole traceability of the agile project course (sprints planning, deliveries, requirements fulfillment, etc.) regarding the developments/deliveries of the system parts and their corresponding coverage rate of requirements.

Regarding the organization constraints, they can be managed for the planning with the period defined in each Sprint, for the human resources by the *Team member* attributes (availability, competencies and wishable competencies) and for the material constraints through the definition of the required material resources, the availability of the human and material resources must be confronted with the sprint period that has been defined (through some constraints between the time period and the respective availabilities).

The *stakeholder* points of view are represented by the set of *values* they define in relation to their expectations elicited into requirements. The *Product Owner* represents their interests by prioritizing the requirements to be included as *Sprint backlog elements* in a given *Sprint*.

The requirements selection to define the *Sprint backlog* within a *Sprint* is based on a *goal* that can be related to a *value* (and related stakeholders), a *function* or a *component*. A function will be allocated to a component, which has a given *level* in the System architecture breakdown. These levels have been defined to be compliant with the proposed graph-based approach to support the requirements dependencies identification.

The nature and constraints of the prototype have been represented in the *Sprint Deliverable* (Done) artefact, in connection with the required material resources.

Focusing more particularly on the requirements, they can be *functional* or *non-functional*, impacting on how they will be tested and satisfied. These will be described through some test cases, in relation to the prototype nature and platform of the Sprint deliverable concerned. Moreover, as in the early phase, mechatronic design evolves in a VUCA (Volatility, Uncertainty, Complexity and Ambiguity) context [36], two other aspects that seem important for agile development are (i) the requirement maturity or flexibility, as it determines either the uncertainty degree of the requirement or how the concerned requirement can be modified or relaxed for future trade-offs, (ii) the consideration of design margins. In the agile context, the requirement flexibility will impact the freedom degree at the end of the sprint when the prototype fails to meet certain requirements, while the design margins related to quantified requirements (e.g., performance) offers a large exploration space to adapt the prototype during the sprint. The confidence degree relates to the expertise of the requirement owner regarding its content, while the risk level is more focused on the novelty/degree of innovation raised by the requirement compared to the current practice.

8.4 Conclusions and Future Work

This chapter aims to provide some guidance and approaches to manage requirements engineering activities in an agile context required for mechatronic design. Our contributions relate to the REMAC (Requirement Engineering for Mechatronics in an Agile Context) framework, consisting of key instruments to tackle requirements related activities in an agile context. The benefits of such a framework, regarding related previous existing work are to propose a global set of approaches to consistently support RE activities, adapted to mechatronic systems, where concurrent and multiple design views are required, when they are developed within an agile context. The REMAC allows users to take into account: aspect-oriented development, with a multi-level design definition and Verification & Validation process, in accordance with the requirements management of agile design methodology, to ensure requirements consistency and traceability, in an agile context driven by a VUCA environment. This first preliminary work will be enhanced by the implementation (software development) of the REMAC framework modules and their experiment on some agile projects on mechatronic use cases.

Acknowledgements The presented considerations and results were conducted within the SUNISWELL-cooperation between ISAE-Supméca (Paris, France), University of Applied Sciences Upper Austria (Wels, Austria) and University of Naples Federico II (Naples, Italy). The authors thank all partners for valuable discussions and contributions.

References

1. Akao Y (1992) QFD: quality function deployment; Wie Die Japaner Kundenwünsche in Qualität Umsetzen. Verlag Moderne Industrie
2. Barbedienne R, Penas O, Choley J-Y et al (2015) TheReSE: SysML extension for thermal modelling. In: Systems conference (SysCon), 2015 9th annual IEEE international, Vancouver,

BC, 13 April 2015, pp 301–308. IEEE. Available at: http://ieeexplore.ieee.org/lpdocs/epic03/wrapper.htm?arnumber=7116768. Accessed 15 December 2015
3. Batres R (2017) Ontologies in process systems engineering. Chemie Ingenieur Technik. Epub ahead of print
4. Bennett N, Lemoine GJ (2014) What a difference a word makes: understanding threats to performance in a VUCA world. Bus Horiz 57(3):311–317
5. Böhmer A, Schweigert S, Devecka J et al (2017) Towards agile development of physical products a startup case study. In: 2017 International conference on engineering, technology and innovation (ICE/ITMC), Funchal, June 2017, pp 78–85. IEEE. Available at: http://ieeexplore.ieee.org/document/8279872/. Accessed 20 July 2021
6. Chan L-K, Wu M-L (2002) Quality function deployment: a literature review. Eur J Oper Res 143(3):463–497
7. Cho J (2008) Issues and challenges of agile software development with SCRUM. Issues Inform Syst 9(2):188–195
8. Cohen D, Lindvall M, Costa P (2004) An introduction to agile methods. In: Advances in computers. Elsevier, pp 1–66. Available at: https://linkinghub.elsevier.com/retrieve/pii/S0065245803620012. Accessed 19 July 2021
9. Deo N (2017) Graph theory with applications to engineering and computer science. Courier Dover Publ
10. DoD D of D (2001) Systems engineering fundamentals. Syst Manag College
11. El Ghosh M, Naja H, Abdulrab H et al (2016) Towards a middle-out approach for building legal domain reference ontology. Int J Knowl Eng 2(3):109–114
12. Elghariani K, Kama N (2016) Review on agile requirements engineering challenges. In: 2016 3rd International conference on computer and information sciences (ICCOINS), pp 507–512
13. Estefan JA (2008) Survey of model-based systems engineering (MBSE) methodologies. INCOSE-TD-2007–003–01, 10 June. INCOSE
14. Gérard S, Selic B (2008) The UML–marte standardized profile. IFAC Proceed Vol 41(2):6909–6913
15. Gräßler I, Hentze J, Bruckmann T (2018) V-models for interdisciplinary systems engineering. In: DS 92: Proceedings of the DESIGN 2018 15th international design conference, pp 747–756
16. Heck P, Zaidman A (2018) A systematic literature review on quality criteria for agile requirements specifications. Softw Qual J 26(1):127–160
17. Hehenberger P, Bradley D (2016) Mechatronic futures: challenges and solutions for mechatronic systems and their designers. Springer
18. Hehenberger P, Bradley D, Dehghani A et al (2019) Mechatronic and cyber-physical systems within the domain of the internet of things. In: Systems engineering in research and industrial practice. Springer, pp 177–207
19. Inayat I, Salim SS, Marczak S et al (2015) A systematic literature review on agile requirements engineering practices and challenges. Computers in human behavior 51. Computing for human learning, Behaviour and Collaboration in the Social and Mobile Networks Era, pp 915–929
20. ISO/IEC/IEEE 29148: 2018 (2018) Systems and software engineering—life cycle processes—requirements engineering, https://www.iso.org/standard/72089.html
21. Johnson N, XtremeEDA Corp (2012) Agile hardware development—nonsense or necessity? In: EDN. https://www.edn.com/agile-hardware-development-nonsense-or-necessity/
22. Kasauli R, Knauss E, Horkoff J et al (2021) Requirements engineering challenges and practices in large-scale agile system development. J Syst Softw 172:110851
23. Kharrat M, Penas O, Plateaux R et al (2021) Integration of electromagnetic constraints as of the conceptual design through an MBSE approach. IEEE Syst J 15(1):747–758
24. Kiczales G, Hilsdale E (2001) Aspect-oriented programming, In: ESEC/FSE-9: Proceedings of the 8th European software engineering conference held jointly with 9th ACM SIGSOFT international symposium on Foundations of software engineering, September 2001
25. Kuhn M, Franke J (2021) Data continuity and traceability in complex manufacturing systems: a graph-based modelling approach. Int J Comput Integr Manuf 34(5):549–566. https://doi.org/10.1080/0951192X.2021.1901320

26. Lakhoua MN, Salem JB, Battikh T et al (2020) Review on modelling and design of mechatronic systems. Int J Mechatron Autom 7(2):57–63
27. Mhenni F, Choley J-Y, Penas O et al (2014) A SysML-based methodology for mechatronic systems architectural design. Adv Eng Inform 28(3):218–231
28. Mule S, Hehenberger P (2020) An empirical study on the usage of agile methods in the mechatronics industry. In: 2020 21st International conference on research and education in mechatronics (REM), December 2020, pp 1–6. Available at: https://ieeexplore.ieee.org/abstract/document/9313905. Accessed 31 January 2024
29. Mule S, Plateaux R, Hehenberger P et al (2020) A new agile hybridization approach and a set of related guidelines for mechatronic product development. In: Product lifecycle management enabling smart X (eds F Nyffenegger, J Ríos, L Rivest, et al.), Cham, 2020, pp 618–633. IFIP Advances in Information and Communication Technology. Springer International Publishing
30. Mule S, Hehenberger P, Plateaux R et al (2021) An approach and an illustrative case study for a hybrid development process in mechatronic system design. Int J Product Lifecycle Manag 13(3):265–289
31. Murtazina MS, Avdeenko TV (2019) An ontology-based approach to support for requirements traceability in agile development. Proced Comput Sci 150:628–635
32. OpenModelica (2024) Modelica modelling language (ModelicaML). Available at: https://openmodelica.org/free-and-open-source-software/modelica-modelling-language-modelicaml/. Accessed 1 February 2024
33. Patalano S, Vitolo F, Lanzotti A (2013) A graph-based software tool for the CAD modelling of mechanical assemblies. GRAPP/IVAPP 2013:60–69
34. Patalano S, Vitolo F, Gerbino S et al (2018) A graph-based method and a software tool for interactive tolerance specification. Proced CIRP 75:173–178
35. Plateaux R, Penas O, Barbedienne R et al (2017) Use of technologically and topologically related surfaces (TTRS) geometrical theory for mechatronic design ontology. Comput Aided Des Appl 14(5):595–609
36. Plateaux R, Penas O, Mule S et al (2020) SCRUM++ framework concepts. In: 2020 IEEE International symposium on systems engineering (ISSE), October 2020, pp 1–8
37. Rempel P, Mäder P (2015) Estimating the implementation risk of requirements in agile software development projects with traceability metrics. In: Requirements engineering: foundation for software quality: 21st International working conference, REFSQ 2015, Essen, Germany, March 23–26, 2015. Proceedings 21. Springer International Publishing, p 81–97
38. Schön EM, Thomaschewski J, Escalona MJ (2017) Agile requirements engineering: a systematic literature review. Comput Stan Interfaces 49:79–91
39. Schön E-M, Sedeño J, Mejías M et al (2019) A metamodel for agile requirements engineering. J Comput Commun 07(02):1–22
40. Spijkman T, Molenaar S, Dalpiaz F, Brinkkemper S (2021) Alignment and granularity of requirements and architecture in agile development: a functional perspective. Inf Softw Technol 133:106535
41. Spivak DI, Kent RE (2012) Ologs: a categorical framework for knowledge representation. PLoS ONE Mavergames C (ed) 7(1):e24274
42. Suh NP (1990) The principles of design. 6. Oxford University Press on Demand
43. Sutherland J, Schwaber K (2007) The scrum papers: nut, bolts, and origins of an agile framework. Nuts, Bolts and Origins of an Agile Process. Epub ahead of print 2007
44. Takeuchi H, Nonaka I (1986) The new new product development game. Harv Bus Rev 64(1):137–146
45. Ullah I, Tang D, Wang Q et al (2017) Managing engineering change requirements during the product development process. Concurrent Eng Res Appl, 26
46. VDI-Guideline2206 (2003) Design methodology for mechatronic systems. VDI-Publishing Group
47. VDI-Guideline2206 (2021) Design methodology for mechatronic systems. VDI-Publishing Group

48. Vogel-Heuser B, Bi F (2021) Interdisciplinary effects of technical debt in companies with mechatronic products—a qualitative study. J Syst Softw 171:110809
49. Walden DD, Roedler GJ, Forsberg K (2015) INCOSE systems engineering handbook version 4: updating the reference for practitioners. INCOSE Int Symp 25(1):678–686
50. Wibowo A, Davis J (2020) Requirements traceability ontology to support requirements management. In: Proceedings of the Australasian computer science week Multiconference, Melbourne VIC Australia, 4 February 2020, pp 1–9. ACM. Available at: https://doi.org/10.1145/3373017.3373038. Accessed 31 January 2024
51. Žužek T, Kušar J, Rihar L et al (2020) Agile-concurrent hybrid: a framework for concurrent product development using Scrum. Concurr Eng 28(4):255–264

Chapter 9
Design Processes of Mechatronic Systems

Matthieu Bricogne, Julien Le Duigou, and Benoît Eynard

Abstract Nowadays, to design innovative and multidisciplinary products such as mechatronics systems, product development process needs to be rethought or at least adapted. These current processes are based on organizational models, business standards IT solutions, etc. They must consider that electronics and software represent an ever-increasing part of the final product. In this context, traditional mechanical engineering, electrical/electronic, and software methods cannot just be coordinated, and common interdisciplinary design approaches have to be proposed. In this chapter, we will explore current mechatronic systems design processes and present the work of convergence and synergy between mechanical, electronic, and software design processes, particularly with the use of system engineering and model-based system engineering but also with the growing integration of IoT and related data. Then we will explore new trends in mechatronic systems design, by focusing on the integration of sustainability aspects at the mechanical, energy, and software levels, the management of the obsolescence of components and software in mechatronic systems, and the application of AI to use data from mechatronic systems to generate new sources of value. We will conclude this work by exploring the possible links between these different trends and how they can complement and feed each other.

9.1 Introduction

Since the end of the twentieth century, many changes have occurred in the systems' design within the manufacturing industries. Whatever you call them, mechatronic systems, cyber-physical systems or smart/intelligent systems, product development

M. Bricogne (✉) · J. Le Duigou · B. Eynard
Roberval, Université de technologie de Compiègne, Compiègne, France
e-mail: matthieu.bricogne@utc.fr

J. Le Duigou
e-mail: julien.le-duigou@utc.fr

B. Eynard
e-mail: benoit.eynard@utc.fr

P. Hehenberger and D. Bradley (eds.), *Mechatronic Futures*,
https://doi.org/10.1007/978-3-031-83571-1_9

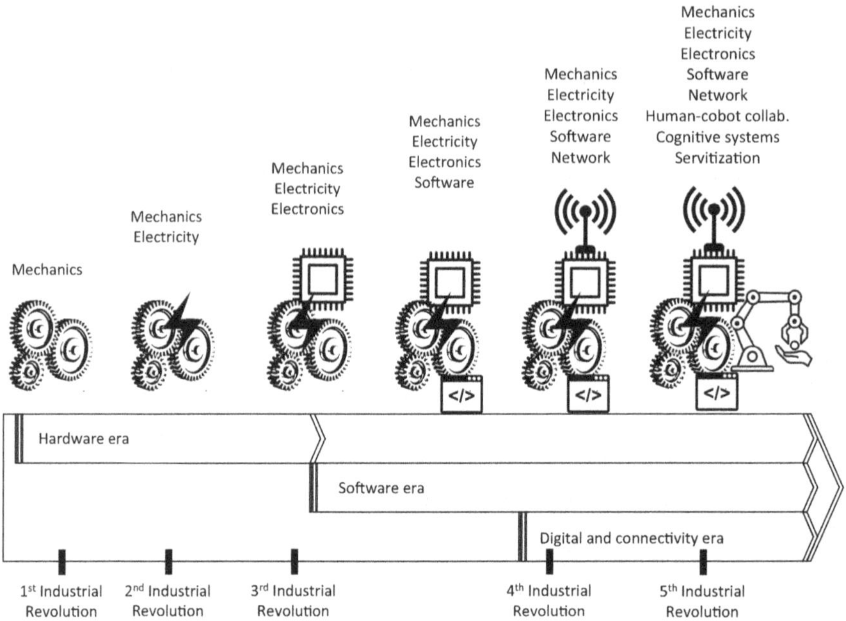

Fig. 9.5.1 Product evolution stages with technologies breakthroughs, related to disciplines involved in mechatronic systems design (adapted from [3])

process has been, and continues to be, rethought or at least adapted to the integrated expertise required for such multidisciplinary products. These current processes are especially based on organizational models and business standards IT solutions. They must consider that electronics, software, and connectivity solutions represent an ever-increasing part of the final product [1, 2]. As illustrated in Fig. 9.5.1, [3] summarize the interrelation between product evolution stages with technology breakthroughs, related to disciplines involved in mechatronic systems design. We are adding to this framework the 5th industrial revolution, which now requires companies to consider sustainability, resilience, and human integration. The impact of these trends on the design of mechatronic systems will be detailed in the final part of this chapter.

In this context, traditional mechanical engineering, electrical/electronic, and software/network methods cannot just be coordinated, and common interdisciplinary design processes have to be proposed. These processes will be influenced by several approaches such as agility or systems engineering to respond to actual trends such as servitization[1] of products [4], Internet of Things, or increasing demand for mass-personalized products [5].

This chapter is divided into two main sections. The first reviews some of the current design processes of mechatronic systems to point out the heterogeneity of

[1] Servitization (also found as servicization or servification), refers to a paradigm of transition of a product centric offer to a combined product-service offer, underpinning a change of the business model for the company.

practices and the gap remaining in integrating these discipline-specific practices. The second section then presents several industrial trends that will, from our point of view, greatly influence the future of mechatronic systems design.

9.2 Current Design Processes of Mechatronic Systems

Mechatronic systems have evolved from electro-mechanical systems with discrete electrical and mechanical parts to connected systems, based on integrated electronic–mechanical systems with sensors, actuators, and digital micro-electronics driven by sophisticated software and network modules [2, 6]. This evolution is driven by changes in market conditions, expressed as new demands such as integrating more functions into the product, reducing the weight and size of the product, increasing reliability, and so on. Connectivity also "expands the boundaries of products and offers new possibilities" leading to the observation mechatronic "products are now part of a larger system as the system's boundaries are expanded [3]. Such evolutions necessitate a re-evaluation of traditional development strategies to address challenges across diverse disciplines [7]. Consequently, companies are urged to adapt their product development strategies, often based on tailored processes and methodologies, supported by appropriate tools [8].

9.2.1 System Development Process and Mechatronic Systems Specificities: Focus on Multidisciplinary Integration

Product development process models have mainly emerged from the mechanical communities [9–11], even if specific models also come from the software community [12]. Despite this historical connection with mechanical engineering, the titles of these models and the step names illustrate the fact that the originators want to present these process models as generic, whatever the type of product developed, or the type of technology used. The authors usually speak of "engineering design" instead of "mechanical design", encompassing all types of product designs. Here, only one product development process model, which is presented in Fig. 9.5.2 [10], has been chosen as it is a questionable standard, but a more general summary is also presented by Howard et al. [13]. In this section, we describe the main steps of this development process model, or at least the most important in this chapter context, without giving much detail about the methods and tools used within these steps, as they will be covered in the following sections. Here, our focus is on outlining the general stages of the process model.

Needs Clarification and Conceptual Design Phases [13]

The design of a mechatronic system requires multi-disciplinary collaboration. To deal with such multi-disciplinary design issues, systems engineering has been

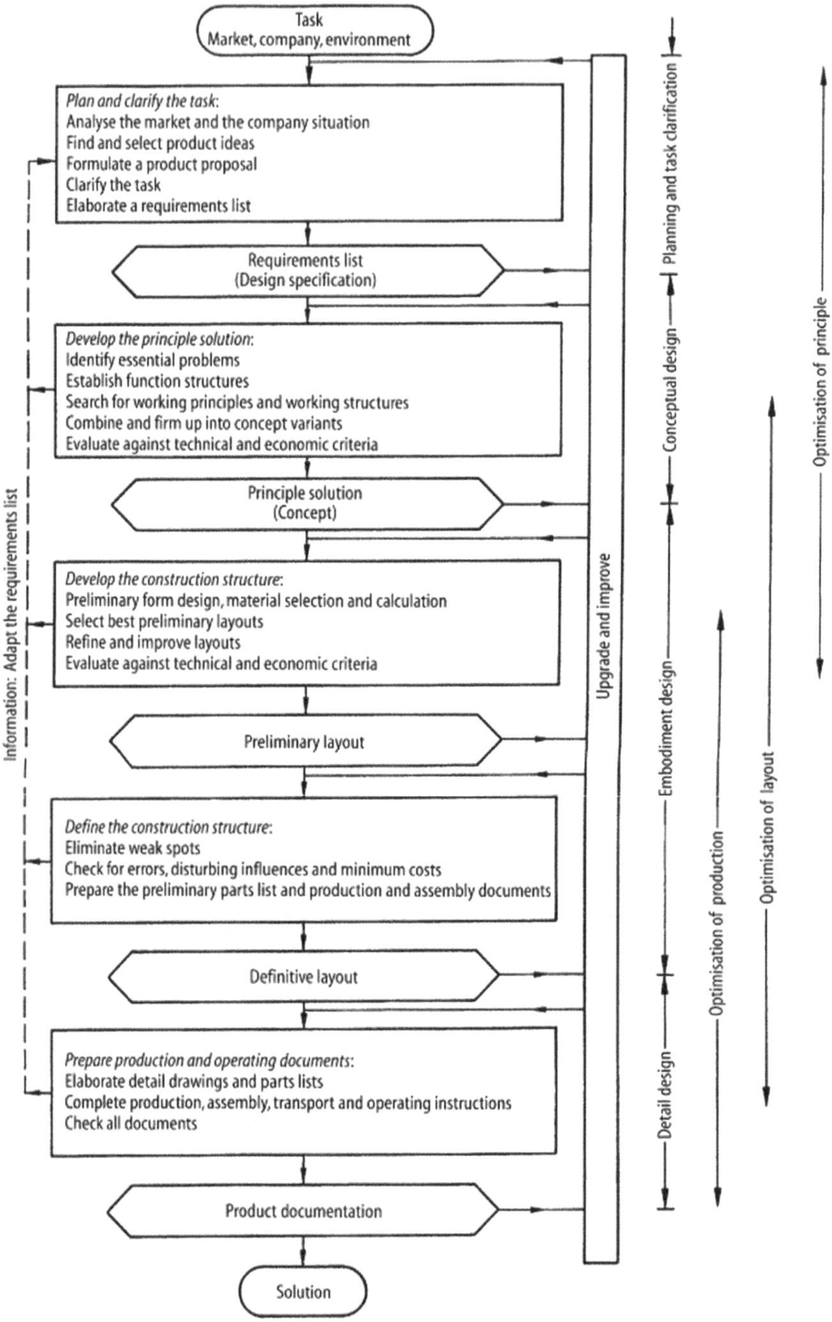

Fig. 9.5.2 Steps in the planning and design process [10]

proposed as an interdisciplinary approach [14]. In the mechatronic context, systems engineering is mostly used to consolidate a requirements list, establish function structures, and propose preliminary product architecture. For instance, SysML [15] and associated methodologies [16] are often presented as a solution to support the need establishment phase whereas Model-Based Systems Engineering (MBSE) activities [17] and Design Structure Matrix [18] are used to support system architecture elaboration.

Embodiment Design and Detailed Design Phases [13]

During the embodiment design and detailed design phases, one of the biggest remaining challenges is data management. For these phases, most scientific efforts in the mechatronics field concern product models dedicated to mechatronics to store information, unique Bill Of Material (BOM) to federate expert contributions, standards creation/usage to support data exchange [19, 20], consistency management between expert's models [21] and so forth. This focus on product data management is mainly because methodologies remain specific for each discipline. Some of these discipline-specific methods are presented in the next section.

While this process model, referred to as a systematic approach, remains standard, different approaches, sometimes in opposition to each other, are gradually changing the way mechatronic systems are designed today. These approaches are presented below.

9.2.2 Hardware/Software Design Methods Convergence

Many software development practices and Information Technology system administration methods inspire or are gradually being adopted, in mechatronic systems design. Among these practices, we can mention agile processes and DevOps.

Agile processes

Agile processes promote adaptive planning, collaboration between team members, and customer involvement throughout the development process. These approaches emphasize delivering working components frequently and responding to changing requirements in an agile manner. They are generally based on several tools such as sprints, burndown charts, user stories, scrum, and backlogs. Early attempts to introduce this agile approach were described as hybridization between "stage-gate" and agile [22, 23], considering plan-driven and agile approaches differ in their handling of requirements, development process, and team organization [24]. Other full agile efforts have been achieved, from a scientific and industrial point of view [25, 26] leading to highlighting some concerns linked to the project's objectives, the characteristics of the mechatronic system to be developed, and the company's culture. The limitations ("large scale systems, traceability, and safety and distributed environment") mentioned a few years ago [27] now seem to be under control, but another point concerning expectations regarding uncertainty management has been raised

[28], leading to the conclusion agile approach has to be integrated or mixed with a system engineering approach with the idea of capitalizing on the best of each.

DevOps

DevOps is a set of practices that combines software development (Dev) and IT operations (Ops). It aims to shorten the development cycle, increase collaboration between development and operations teams, and improve the reliability, security, and performance of systems. DevOps emphasizes infrastructure automation, continuous integration, continuous delivery, monitoring, and feedback loops. When it comes to applying or adapting this process to mechatronics, continuous integration, and continuous delivery seem to be the most important elements to consider, in particular to ensure Verification & Validation (V&V) objectives [29].

9.2.3 Systems Engineering in Mechatronics Design

Systems Engineering (SE) plays a significant role in supporting mechatronics design by providing a systematic approach to the development of complex systems. Generally based on the V-model, this approach relies on several methodological elements, such as system modeling with Model-Based Systems Engineering (MBSE) [30, 31], system analysis and simulation, requirements management [31, 32], system integration and testing and design verification and validation [29]. System Engineering (SE) plays a significant role in supporting mechatronics design by providing a systematic approach to the development of complex systems. Generally based on the V-cycle, this approach relies on several methodological elements, such as system modeling with Model-Based Systems Engineering (MBSE) [30, 31], system analysis and simulation, requirements management [31, 32], system integration and testing and design verification and validation [29].

Model-Based Systems Engineering (MBSE)

MBSE has attracted numerous researchers' attention recently. It is considered a significant methodology for the design of mechatronic systems with increasing complexity [33]. The Object Management Group's Unified Modeling Language and System Modeling Language (SysML) has been recently widely used to support the MBSE [34]. Some extensions have been developed for SysML to support the specific requirements of the design of mechatronic systems, such as automatic simulation [35], design-making process [36], etc. MBSE improves the exchange of design models and data, multi-disciplinary modeling, simultaneous consideration of design from different disciplines, tools, and methods supporting multi-disciplinary design and support of the design of control software. However, current studies on MBSE mainly focus on early design phases and seldom involve the detailed design phases. Moreover, MBSE are more adapted to complex systems but it is still missing effective and user-friendly tools for such development.

Systems Engineering has been proposed as a multidisciplinary approach to enable the realization of complex systems. Bathelt et al. [37] Apply extensions for every design phase based on the VDI guideline 2206 to improve the development of systems controlled by a Programmable Logical Controller. The Requirement—Functional—Logical—Physical (RFLP) approach is a specific V-model derived method, particularly adapted to mechatronic systems design and formed of four phases (R-F-L-P) which are each supported by different technical tools aiding the designers [27]. Zheng et al. [38] propose a two-level process based on V-model and hierarchical modeling, focusing on interfaces to improve multidisciplinary integration in detail design. System engineering improves cooperative work and communication among the design engineers at the early stages of the design process, multi-disciplinary modeling, early testing, and verification and provides a set of tools and methods supporting multi-disciplinary design. Nevertheless, the detailed design is still very domain-specific and needs further development.

9.3 Further Challenges for Integration of Mechatronic Systems: Current Research Approaches

In the previous sections, several scientific approaches attempting to address organizational challenges for mechatronic system design have been presented. But other challenges, related to mechatronic systems design, have to be taken into consideration. These challenges will relate to several trends, such as the ever-increasing software element in products and the necessary collaboration between systems.

9.3.1 Hardware/Software Design Methods Convergence: Cyber-Physical Systems Versus Mechatronic Systems

The science of Cybernetics has given birth to a type of system referred to as a Cyber-Physical System (CPS). These CPSs refer, according to Rajkumar et al. [39], to the next generation of systems that require close integration of information, communication, and control technologies to achieve stability, performance, reliability, robustness, and efficiency in the management of physical systems of many application areas [39]. Although CPS development brings up specific issues and challenges, a large majority of integration problems encountered are similar to those identified in the design of systems involving multidisciplinary teams as required for mechatronic systems. Despite this observation, the mechatronics and CPS communities continue to seem to only slightly interact and share their expertise. The explanation proposed here is that mechatronic products are historically electromechanical products introducing more and more electronics and software while CPSs are historically the "Cyber-Systems" [40], which are more and more interacting with physical reality.

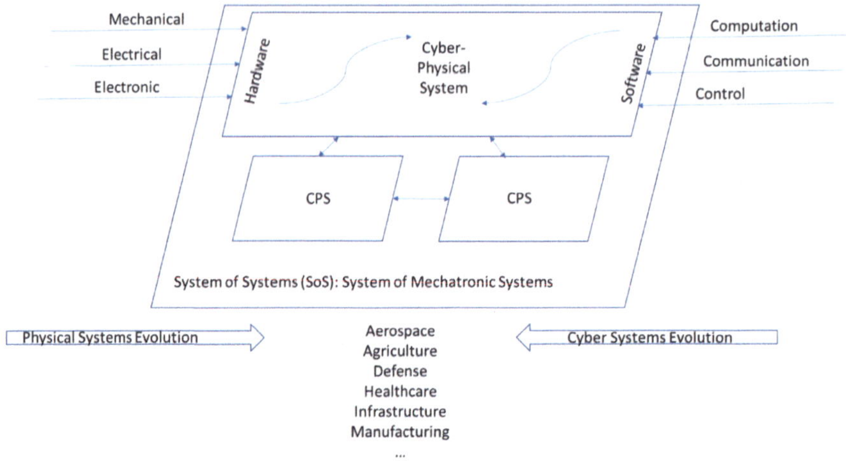

Fig. 9.5.3 Co-influence of mechatronic system and CPS (adapted from [27] and [40]) and extension to system of mechatronic systems

As can be seen, mechatronics and CPS challenges and research issues are complementary. While mechatronics is particularly focused on the hardware part of a system, CPSs remain focused on the software parts. Nevertheless, these two areas share a common goal which is to design and produce integrated systems. Figure 9.5.3 summarizes this convergence between CPS and mechatronic systems and extended it to system of mechatronic system (see 5.3.3).

9.3.2 *Mechatronics in the Internet of Things*

This convergence between mechatronic systems and CPSs is also subject to the pressure of the Internet of Things (IoT) trend, "*The connection of physical things to the Internet makes it possible to access remote sensor data and to control the physical world from a distance*" [41]. Thanks to this definition, we can understand the future role of mechatronic systems in this (r)evolution. IoT is based on smart objects, which can be considered as a mechatronic system connected to the Internet.

The need for new devices or the integration of sensors or/and actuators in the new generation of existing products will probably greatly increase the development of the mechatronics field. Whereas mechatronics applications were sometimes considered for specific and complex systems, this pervasive deployment of smart objects [41] will greatly amplify the need for mechatronic systems. However, some authors argue that "*there will need to be significant changes to the way mechatronic, and related, systems are designed and configured*" [42] to be fully integrated into the IoT trend.

They underline that "*increasing complexity while managing the transfer of functionality, particularly from the mechanical domain to the information technology and electronics domains, has long been an issue facing system designers*" [42, 43]. This state will force "*practitioners and educators to further review the ways in which mechatronic systems and components are perceived, designed and manufactured. In particular, the role of mechatronic smart objects as part of an IoT based system in which the structure is defined by context is resulting in an increased and increasing emphasis on issues such as machine ethics, user interaction, complexity and context as well as with issues of data and individual security*" [42].

From a technical point of view, if every mechatronic system can be individually identified and can conform to a standard protocol, the interoperability of the systems will increase and make systems even more autonomous and intelligent. New applications can then be envisioned. Another perspective relies on new synergistic services offered by mechatronic products compared to an isolated embedded system. Both these new opportunities are detailed in the next sections.

9.3.3 System of Mechatronic Systems

The role of mechatronic systems in the System of Systems (SoS) can be defined as "*large-scale integrated systems that are heterogeneous and independently operable on their own but are networked together for a common goal*" [44]. Systems participating in a system of systems are independently designed and can operate autonomously [45]. Following this definition, mechatronic systems and CPS can be considered components of SoS (see Chap. 10).

The main characteristics of SoS are defined in [45]. An SoS performs functions that are not achievable by single systems (so-called emergent behavior). SoS are geographically distributed, such that all participating systems need to exchange data in a distributed and remote way. This implies the emphasis of previous sections on CPS and IoT trends and the next generation of connected mechatronic systems. Another characteristic is the evolutionary development. This means that the SoS offered services and purposes change over time and mechatronic systems must be able to enter or leave the SoS during run-time. Managerial independence is another characteristic of a SoS. The mechatronic systems are designed independently and manage their own goals.

This implies new methods and tools to develop SoS that could constrain mechatronic systems design. SoS engineering deals with planning, analyzing, organizing, and integrating the capabilities of a mix of existing and new systems into an SoS capability greater than the sum of the capabilities of the constituting parts [46]. New mechatronics systems should enable this SoSE, with standardized interfaces and cooperation enhancement.

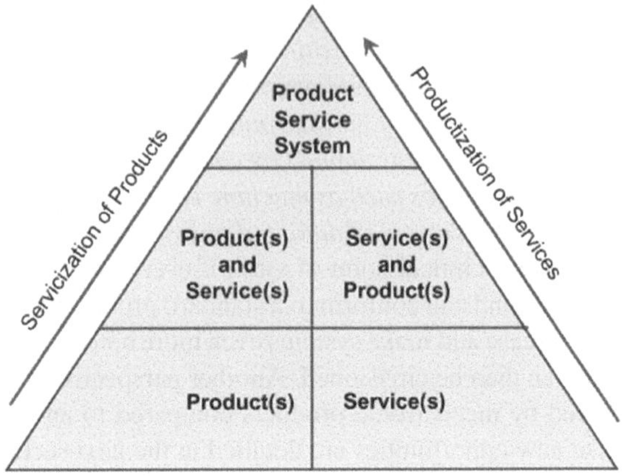

Fig. 9.5.4 Evolution of the PSS model [4]

9.3.4 Servitization of the Mechatronic Systems

To improve customer loyalty and to optimize the balance between the offer and the customer's requirements, significant numbers of industries are shifting from a product-centric approach to a bundle of products associated with and to services. This is presented as a way to achieve objectives such as risk reduction, competitiveness exposure reduction, and sustainability through business model evolution. The shift toward an integrated offer of products and services is illustrated by both concepts of productization and servicization, in a paradigm of a transition from a service or a product to an "integrated bundle of products and services" called a Product-Service-System (PSS) [47]. These evolutions are presented in Fig. 9.5.4.

In this context, mechatronics systems have a great role in transforming products into platform supporting services. Even if the system is neither connected all the time (IoT) nor able to interact autonomously with other systems (SoS), the fact mechatronic systems can transform physical products into intelligent systems allows us to envision new perspectives. Some authors recently presented some approaches to consider the service during the mechatronic system design [48].

9.3.5 Artificial Intelligence in the Design Process of Mechatronic Systems

The rapid evolution of Artificial Intelligence (AI) has revolutionized the design and development of systems across various domains. From software design to complex infrastructures, AI offers unique capabilities that transcend traditional design methods. This is particularly true in mechatronics systems design.

In the field of software design, AI is extensively employed to automate repetitive tasks such as bug detection, code optimization, and even the generation of new features. Techniques like machine learning and natural language processing are applied to enhance the efficiency and quality of produced software [49]. This facilitates the generation of embedded code and its update within mechatronics systems.

In the electronics domain, AI is utilized to automatically generate circuit schematics and layouts based on systems specifications. AI algorithms effectively explore solution spaces to identify optimal electronics architectures, considering multiple criteria such as power consumption, latency, and component density [50].

Finally, on the mechanical part, AI is leveraged to explore design spaces, identify optimal configurations, and resolve multi-physics complex optimizations. An exemplary work by Soni et al. [51] demonstrated this using both artificial neural networks and genetic algorithms to develop a basic design formulation for aesthetic products, Gu et al. [52] used both convolutional neural networks and generative adversarial networks (DCGAN) to produce new 3D spring models and Gao et al. [53] optimized the combustion chamber geometry of a diesel engine using artificial neural network and genetic algorithm (NSGA III). By employing predictive models based on AI, it becomes possible to simulate the behavior of mechanical systems under various conditions, thereby facilitating an efficient evaluation of different design configurations.

These three domains are further augmented by AI-enabled multidisciplinary collaboration, where AI facilitates collaboration between engineers and designers in the design process of mechatronic systems. Chen et al. [54] used rule-based analysis to determine the interface problems and enhance experts from different fields to solve them in the earliest stage. AI systems can analyze user needs, interpret design specifications, and generate suggestions to enhance existing concepts. This enables a more creative and exhaustive exploration of the design space, leveraging both human expertise and AI computational power.

Regarding systems of systems such as autonomous vehicles or drone networks, AI plays a central role. For example, Obbati et al. [55], use a multi-agent deep reinforcement learning method to jointly optimize trajectories, and data collection delay, maximize the throughput of IoT devices, and minimize the energy utilization of unmanned aerial vehicles. These systems integrate AI algorithms to interpret sensor data, make real-time decisions, and autonomously adapt to changing environments. This capability enables mechatronic systems to operate autonomously and effectively in diverse contexts.

Despite the progress made, the integration of AI into mechatronic system design still presents challenges, including the increasing complexity of AI models, which can make system validation more difficult. Additionally, the interpretability of decisions made by AI algorithms remains a challenge, particularly in sensitive domains such as healthcare or security. However, with ongoing research in the field, AI will continue to play an essential role in innovating and improving mechatronic systems, paving the way for new possibilities in optimization and performance.

9.3.6 Integration of Sustainability in Mechatronic Design

Regarding the reality of climate change and after the United Nations framing an agenda with Sustainable Development Goals [56] and road mapping for the European Green Deal [57], integrating sustainability issues in mechatronic design is an unquestionable challenge [58]. Fifteen years ago, the main debate was the compliance with Reach [59] and RoHS [60] standards for DEEE waste management [61].

As the principles of Life Cycle Assessment (LCA) and eco-design are now largely integrated into the product development process [62, 63], the perspective for mechatronic designers has changed. A new expert profile "the eco-designer" is joining the product development team. The viewpoint balancing and trade-off merging between design team members are then strongly discussed sustainability issues [64, 65]. The challenge is now to keep the alignment with satisfactory technical performances according to end user or customer requirements and be compliant with an eco-friendly impact on the environment.

For addressing such a challenge, numerous methods and tools exist for goods mass production [66, 67] but due to the technical complexity of mechatronic systems, the needed methods and tools should be carefully chosen [58]. The choices and decision-making of the electronic components, the power supply, the technical architecture of system assembly, the component material, etc. are numerous questions to be addressed [68]. Sustainability issues are also coming with the sourcing of primary material, the production engineering and manufacturing technology, the supply chain, and shipping. The ecological impact of product usage and maintenance should also be taken into account. And of course, last but not least, the product end of life requires paid attention and it is not only a question for material recycling but also dismantling and disassembly for reusing, repairing, and refurbishment according to R-strategies for circularity [69].

According to all the above-mentioned issues, the challenge is nowadays to think about new strategies for design processes for mechatronic systems. Innovation in methods and tools is on the way but definitively requires to be deeply intensive. The integration of novel sustainability expertise in the design processes, management, and organization is also a key perspective to creating eco-friendly mechatronic products and systems for the gain of circularity, the benefit of an ecological environment, and a better future.

9.3.7 Management of Obsolescence for Long Lifecycle Perspective of Mechatronic Systems

Mechatronic systems are typical products where obsolescence is a major issue. Obsolescence is defined by IEC as the "transition of a required item still in use from available to unavailable from the manufacturer" [70]. This definition introduces one of the obsolescence causes, which is linked to consumer's wishes. However, obsolescence can also be derived from many other causes. According to Trabelsi et al. 2021 [71], obsolescence causes can be categorized by technological innovation; the loss of original suppliers from the market for a variety of reasons; the lack of support and/or spare parts from suppliers; the directives, rules and other laws imposed by governments; and the fashion effects and consumer psychology.

In the field of technological innovation, Weerasuriya and Wijayanayake [72] distinguish three categories of obsolescence, i.e., hardware obsolescence, software obsolescence, and liveware (people) obsolescence: hardware obsolescence generally depends on the end of the supply and availability of a product [73]; software obsolescence occurs when the software becomes obsolete for several reasons independently of the hardware, according to Abili et al. [74]; Liveware obsolescence occurs in the form of skills and knowledge of hardware and software becoming obsolete. The extent of people's obsolescence can be seen by the amount of training and certification available for people to learn new hardware and software [75].

The business strategy for managing obsolescence is nowadays an unquestionable subject and should be linked with the trade-off in technological innovation and the sustainable lifecycle of mechatronic systems.

9.4 Conclusion

The chapter has dealt with new models for design processes of mechatronic systems and their future trends with novel applications. After a detailed presentation of the current models and standards of development processes for mechatronic systems engineering, the future trends for mechatronic system design are discussed. First, the new developments of mechatronics in the field of cyber-physical systems were considered. Second, the added value of mechatronic systems for the implementation of the Internet of Things has been detailed. Third, the role of mechatronics in the design and integration of the system of systems was presented as the last research trend.

Regarding the applications, two very interesting topics were considered with the integration of mechatronic systems in the servitization of products on the one hand. On the other hand, the generalization of mechatronics as the backbone of the factories of the future was discussed.

The future trends and models for the design processes of mechatronic systems have to consider the unquestionable issues derived from artificial intelligence usage in engineering design, sustainability, and circularity challenges, and lastly obsolescence management. These design processes for mechatronic engineering have to support the development of new services or the implementation of the industrial internet to benefit from proactive support for the cyber-physical systems.

References

1. Isermann R (2002) Mechatronic design approach. In: Bishop RH (ed) The mechatronics handbook. CRC Press LLC
2. Zheng P, Xu X, Chen C-H (2020) A data-driven cyber-physical approach for personalized smart, connected product co-development in a cloud-based environment. J Intell Manuf 31:3–18
3. Guérineau J, Bricogne M, Rivest L, Durupt A (2022) Organizing the fragmented landscape of multidisciplinary product development: a mapping of approaches, processes, methods and tools from the scientific literature. Res Eng Des 33:307–349. https://doi.org/10.1007/s00163-022-00389-w
4. Baines TS, Lightfoot HW, Evans S, Neely A, Greenough R, Peppard J, Roy R, Shehab E, Braganza A, Tiwari A, Alcock JR, Angus JP, Bastl M, Cousens A, Irving P, Johnson M, Kingston J, Lockett H, Martinez V, Michele P, Tranfield D, Walton IM, Wilson H (2007) State-of-the-art in product-service systems. Proceed Inst Mech Eng Part B J Eng Manuf 221:1543–1552. https://doi.org/10.1243/09544054JEM858
5. Fogliatto FS, da Silveira GJC, Borenstein D (2012) The mass customization decade: an updated review of the literature. Int J Prod Econ 138:14–25. https://doi.org/10.1016/j.ijpe.2012.03.002
6. Isermann R (2007) Mechatronic design approach. The mechatronics handbook. CRC Press, Boca Raton, USA, pp 2–3
7. Kernschmidt K, Feldmann S, Vogel-Heuser B (2018) A model-based framework for increasing the interdisciplinary design of mechatronic production systems. J Eng Des 29:617–643. https://doi.org/10.1080/09544828.2018.1520205
8. Clarkson J, Eckert C (2005) Design process improvement: a review of current practice. Springer, London
9. Hyman BI (2003) Fundamentals of engineering design, 2nd edn. Prentice Hall
10. Beitz W, Pahl G, Wallace K (2003) Engineering design: a systematic approach. Springer
11. Ullman DG (2010) The mechanical design process. McGraw-Hill Higher Education
12. Shylesh S (2017) A study of software development life cycle process models
13. Howard TJ, Culley SJ, Dekoninck E (2008) Describing the creative design process by the integration of engineering design and cognitive psychology literature. Des Stud 29:160–180. https://doi.org/10.1016/j.destud.2008.01.001
14. Blanchard BS (2012) System engineering management. John Wiley & Sons
15. OMG Group (2007) OMG Systems modeling language (OMG SysML ™)
16. Plateaux R, Choley JY, Penas O, Riviere A (2009) Towards an integrated mechatronic design process. In: 2009 IEEE International conference on mechatronics. IEEE, pp 1–6
17. Graignic P, Vosgien T, Jankovic M, Tuloup V, Berquet J, Troussier N (2013) Complex system simulation: proposition of a MBSE framework for design-analysis integration. Proced Comput Sci 16:59–68. https://doi.org/10.1016/j.procs.2013.01.007
18. Ye Y, Jankovic M, Bocquet JC (2012) Managing semantic consistency in model-based systems engineering using a matrix structure. In: Volume 2: 32nd Computers and information in engineering conference, Parts A and B. ASME, pp 1193–1204
19. Zheng C, Bricogne M, Le Duigou J, Eynard B (2014) Survey on mechatronic engineering: a focus on design methods and product models. Adv Eng Inform 28:241–257. https://doi.org/10.1016/j.aei.2014.05.003
20. Chen K, Bankston J, Panchal JH, Schaefer D (2009) A Framework for Integrated Design of Mechatronic Systems. In: Wang L, Nee AYC (eds) Collaborative design and planning for digital manufacturing. Springer, London, London, UK, pp 37–70
21. Hehenberger P (2014) Perspectives on hierarchical modeling in mechatronic design. Adv Eng Inform 28:188–197. https://doi.org/10.1016/j.aei.2014.06.005
22. Cooper RG, Sommer AF (2016) The agile-stage-gate hybrid model: a promising new approach and a new research opportunity. J Prod Innov Manag 33:513–526. https://doi.org/10.1111/jpim.12314

23. Mule S, Plateaux R, Hehenberger P, Penas O, Patalano S, Vitolo F (2020) A New Agile Hybridization Approach and a Set of Related Guidelines for Mechatronic Product Development. In: Nyffenegger F, Ríos J, Rivest L, Bouras A (eds) Product lifecycle management enabling smart X. Springer International Publishing, Cham, pp 618–633
24. Goevert K, Heimicke J, Lindemann U, Albers A (2019) Interview Study on the agile development of mechatronic systems. Proceed Des Soc Int Conf Eng Des 1:2287–2296. https://doi.org/10.1017/dsi.2019.235
25. Plateaux R, Penas O, Mule S, Hehenberger P, Patalano S, Vitolo F (2020) SCRUM++ Framework concepts. In: 2020 IEEE International symposium on systems engineering (ISSE). IEEE, pp 1–8
26. Groupe de travail de la Communauté d'Experts Fiabilité—Qualité—Sécurité (2023) Guide sur l'application des approches AGILES dans le domaine automobile
27. Bricogne M, Le Duigou J, Eynard B (2016) Design Processes of Mechatronic Systems. In: Bradley D, Hehenberger P (eds) Mechatronic futures. Springer International Publishing, Cham, pp 75–89
28. Heimicke J, Niever M, Zimmermann V, Klippert M, Marthaler F, Albers A (2019) Comparison of existing agile approaches in the context of mechatronic system development: potentials and limits in implementation. Proceed Des Soc Int Conf Eng Des 1:2199–2208. https://doi.org/10.1017/dsi.2019.226
29. Sima LM (2022) Testing mechatronic systems and their integration into the digital enterprise. IOP Conf Ser Mater Sci Eng 1268:012011. https://doi.org/10.1088/1757-899X/1268/1/012011
30. Mhenni F, Choley J-Y, Penas O, Plateaux R, Hammadi M (2014) A SysML-based methodology for mechatronic systems architectural design. Adv Eng Inform 28:218–231. https://doi.org/10.1016/j.aei.2014.03.006
31. Vidal EJ, Villota ER (2018) SysML as a tool for requirements traceability in mechatronic design. In: Proceedings of the 2018 4th international conference on mechatronics and robotics engineering. Association for computing machinery, New York, NY, USA, pp 146–152
32. Zheng C, Eynard B, Qin X, Li J, Bai J, Gomes S, Zhang Y (2019) A requirement-driven architecture definition approach for conceptual design of mechatronic systems. Integr Comput Aided Eng 26:361–382. https://doi.org/10.3233/ICA-180595
33. Fisher J (1998) Model-based systems engineering: a new paradigm. Incose Insight Incose Insight 1:3–16
34. Object Management Group Systems modeling language specification. http://www.omg.org/spec/SysML/1.2/PDF/
35. Cao Y, Liu Y, Paredis CJJ (2011) System-level model integration of design and simulation for mechatronic systems based on SysML. Mechatronics 21:1063–1075. https://doi.org/10.1016/j.mechatronics.2011.05.003
36. Kerzhner AA, Paredis CJJ (2012) A SysML-based language for modeling system-level architecture selection decisions. In: ASME 2012 International design engineering technical conferences and computers and information in engineering conference. Chicago, USA
37. Bathelt J, Jonsson A, Bacs C, Dierssen S, Meier M (2005) Applying the new VDI Design Guideline 2206 on mechatronic systems controlled by a PLC. In: International conference on engineering design (ICED 05). Melbourne, Australia
38. Zheng C, Le Duigou J, Bricogne M, Hehenberger P, Eynard B (2015) Multidisciplinary interface modelling: a case study on the design of 3D measurement system. In: IFIP WG 5.1, 12th International conference on product lifecycle management (PLM15). Doha, Qatar
39. Rajkumar R (Raj), Lee I, Sha L, Stankovic J (2010) Cyber-physical systems: the next computing revolution. In: Proceedings of the 47th design automation conference on—DAC'10. ACM Press, New York, New York, USA, p 731
40. Rajkumar R (2012) A cyber-physical future. Proc IEEE 100:1309–1312. https://doi.org/10.1109/JPROC.2012.2189915
41. Kopetz H (2011) Real-time systems. Springer, US, Boston, MA
42. Bradley D, Russell D, Ferguson I, Isaacs J, MacLeod A, White R (2015) The internet of things—the future or the end of mechatronics. Mechatronics 27:57–74. https://doi.org/10.1016/j.mechatronics.2015.02.005

43. Vasi VS, Lazarevic MP (2008) Standard industrial guideline for mechatronic product design. FME Trans 36:103–108
44. Jamshidi M (2008) Systems of systems engineering: principles and applications. Boca Raton, USA
45. Maier MW (1996) Architecting principles for systems-of-systems. INCOSE Int Symp 6:565–573. https://doi.org/10.1002/j.2334-5837.1996.tb02054.x
46. Department of Defense (2004) Systems engineering. In: Defense Acquisition Guidebook. Washington, D.C., USA, p 520
47. Mahut F, Daaboul J, Bricogne M, Eynard B (2015) Survey on product-service system applications in the automotive industry. IFAC-PapersOnLine 48:840–847. https://doi.org/10.1016/j.ifacol.2015.06.188
48. Pezzotta G, Pirola F, Pinto R, Akasaka F, Shimomura Y (2015) A service engineering framework to design and assess an integrated product-service. Mechatronics 31:169–179. https://doi.org/10.1016/j.mechatronics.2015.05.010
49. Mezouar H, Afia AE (2022) A systematic literature review of machine learning applications in software engineering. In Proceedings of the 5th international conference on big data and internet of things. BDIoT 2021. Lecture notes in networks and systems, Springer, Cham. Vol 489. https://doi.org/10.1007/978-3-031-07969-6_24
50. Nti IK, Adekoya AF, Weyori BA et al (2022) Applications of artificial intelligence in engineering and manufacturing: a systematic review. J Intell Manuf 33:1581–1601. https://doi.org/10.1007/s10845-021-01771-6
51. Soni S, Khanna P, Tandon P (2013) Knowledge support system for aesthetics in product design. ASME J Comput Inform Sci Eng 13(1):011006. https://doi.org/10.1115/1.4023355
52. Gu Z, Hou X, Saafi M, Ye J (2022) A novel self-updating design method for complex 3D structures using combined convolutional neurons and deep convolutional generative adversarial networks. Adv Intel Syst 4(6):2100186. https://doi.org/10.1002/aisy.202100186
53. Gao S, Zhang Y, Zhang S, Tan D, Li J, Yin Z, Hu J, Zhao Z (2023) Multi-objective optimization of the combustion chamber geometry for a highland diesel engine fueled with diesel/n-butanol/PODEn by ANN-NSGA III. Energy 282:128793. https://doi.org/10.1016/j.energy.2023.128793
54. Zheng C, Le Duigou J, Bricogne M, Eynard B (2016) Multidisciplinary interface model for design of mechatronic systems. Comput Ind 76:24–37. https://doi.org/10.1016/j.compind.2015.12.002
55. Oubbati OS, Atiquzzaman M, Lim H, Rachedi A, Lakas A (2022) Synchronizing UAV teams for timely data collection and energy transfer by deep reinforcement learning. IEEE Trans Veh Technol 71(6):6682–6697. https://doi.org/10.1109/TVT.2022.3165227
56. United Nations (2015) Transforming our world: the 2030 Agenda for sustainable development. United Nations: New York, NY, USA
57. Fetting C (2020) The European Green Deal, European sustainable development network report, ESDN Office, Vienna, Austria
58. Merschak S, Diallo TML, Hehenberger P (2022) Life cycle sustainability assessment as a decision-making tool for the design of mechatronic systems. Int J Prod Lifecycle Manag 14(2/3):142–173. https://doi.org/10.1504/IJPLM.2022.10050745
59. De Tandt I, Demuytere C, Van Asbroeck E, Moerman H, Mys N, Vyncke G, Delva L, Vermeulen A, Ragaert P, De Meester S, Ragaert K (2021) A recycler's perspective on the implications of REACH and food contact material (FCM) regulations for the mechanical recycling of FCM plastics. Waste Manag 119:315–329. https://doi.org/10.1016/j.wasman.2020.10.012
60. Zhou CH, Eynard B, Roucoules L (2009) Interoperability between PLM and RoHS compliance management based on XML and smart client, J Comput Inform Sci Eng 9(3). https://doi.org/10.1115/1.3184601
61. Baxter J, Lyng KA, Askham C, Jørgen Hanssen O (2016) High-quality collection and disposal of WEEE: environmental impacts and resultant issues. Waste Manag 57:17–26. https://doi.org/10.1016/j.wasman.2016.02.005

62. Vallet F, Eynard B, Millet D, Mahut-Glatard S, Tyl B, Bertoluci G (2013) Using eco-design tools: an overview of experts' practices. Des Stud 34(3):345–377
63. Kim H, Cluzel F, Leroy Y, Yannou B, Yannou-Le Bris G (2020) Research perspectives in ecodesign. Des Sci 6:e7. https://doi.org/10.1017/dsj.2020.5
64. Bhamra T, Hernandez RJ (2021) Thirty years of design for sustainability: an evolution of research, policy, and practice. Des Sci 7:e2. https://doi.org/10.1017/dsj.2021.2
65. Arena M, Ciceri ND, Terzi S, Bengo I, Azzone G, Garetti M (2009) A state-of-the-art of industrial sustainability: definitions, tools, and metrics. Int J Prod Lifecycle Manag 4(1–3):207–251
66. Mandolini M, Marconi M, Rossi M, Favi C, Germani M (2019) A standard data model for life cycle analysis of industrial products: a support for eco-design initiatives. Comput Ind 109:31–44. https://doi.org/10.1016/j.compind.2019.04.008
67. Rossi M, Germani M, Zamagni A (2016) Review of ecodesign methods and tools. Barriers and strategies for effective implementation in industrial companies. J Clean Prod 129:361–373. https://doi.org/10.1016/j.jclepro.2016.04.051
68. Marconnet B, Gaha R, Assuad C, Martinsen K, Eynard B (2024) Selecting sustainability indicators for smart product design based on industry 4.0/5.0 technologies: analysis and proposal of a methodological framework. Proceed Des Soc 4:1369–1378
69. Rodrıguez NB, Favi C (2023) Disassembly and repairability of mechatronic products: insight for engineering design. J Mech Des 146(2):020906. https://doi.org/10.1115/1.4064075
70. IEC62402 (2019) Obsolescence management, Int Electrotech Comm
71. Trabelsi I, Zoghadri M, Zeddini B, Barkallah M, Haddar M (2021) Prediction of obsolescence degree as a function of time: a mathematical formulation. Comput Ind 129:103470. https://doi.org/10.1016/j.compind.2021.103470
72. Weerasuriya G, Wijayanayake W (2014) An evaluation of factors affecting information systems obsolescence. J Emerg Trends Comput Inform Sci 5:158–164
73. Sandborn P (2007) Designing for technology obsolescence management, Industrial engineering research conference. Proceedings of the institute of industrial and systems engineers (IISE)
74. Abili N, Onwuzuluigbo R, Kara F (2013) Subsea controls future-proofing: A systems strategy embracing obsolescence management. Underwater Technol 31:187–201. https://doi.org/10.3723/ut.31.187
75. Whitley JA (2001) A study of computer obsolescence and its impact, M.S. Thesis of Towson University

Chapter 10
Advances in Intelligent Healthcare from e-Health to Self-Care

Jorge Azorin-Lopez, Andres Fuster-Guillo, Nahuel E. Garcia-D'Urso, Pablo Rocamora-Garcia, and Marcelo Saval-Calvo

Abstract This chapter outlines the significant impact of Artificial Intelligence (AI) and assistive technologies in transforming healthcare from e-Health to self-care based on personalized and human-centered smart health technologies. AI has become instrumental in improving diagnostic accuracy, customizing treatment plans, and enhancing healthcare operations over the past decade. Its applications, including advanced medical imaging, predictive analytics, and patient monitoring, have extended healthcare delivery from traditional settings to home and community environments through telehealth and remote monitoring. It evaluates the predictions made 10 years ago and the challenges and lessons learned. Looking ahead, the chapter speculates on the potential of emerging and/or mature technologies putting AI technologies in the center to further revolutionize healthcare.

Keywords Artificial intelligence · Healthcare · Self-care · Elderly · Smart homes

10.1 Introduction

Smart environments, Industry 5.0, and Healthcare 5.0 are keywords defining the current, but in constant evolution, state of society. This evolution has come hand in hand with technological improvements that allowed more sophisticated Artificial Intelligence (AI) systems to reach a level of maturity that has been transferred to impact society. This chapter analyses the progression of technology and AI in smart environments, focusing attention on healthcare [1]. The starting point of this study is our previous chapter [2] where a set of predictions were made.

Section 10.2 identifies the main trends and challenges in home and healthcare technologies over the past decade. In the case of Smart Homes, the main challenge has been to enable diverse devices to communicate seamlessly, evolving toward

J. Azorin-Lopez (✉) · A. Fuster-Guillo · N. E. Garcia-D'Urso · P. Rocamora-Garcia · M. Saval-Calvo
University of Alicante, San Vicente del Raspeig, Spain
e-mail: jazorin@dtic.ua.es

P. Hehenberger and D. Bradley (eds.), *Mechatronic Futures*,
https://doi.org/10.1007/978-3-031-83571-1_10

distributed IoT architectures. In healthcare, the focus has been on adapting to the needs of an aging population by integrating e-Health and mobile technologies.

The evolution of healthcare technology has paralleled industrial advances, moving from basic patient care to sophisticated patient-centric systems (Healthcare 1.0 to 5.0). Section 10.3 discusses different technologies and how they are transforming healthcare into a digitized and interconnected ecosystem, with data-driven and patient-centric approaches.

In Sect. 10.4 we look at how artificial intelligence (AI) is being successfully applied in healthcare to improve diagnostics, personalize treatments, or perform predictive analytics. Machine and deep learning are key in Healthcare 5.0. These paradigms make it possible to classify patients to predict disease progression and facilitate preventive and proactive care, optimizing treatments and follow-up through continuous monitoring and analysis. A notable application of AI is its ability to improve the management of prevalent and chronic high-impact diseases such as obesity, where AI-based tools facilitate comprehensive and personalized treatment plans and continuous patient follow-up.

Section 10.5 discusses how AI will change our homes and healthcare by improving interoperability and integration to enable intuitive and proactive health management using advanced wearable devices and real-time data analytics. Evolving blockchain and encryption technologies will ensure data security in Smart Homes and healthcare ecosystems. AI-based personalized assistants will provide individualized health guidance and federated learning will preserve privacy in collaborative model training. Finally, explainable AI will increase transparency and trust in AI decisions and generative AI will accelerate the discovery of new healthcare innovations.

10.2 Past-Decade Predictions and Trends

In the earlier version of this chapter,[1] it was identified that the implementation of Smart Homes and related technologies involved several systems issues for the short, medium and long terms such as the choice and future-proofing of technologies, ethics, and costs. This chapter now looks at the background, and structure of, Smart Home technologies and systems before progressing to look at one area of application, that of eHealth and mHealth, in more detail.

10.2.1 Smart Home Trends

Regarding the home systems challenges, the most important goal identified was that of enabling individually intelligent devices to effectively communicate and interact

[1] Azorin-Lopez et al. (2016) Home Technologies, Smart Systems and eHealth, **Mechatronic Futures**, Springer.

with the other devices and actors in the environment. A significant issue for the short and medium terms was the **integration of various Smart Home systems**. This included ensuring that different devices could communicate with one another effectively, which was crucial for providing services to the users. However, this comes with a set of challenges. One of the primary concerns is the integration of disparate technologies which can lead to issues in system compatibility and interoperability. Ensuring that different devices and systems can communicate and operate together was a significant technical challenge.

The previous chapter predicted the ongoing integration and future potential of *Smart Home systems that manage everything from entertainment to domestic functions like washing and cleaning*. It predicts that Smart Homes will increasingly contain networked systems that manage these activities smoothly. Moreover, it was envisioned that there would be greater integration of home and medical devices to create more cohesive systems. According to the use of residential gateways, the evolution of technology changed from a centralized structure to a distributed structure trend. The Internet of Things (IoT) was expanding, leading to more connected devices that interact and communicate to provide a cohesive living experience. It envisioned the use of the IoT to provide a standardized architecture to support the development of new services, this is an IoT as a means of integrating a range of smart objects within that infrastructure.

Another important challenge identified is how the services that can be provided in the Smart Home can be used. In other words, **the technology should not be visible to the user**, he or she should perceive the service, not the technology. There is an emphasis on the need for Smart Homes to be user-friendly, allowing individuals of all ages and technological skill levels to interact with systems effectively. First, user interfaces often lack intuitiveness, which can make technology appear inaccessible. Hence, the development of user-friendly interfaces that are intuitive and require minimal learning was seen as crucial. Further, the user interface must emphasize autonomy while learning to perform tasks for their users and providing proactive assistance as necessary. The user interface could observe and monitor the actions performed by the user to learn, suggest, or perform an action. Additionally, the user interface should have the ability to adapt to the user.

Hence, from an automated reactive home system to an intelligent proactive system was envisioned as the trend. Within this aspect, the integration of ubiquitous computing and ambient intelligence was transforming homes into environments that are responsive and sensitive to the presence of individuals. In this context, the previous chapter discussed the evolution toward cyber-physical systems and the Internet of Things (IoT), emphasizing the future of interconnected intelligent devices that not only communicate with each other but also support a wide range of intelligent applications and services, thereby creating more integrated service ecosystems.

10.2.2 Healthcare Trends

Regarding eHealth, the previous chapter highlighted the significant role that eHealth and mHealth were expected to play, especially in response to the aging population. The main issues related to an aging population, firstly, **there is a growing number of older adults**, which increases demand on healthcare systems due to their more complex and frequent healthcare needs. Next, elderly people also prefer to **maintain independence** and live at home for as long as possible, requiring healthcare support that can be provided at home or via technologies like telemedicine, allowing them to live independently while receiving necessary care. Finally, the healthcare system needs to adapt by **expanding services and improving access** to accommodate the increasing number of elderly people. This includes integrating technologies such as eHealth and mHealth to provide continuous monitoring and support, facilitating better chronic condition management and daily health care remotely.

This situation involves the use of advanced monitoring systems tailored to individual needs, predicting a shift toward more pervasive use of technology in healthcare. In particular, the issues discussed above about the lack of device integration related to different technologies and devices and those related to devices being visible to users remain major problems for the development of eHealth. These problems could be more severe given the population they must deal with. The challenges were primarily centered around the integration of health technologies. Ensuring seamless interoperability between diverse health devices and systems was essential yet challenging, requiring standardized communication protocols and platforms. The predictions made included, in addition to advancements in the integration of various Smart Home devices and health technologies to create more cohesive systems, the advancement of integration across different health devices and systems and promoting interoperability through standards such as the ISO/IEEE 11,073 Personal Health Data (PHD). This facilitated seamless communication between devices used in home health monitoring and broader healthcare systems.

An important aspect of the provision of these health services was the use of wearable and implantable sensors which was highlighted as a significant trend to revolutionize health monitoring. These devices must be designed to monitor a range of physiological parameters continuously, supporting both ongoing health management and emergency responses. A very closely related trend is the expansion of mobile health (mHealth). It utilizes mobile devices to extend healthcare beyond traditional environments, offering new methods for disease management and health monitoring. This includes using the previous wearable and implantable sensors that provide real-time health monitoring and connect with mobile applications to offer continuous care and support. Finally related to this problem, it highlighted the importance of avoiding technological lock-in by adopting more flexible and open system architectures adaptable to future devices and services without being hindered by legacy systems.

Data security and privacy were significant concerns because health data are particularly sensitive. Ensuring robust security measures and maintaining patient confidentiality while allowing necessary access to health data for caregiving purposes pose ongoing challenges. Effective data management and the secure handling of health-related information are paramount, requiring advanced data security measures. It was discussed that it would play an important role in developing personal health databases to achieve the necessary levels of user confidence while ensuring appropriate access. Another significant trend was the use of an integrated health informatics system. This system would utilize patient data and health informatics to support the planning, organization, and management of health care. By analyzing trends and patterns within the data, such systems can facilitate earlier medical interventions. The envisioned infrastructure allowed for the free movement of information across different strategic areas involved in healthcare provision, development, and management while maintaining a connection to individual patient data. However, the system faces significant challenges related to security and confidentiality. A key concern was ensuring patient anonymity while still allowing for the strategic use of relevant data.

Finally, the problems and trends toward user-centric design discussed above are valid in these technologies, being essential to ensure they are accessible to all users, thereby facilitating broader adoption and more effective health interventions. The development of user-friendly interfaces in eHealth and mHealth technologies, which should be intuitive and accessible to all users, including the elderly or those with limited tech skills. This approach includes simplifying the interaction with technology to enhance user engagement and compliance.

10.3 Technological Advances in Healthcare

The evolution of healthcare parallels the Industrial Revolution, which transformed production and organization. From mechanization to digitalization, the fourth revolution, or Industry 4.0, emphasizes '*Value Networks*', '*Data-driven Business Models*', '*Customer-centric*' approaches, and '*Global digital ecosystems*' [3, 4].

Industry 4.0 transforms industrial processes using technologies like the Internet of Things (IoT), Cyber-Physical Systems (CPS), Big Data Analytics, Artificial Intelligence (AI) and Machine Learning (ML), Cloud Computing, Augmented Reality (AR) Virtual Reality (VR) and Mixed Reality (MR), Advanced Robotics, Additive manufacturing (3D printing), Cyber Security, Blockchain, Simulation and modelling (Digital Twins) or, 5G/6G communications.

Similarly, healthcare has evolved through stages from Health Care 1.0 to 5.0 [5] (see Fig. 10.1). HC1.0 involved basic patient-specialist encounters, testing, diagnosis, and follow-up. HC2.0 saw the development of diagnostic, monitoring, surgical, and life support devices. HC3.0 introduced electronic medical records (EMR/EHR) and remote care. HC4.0 brought smart and connected systems for personalized, patient-centered care. HC5.0 envisions smart disease control, smart remote monitoring and

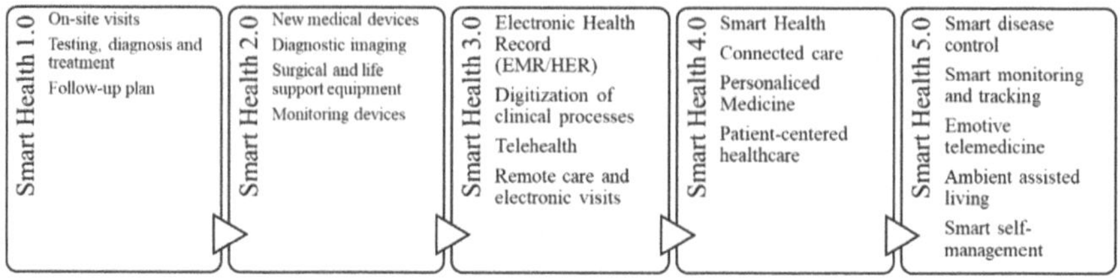

Fig. 10.1 Progressive transformation from health care 1.0 to 5.0

tracking, virtual clinics, ambient assisted living, smart self-management, wellness monitoring and control, smart treatment reminders compliance and adherence, and personalized and connected health care [6]. Healthcare 4.0 and 5.0 focus on patient-centric models, transforming value chains into value networks where all stakeholders (health centers, pharmacies, doctors, hospitals, nursing services, telehealth services, emergency services and so forth) put the patient at the center of their business model.

From a technological point of view, there are several technologies with different degrees of transversality to decision-making, the interconnection of stakeholders, and healthcare processes (see Fig. 10.2). Some of these technologies include artificial intelligence and deep learning, big data analytics and statistical inference, virtual reality, augmented reality, modeling and optimization techniques, IoMT (Internet of Medical Things), cyber-physical systems (CPS), RFID, wearable devices, smart sensors, nanotechnology, robotics, blockchain, additive manufacturing or 5G/6G.

Artificial intelligence is being applied in multiple areas of health care in a transversal way, which is why we will dedicate part of the chapter to this. The huge amount of data available in the healthcare sector is posing new challenges that

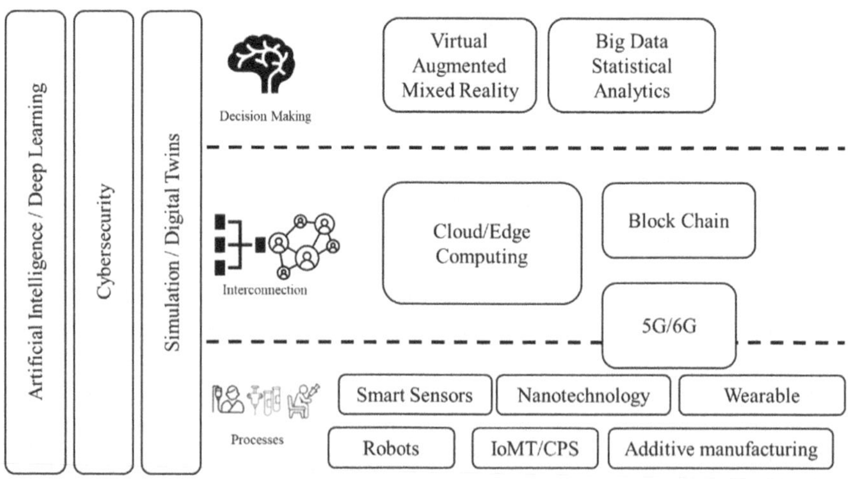

Fig. 10.2 Technologies involved in the health care process

can be addressed with big data analytics and statistical inference techniques. Data from different sources such as electronic health records (EHRs), medical images, and wearable devices with personalized data, provide the oil for big data analytics techniques, which will allow addressing new challenges such as such as personalized medicine or predictive analytics. Virtual, augmented, and mixed reality technologies are being widely applied in the healthcare sector with outstanding applications such as healthcare professional training, surgery planning, and rehabilitation assistants. The revolution generated by the Internet of Things in industrial and domestic environments is having success in the healthcare sector with the idea of IoMT (Internet of Medical Things). These new devices are facilitating remote patient monitoring, personalized medicine, and advanced diagnostic processes.

Healthcare 5.0 technologies bring services into patients' homes. IoMT, CPS, wearables, smart sensors, and robotics have led to body-worn devices, smartwatches, ECG sensors, and assistive robots [7]. Nanotechnology, with materials and devices between 1 and 100 nm, enables new diagnostics and treatments. Nanorobots aid in cancer detection and drug delivery, while nanotechnology improves vaccines and disease prevention [6].

10.3.1 Smart Home Situation

The situation of problems and trends detected in the Smart Home (Table 10.1) are analyzed:

T1. Smart Home systems that manage everything from entertainment to domestic functions. Smart Home systems (SHS) have evolved to integrate domestic devices and external services like utilities, information, and entertainment, moving toward broader smart city integration [8]. SHS integrate more and more various types of devices to manage and automate domestic functions efficiently. Data from sensors (including user inputs) is processed by a home automation hub, which makes decisions based on pre-set rules or machine learning algorithms and sends commands to actuators to take appropriate actions.

Nowadays, sophisticated central software platforms run services such as security, entertainment, energy, and climate control, providing the infrastructure to manage, automate, and control diverse devices and functions seamlessly. These platforms, like Amazon Alexa, Google Assistant, Apple HomeKit, and Samsung SmartThings, are Cloud-based and use usually devices exclusively manufactured for them. IFTTT (If This Then That) is a web-based service that creates chains of simple conditional statements to enable the integration of services and devices that would not normally communicate with each other. To provide compatibility between different manufacturers and security, the Matter protocol is an open-source standard for connecting IoT Smart Home devices [9]. Finally, it is worth mentioning Home Assistant, an open-source platform, integrates various smart objects and services.

Table 10.1 Problems detected and trends in smart homes

Problem	Trends	Achieved
P1.Lack of integration of various smart home systems (services and technologies)	T1. Smart home systems that manage everything from entertainment to domestic functions	Totally
	T2. Homes will increasingly contain networked systems that manage entertainment and automation smoothly	Totally
	T3. Integration of smart home devices and health technologies to create more cohesive systems	Partially
	T4. Evolution from a centralized structure to a distributed architecture of devices within the Internet of Things (IoT)	Totally
P2.Technology and devices are visible to the users	T5. The user interfaces with more autonomy, providing proactive assistance and adapting to the users	Partially (beginning)
	T6. Integration of ubiquitous computing and ambient intelligence to provide environments that are responsive and proactive	Partially
	T7. Evolution toward CPS and IoT, emphasizing the future of intelligent interconnected devices providing intelligent applications	None

T2. Homes will increasingly contain networked systems that manage entertainment and automation smoothly. The integration of sensors, actuators, and services relies heavily on communication protocols like 6LowPAN, Zigbee, and Z-Wave, as low-power wireless technologies [10]. Wi-Fi is also used for devices and audio/video (some Ethernet is also used for TVs), and Bluetooth is often used for short-range direct control of devices without the need for Wi-Fi. However, achieving seamless interoperability between devices from different manufacturers remains challenging.

T3. Integration of Smart Home devices and health technologies to create more cohesive systems. This integration is growing, driven by increased health awareness, aging populations, and the adoption of wearable health technologies [11]. Wearable devices in Smart Homes monitor health in real-time, tracking vital signs like heart rate, sleep patterns, and activity levels to adjust home settings for comfort and health. Air quality monitors detect pollutants and adjust purification systems. Moreover, smart exercise equipment tracks fitness and adjusts goals, while smart mirrors and treadmills link to other home systems for added functionalities. Smart fridges and kitchen appliances manage nutrition and suggest recipes based on health goals. Telehealth services are being integrated into Smart Homes, enabling video calls and remote consultations with healthcare providers. However, much more integration work is needed to provide cohesive systems where many challenges need to be addressed, which will be discussed in the following sections.

T4. Evolution from a centralized structure to a distributed architecture of devices within the Internet of Things (IoT). The IoT paradigm has shifted from centralized systems to more integrated, distributed architectures [12]. Initially, they used a central server or Cloud platform to process and manage data favored for its straightforwardness and ease of management, aligning with the trends that heavily tended to Cloud integration for solutions. However, as IoT devices proliferate and generate vast amounts of data, there is a growing need for real-time processing and efficient data management. This shift has led to the adoption of microservices-based architectures [13, 14]. Distributed IoT architectures enhance scalability and processing efficiency by allowing data to be processed locally, minimizing latency and reducing bandwidth usage. This decentralization optimizes system management, where devices filter data locally before transmitting only necessary information to the central system. The trend toward edge computing further supports distributed architectures by processing data close to where IoT devices are located, reducing latency and improving responsiveness (Cloud, mist, fog, and edge computing).

T5. The user interfaces with more autonomy, providing proactive assistance and adapting to the users. The most widespread user interfaces are voice-based, such as Amazon Alexa, Google Assistant, and Siri. These interfaces handle routine tasks like scenes, reminders, alarms, weather alerts… but lack proactive assistance capabilities [15]. Although routine learning and contextual understanding are improving, the full vision of proactive AI remains a work in progress. Advances in machine learning and user data integration are expected to enhance these capabilities. However, certain smart devices are designed for specific proactive assistance. For example, the Nest Learning Thermostat adjusts heating and cooling based on the resident's habits, optimizing climate control for comfort and energy efficiency. Similarly, the iRobot Roomba cleans homes autonomously, adapting their cleaning routes but not predicting new needs.

T6. Integration of ubiquitous computing and ambient intelligence to provide environments that are responsive and proactive. Recent research has demonstrated significant advancements in integrating distributed IoT devices using edge and Cloud computing, emphasized by the potent application of artificial intelligence. These proofs of concept demonstrate how AI can augment ubiquitous computing and ambient intelligence, enabling systems to not only respond but also proactively adapt to user needs through advanced data analysis and machine learning. Such systems are designed to learn from user interactions and context changes without direct user input. However, despite the promising potential demonstrated in research labs, the practical application of these advanced AI-driven systems in everyday environments is waiting.

10.3.2 Healthcare Situation

The problems and trends set out in Table 10.2 are analyzed in the following paragraphs.

Table 10.2 Problems detected and trends in healthcare

Problem	Trends	Achieved
P3.Lack of integration of health technologies (including services)	T8.Promoting interoperability through standards such as the ISO/IEEE 11,073 personal health data (PHD)	Totally
	T9.The use of wearable and increasingly implantable sensors is highlighted as a significant trend in providing healthcare	Partially
	T10. Expansion of mobile health (mHealth) to extend healthcare beyond traditional environments	Partially
	T11. Avoiding technological lock-in by adopting more flexible and open system architectures	Partially (beginning)
P4.Data security and privacy	T12. Achieve the necessary levels of user confidence while ensuring appropriate access to develop personal health databases	In progress (beginning)
	T13. Use of an integrated health informatics system to support the planning, organization, and management of health care ensuring patient anonymity while still allowing for the strategic use of relevant data	In progress

T7. Evolution toward CPS and IoT, emphasizing the future of intelligent interconnected devices providing intelligent applications. Cyber-physical systems (CPS) and IoT are revolutionizing healthcare by integrating intelligence and autonomy into physical processes. These systems create interconnected environments where intelligent objects communicate and operate autonomously. Research in this area is pushing the boundaries of what is possible, suggesting a future where CPS can form autonomous ecosystems without central oversight. This vision of interconnected, intelligent objects that manage themselves in a decentralized manner is highlighted in recent academic discussions and conceptual frameworks. However, practical implementation is still in its infancy, relying on centralized decision-making systems, which coordinate the actions of interconnected devices.

T8. Promoting interoperability through standards such as the ISO/IEEE 11,073 Personal Health Data (PHD). The ISO/IEEE 11,073 standards ensure efficient communication and interoperability among medical devices handling Personal Health Data, allowing devices from different manufacturers to exchange data seamlessly. This enhances health monitoring across various settings. The latest updates in 2022 have proven effective in pilot studies, improving health management in multi-user environments like assisted living facilities. Adopting the ISO/IEEE 11,073 standards significantly advances the management and continuity of care, particularly in settings requiring close monitoring and integration of diverse health technologies.

T9. The use of wearable and increasingly implantable sensors is highlighted as a significant trend in providing healthcare. In the T3 trend assessment, we talked about wearable devices like smartwatches and fitness trackers integrated into Smart Homes. These devices support health and wellness monitoring, offering insights that aid overall health management. They monitor health and wellness, providing valuable insights. They track physical activity, heart rate, and sleep patterns, and some can perform ECGs and detect irregular heart rhythms like atrial fibrillation. Increasingly, wearables can detect falls, send emergency alerts, and offer activity reminders to promote a healthier lifestyle. However, these devices are not substitutes for medical devices, as they often lack regulatory approval. They complement health services rather than replace them. Medical wearables, designed for clinical monitoring and chronic disease treatment, provide more accurate data and are regulated by health authorities. There are two types: portable biosensors and implantable devices, both crucial for real-time monitoring and data collection [16].

On one hand, wearable biosensors monitor physiological parameters like glucose and heart rate, aiding chronic disease management for patients and healthcare professionals. They are used in fitness tracking, hydration assessment, and electrolyte balance assessment, making them versatile tools for personalized healthcare and sports efficiency. On the other hand, implantable devices operate within the body, enabling long-term monitoring and therapeutic intervention. They are used in treating cardiac disorders with implantable defibrillators (ICD), neurostimulators for pain treatment, and continuous glucose monitoring.

T10. Expansion of mobile health to extend healthcare beyond traditional environments. This expansion is being realized partially, with successful case studies demonstrating its potential. mHealth apps are designed to be accessible, low-cost, and user-friendly, promoting healthy behaviors in fitness, nutrition, and mental well-being [17]. Important areas of study include remote monitoring and management of chronic diseases like diabetes and cardiovascular conditions. For example, the work of [18] allows patients to receive timely care and adhere to treatment or personalized care [17]. mHealth has also shown beneficial effects on pain intensity, enhancing quality of life [19]. Additionally, mHealth extends healthcare access to remote and underserved areas, reducing the need for in-person visits [20]. It is a vital tool for improving healthcare efficiency, public health monitoring, and patient outcomes, particularly in patient rehabilitation [21].

T11. Avoiding technological lock-in by adopting more flexible and open system architectures. Today, several steps are being taken to make this trend a reality. The idea of open and flexible system architectures that can be adopted to avoid technological lock-in is completely attached to the concept of the Internet of Medical Things (IoMT). IoMT ensures collaboration and data sharing among diverse medical devices. It enables the integration of data from wearable devices, implantable sensors, and electronic medical records, supporting personalized health. For healthcare organizations, IoMT means the use of new technologies without being tied to a specific vendor system [22]. Open architectures address issues necessary to remove e-Health obstacles like security, privacy, and interoperability [23, 24].

T12. Achieve the necessary levels of user confidence while ensuring appropriate access to develop personal health databases. Achieving user trust while ensuring adequate access to personal health databases involves balancing data privacy and accessibility. Many factors determine users' confidence in using digital healthcare, such as familiarity with eHealth systems and trust in the system; willingness to share personal health data, digital literacy, and attitude toward information systems [25]. Improving security measures and transparent data management can enhance patient trust in digital health systems [26]. Another critical aspect of this trend is the need to increase user confidence with data-sharing practices [27]. Finally, context-aware data access policies will help implement the trade-off between user confidence and appropriate access [28].

T13. Use of an integrated health informatics system to support the planning, organization, and management of health care ensuring patient anonymity while still allowing for the strategic use of relevant data. The current situation for this trend is that it is in progress and the challenges remain to be solved. Integrated health informatics systems are increasingly being adopted across healthcare organizations, driven by the need for efficient data management and improved patient outcomes. According to Yogesh & Karthikeyan [29], integration plays a significant role in modern healthcare, aiming to ensure efficient planning, organization, and management of healthcare. In this context, the use of Cloud computing has enhanced the availability and accessibility of patient data for healthcare providers [30]. This improves care coordination and decision-making but also requires strong security measures to protect patient data. The adoption of the latest standard in health informatics, Fast Healthcare Interoperability Resources (FHIR) architecture, has facilitated the exchange of health data across various subsystems, helping to reorganize operations and support patient care. The growing use of standardized frameworks like HL7 FHIR speeds up the healthcare industry a better integration of several health informatics systems and leverages data for better planning and management. However, challenges such as maintaining privacy, ensuring interoperability, and developing adaptable standards remain crucial to achieving full integration.

10.4 Advances in Healthcare Through Artificial Intelligence

10.4.1 AI's Transformative Role in Healthcare

Healthcare 5.0 is introducing smart and connected systems using artificial intelligence to provide personalized and patient-centered self-care and AI is now present in most of the levels and applications in healthcare, including diagnosis, robot guidance in surgical environments or home care, monitoring, etc. There is a strong focus on machine learning, and particularly on deep learning, for the decision-making that is involved in most of the functionalities in Healthcare 5.0, e.g., pre-diagnosis for expert assistance is based on AI and machine learning, or surgical robot's commandment is based on learning techniques. But, as was discussed before, special attention must be given to misdiagnosis by unrepresented or incomplete data.

If we pay attention to self-care, the different applications share some common functionalities that the system shall provide. Schork states that personalized medicine and health care include *diagnoses*, *treatment,* and follow-up *monitoring* of individual patients in a coordinated way with a smooth transition between stages [31]. Ali et al. have reviewed several AI-based healthcare proposals and some general functionalities have been found [32].

The **classification of patients** provides information, feedback, and warnings about a patient's status, either by metrics or in general. Different approaches have been used here including classical machine learning methods, such as the case of An et al. [33] who used Principal Component Analysis (PCA), K-Means clustering-based classification strategy, or partial least squares (PLS). Others such as Brom et al. and Salgado and Vieira [34, 35] used Decision Trees to classify the risk of readmission using EHR data.

Based on the patient classification, the **predictive analytics** provide accurate forecasts of disease development that can be made to aid diagnosis and prognosis. Using historical data, statistics, and AI algorithms, predictive analytics can provide significant forecasting information to estimate future indicators [36]. Different methodologies have been used for this purpose, such as the case of Kibiria et al. [37] which presents a diabetes prediction using soft voting classification with ensemble learning, similar to the work of Rastogi and Bansal [38]. Novel deep learning architectures have also been used for prediction in healthcare. To forecast health parameters, time series of data and temporal networks are ideal, being Long Short-Term Memory (LSTM) one of the most common approaches among other techniques. DeepSigN [39, 40] is based on LSTM presented for health deterioration forecasting. Similarly, Sathayanarayana et al. used time-batched LSTM for quality sleeping prediction [41].

Other functionalities are **preventive and proactive care.** The results of predictive analytics can be used to create preventive and proactive care plans [42]. Preventive healthcare focuses on providing relevant information as soon as possible using historical data and general knowledge of the diseases. Proactive, on the other hand, goes a step further by actively managing health to improve personalized treatments. Currently, health systems are suffering overloads due to higher demand and limited revenues. Proactive smart management can help reduce queuing times and budget. Using AI-based systems it will be possible to prioritize patients automatically, both before and during the hospital arrival. Also, a proactive approach will help individuals to have a continuous monitoring of their health, which could result in better adherence to treatments and reducing health visits.

Lastly, **optimal intervention, treatment, and follow-up** provide continuous monitoring, analysis, and prediction of the patient's condition that are necessary to dynamically update interventions and optimal treatment plans. The use of AI for modeling the evolution and continuous monitoring is a key aspect of disease forecasting [43]. Regression models and other statistical proposals are commonly used. These functionalities are the result of several previous steps (predictions, classification, stratification, etc.) that provide relevant information to take action. AI can be used to allow the follow-up of patients with smart devices that provide remote and self-monitoring [44, 45].

10.4.2 Challenges and Lessons Learned

We are currently in one of the most disruptive historical transitions that humans have ever experienced. This produces both excitement and discomfort at the same time as the current state-of-the-art technology will improve the quality of life but, at the same time, it is perceived as an intrusive entity in people's lives. The expansion of assistive technologies in our everyday lives is technologically, regulatory, and socially limited. Most Artificial Intelligence-based systems greatly depend on large datasets and computational power as well as ethical, legal and social limitations. Assistive technologies imply several decision-making systems working simultaneously and developing Artificial General Intelligence (AGI) is still far from being possible, but novel proposals emerge every day to cover unsolved problems and improve the effectiveness of existing solutions.

Currently, in 2024, we have reached the point where some assistive technologies are eligible to be at home, such as Siri, Alexa, and Google Assistant [46]. However, these provide basic information based on simple indications, which limits the efficacy as in healthcare a patient's condition includes several aspects. Recently, more complex tools, such as ChatGPT, are providing more elaborate information. Particularly, ChatGPT has been studied for its advantages and disadvantages in medicine by Dave et al. [47]. However, these technologies run on top of highly expensive hardware and a large amount of training data [48].

This data concern raises various problems, on the one hand, the problem of time-consuming data collection that covers all possible cases, and, on the other hand, the information supplied to the models for training them. Selecting non-sensitive data in the training process will produce a biased model with a narrow understanding of the topic, which might be critical for a reliable diagnosis [49]. In the technical area, the authors found three main sources of bias. Data drive, created by data gaps or biased samples such as the case of removed personal information that can be relevant for diagnosis; algorithmic bias, the result of an imbalance in classes (mainly in rare diseases) or training bias; and human bias, produced by the algorithm design. Biases have other effects such as *hallucination*, which is defined as the inference of information as nonsensical or unfaithful to the provided source content [50]. This might create a diagnosis that might look accurate but in essence, is non-realistic and might mislead to the wrong treatments.

In terms of social limitations, some papers have recorded hesitancy in elderly people to accept technology at home. Assistive technology mainly benefits older people because their health condition tends to be more delicate, and their ability to go to hospitals or care centers is more limited. As Tian et al. have observed, some Smart Home Health Technologies (SHHTs) are too difficult for older persons to learn and tend to be worried about privacy or gossip [51].

Apart from the technical and social limitations, the legal frameworks that apply to artificial intelligence and health bound the appliance of the technology to certain levels and under specific conditions. From a general point of view, medical recordings are composed of personal information that in most cases needs to be treated privately

[52]. The main regulations in this aspect in the EU are established in the General Data Protection Regulation (GDPR), which states the protection measurements of personal data usage and transference. Among the different articles, GDPR includes that subjects must give their explicit consent for the data collected and their usage purpose. Also, data subjects would be able to withdraw their consent at any time. GDPR does not apply directly to AI systems, however, it might add bias as we have mentioned before, as some data that might be relevant to the decision-making process is not allowed to be used.

A complementary directive to the GDPR in the EU is the ePrivacy Directive, which regulates cookies, email marketing, data localization, and other aspects. The main aspect that impacts AI in this case is the data removal when it is not used anymore. However, using historical data might help create more reliable predicting models, and currently old data is starting to be used for this purpose.

Recently, a new law has been introduced in Europe for the use of Artificial Intelligence, the AI Act. This legal framework to '*guarantee the safety and fundamental rights of people and businesses, while strengthening AI uptake, investment and innovation across the EU*'[8]. Among the different aspects that the law covers and given that AI is currently embedded in almost everything that runs on top of digital information, it has an impact on eHealth, such as the machinery regulation, cross-border exchange of information, and mainly data governance. Many AI applications in eHealth are likely to be classified as high-risk by the Act, incrementing the requirements, such as risk assessments, levels of accuracy, robustness, etc. In cases of monitoring for self-care purposes, the applications will need to maintain high standards during their lifecycle.

If we move to the worldwide AI legal regulations, we find the Data Protection Act in the UK that derives from the GDPR, the HIPPA (Health Insurance Portability and Accountability Act) in the USA and the PIPEDA (Personal Information Protection and Electronic Documents Act) in Canada. All these regulations cover similar issues as the GDPR, focusing on protecting patients' data, from the collection, storage, use, and disclosure of personal data.

10.4.2.1 Lessons Learned

The new era of health care with the current capabilities of AI has several benefits, as we will discuss later, but it also has limitations as we have seen. Those limitations have an impact on the actual of ideas to real-world applications. That is why future predictions made on this topic might not have been yet fully developed. Currently, with the IoT, eHealth, mHealth, and all the devices and intelligent modules that coexist in our environment, the quality of life has improved dramatically, but a general integrated system where personalized self-care can be developed is still not on the near horizon. If we can learn a lesson for future forecasting of the trends and implementation limitations is that information is the key, but creating standards will be the key factor. Standards in data form but also standards in legal frameworks allow systems to acquire, store, and interchange information in the right manner and prevent risks.

10.5 Moving into the Future

10.5.1 The Future of Smart Homes and Healthcare

Over the next decade, artificial intelligence (AI) will significantly reshape healthcare and Smart Home, transforming how we monitor, diagnose, and maintain our health. Interoperability and advanced integration will be fundamental as IoT and healthcare technologies standardize, enabling seamless integration between devices and platforms. Open IoT systems, including Medical IoT (MIoT), will ensure Smart Home works with a wide range of devices and services.

Smart Homes will become more intuitive and responsive, with AI systems proactively managing daily activities and health tasks. AI-based predictive analytics will allow homes to anticipate and manage routine tasks and health issues through real-time data analysis. Personalized recommendations and alerts will help users take preventive measures, improving care quality. In addition to the increase in the number and capacity of smart devices in the home, Smart wearable and implantable devices will offer medical-grade accuracy and integrate with home systems, continuously monitoring health and intervening in critical situations, providing more personalized healthcare.

Service provision will transcend the home, with ambient intelligence adapting to different environments in which the individuals are (car, city, etc.) and their needs. This will involve distributed Cloud management and secure personal data handling. Advanced encryption and blockchain technologies will ensure data protection, giving users full control of their data, and keeping privacy and security.

Personalized AI assistants will act as health coaches and life managers, offering individualized advice on diet, exercise, lifestyle, and daily tasks. They will be virtual companions and will maintain natural conversations, learning from personal data, health trends and daily habits to provide personalized and appropriate recommendations. They will manage routine tasks like scheduling appointments and setting reminders, enhanced by advances in large language models keeping a voice-based interface.

AI will be able to establish relationships and trends that are not possible in traditional analysis, providing more insights into health data. AI systems will continuously process and learn from longitudinal health data evolving predictions and recommendations. The same predictive AI would be useful in detecting early signs of mental health problems or chronic diseases and preventing them. Also, it will serve pharmaceutical management to adjust doses based on the patient's responsiveness to treatments reducing side effects and ensuring effectiveness.

Rehabilitation will change fundamentally, with AI aligning exercises and recovery plans to individual progress. By analyzing movements and behaviour, AI will personalize care for optimal recovery. Virtual and augmented reality will become integral to home health, turning Smart Homes into therapy and training environments.

Finally, to provide that, it will be important for international organisms and governments to address the lack of more general international standards and legislation. Current regulations are often tailored to local circumstances, creating fragmentation that hinders the global scalability of AI and IoT in healthcare.

10.5.2 Emerging Innovations

10.5.2.1 Federated Learning

Federated learning (FL) will revolutionize medical care by allowing training in collaborative models and, at the same time, solving one of the big problems of preserving patient privacy. This is a problem that has been ongoing for many years for the definitive introduction of technology in the sector. Medical institutions around the world will use FL to collectively train diagnostic models without sharing raw patient data. This will lead to the development of **universally applicable diagnostic tools** that can accurately detect diseases such as cancer, cardiovascular diseases, and rare genetic disorders. By sharing knowledge and maintaining privacy, diagnoses will be improved, helping patients and healthcare providers.

This FL application extends to **personalized treatment recommendations**. It will analyze patient data locally so that the AI model learns from aggregated patterns and offers a personalized treatment plan. These include personalized medication doses, treatments, and lifestyle changes that would fit each person's unique health profile to optimize patient care. For example, in **remote health monitoring**, FL will drive predictive analytics in wearable devices and home health monitoring systems. These AI-powered devices will easily detect even the smallest changes in people that could indicate a health problem such as arrhythmias, blood sugar irregularities, or dehydration. Based on data collected from many individual networks in completely anonymized data, predictive models will be highly accurate to enable prevention or early intervention for many health conditions.

Predictive maintenance of medical equipment will be optimized. AI models trained using FL will analyze usage data from multiple institutions to make predictions about equipment failures and the need for maintenance, to enable timely repairs. This results in minimizing downtime and maximizing the accuracy of medical devices, thereby improving patient care. This FL of public health agencies will help analyze data on population health across regions while ensuring privacy regulations are respected. Additional knowledge to identify emerging health trends and forecast outbreaks would allow better allocation of resources for disease prevention and control. This would make collaborative analysis an invaluable tool for **public health management**.

AI health assistants will be more precise in the advice they offer and more personalized concerning **self-care**. Integrated virtual assistants will help deliver diet, exercise, sleep, mental well-being, and other recommendations tailored to personal needs and desires by combining local data with global usage patterns. AI models are proving to be very effective in guiding users toward healthy lifestyles. Moreover, it could be used as a triage system achieving significant savings in terms of the time for interacting with patients rather than eliciting information.

Drug discovery and development in the **pharmaceutical industry** will be improved. Companies will share insights derived from distributed datasets without leaking sensitive information, where they will identify promising drug candidates and evaluate their effectiveness much faster to rapidly develop new treatments.

Finally, it will affect **health cybersecurity**. AI models could be fine-tuned across multiple healthcare institutions to analyze attack patterns and threats, learning how to detect and respond to security breaches. This will enhance sensitive patient data to ensure healthcare systems are robust in the face of rapidly changing cyber threats.

10.5.2.2 Explainable AI

Explainable AI (XAI) will significantly improve healthcare by making AI decisions more transparent and understandable. It will also help improve the adoption of AI in healthcare by eliminating the 'black box' status of current AI systems. The ability of AI to be adopted in healthcare has been continually obstructed by this *black box* that is shunned by most healthcare professionals and regulatory bodies. XAI will solve this by providing information about how the AI reaches its conclusions, allowing healthcare providers to verify decisions and justify their use.

XAI will explain the processes behind model decisions by performing specific diagnostics. This transparency will empower doctors to understand the rationale behind each decision, increasing confidence in AI-based diagnoses. Medical staff will be able to confirm AI results with their clinical judgment. For example, XAI will highlight medical characteristics that lead to specific diagnoses, helping doctors validate and ensure the correct course of treatment.

Therefore, this will be in the **personalized medicine recommendations** for the treatment of certain patients. By explaining what underlies the automated recommendations (e.g., genetic markers, lifestyle, previous responses to treatments), XAI is likely to help doctors develop treatments of a more targeted nature. This will also make patients more confident in decisions made with the help of AI when it is clearly explained to them how their data contributes to treatment.

In clinical research, it will provide **virtual laboratories**, allowing researchers to uncover insights from clinical data, identify biases, and adjust algorithms for more accurate results. This will enhance predictive models to identify potential clinical trial candidates, predict disease progression, and design new treatment protocols. Understanding AI decision-making will help researchers interpret results better and align the system with clinical experience.

10.5.2.3 Generative AI

Generative AI (GenAI) focuses on creating new data from learned patterns. It will enable the generation of highly detailed medical data models (e.g., images, clinical data) that can simulate various conditions. This synthetic data will augment training datasets allowing the improvement of diagnosis by learning from a broader range of medical scenarios without the need for extensive manual labeling. It also will enhance the ability of healthcare professionals in the early detection of diseases.

Generating molecular structures and interactions will lead to faster and more efficient identification of promising treatments, significantly reducing the timeline **for drug discovery and development**. The simulation of this kind of data will identify potential drug candidates with specific therapeutic effects, reducing the need for costly and time-consuming laboratory experiments. Regarding **personalized medicine**, the trial-and-error approach, that often characterizes current medical practice, could be reduced by generating virtual patients with individual responses to different treatment protocols. The synthetic responses could consider many individual characteristics including genetic, environmental, and lifestyle data.

Furthermore, this technology will also revolutionize **medical records** by synthesizing or summarizing data from diverse sources (e.g., patient histories, treatment plans, medical images) to generate comprehensive patient profiles that will support more informed clinical decisions and improve patient outcomes by helping healthcare providers to understand the broader context of a patient's health. Taking this point further, it is possible to image **virtual assistants** with personalization capabilities that can dramatically improve the healthcare sector. These assistants could generate personalized responses and recommendations based on the individual's medical history, current health status, and genetic predispositions. This level of personalization will make interactions more intuitive and personalized. Furthermore, virtual assistants could learn from the interaction detecting subtle cues in the user's voice or word choice that indicate emotional states providing empathic responses and support. It will provide not only medical support but also emotional stimulation. Finally, these AI systems could generate accurate, culturally relevant health information and advice in multiple languages, increasing accessibility and inclusivity in healthcare services.

Today GenAI is very well known for its realistic capabilities in providing images and videos. In our opinion, in the coming years, it will be developed much more, obtaining impressive results. In consequence, together with virtual and augmented reality systems, highly realistic environments can be produced for **medical training**. It will allow students and professionals to practice diagnostic skills in a risk-free environment able to simulate patient interactions, surgical procedures, emergency responses, etc.

Finally, creating personalized and realistic videos that explain complex medical conditions and treatments to patients is of enormous importance. These videos can mimic human-like interactions, using virtual avatars to convey empathy and support, making medical information more accessible and understandable. For example, in physiotherapy and rehabilitation exercises adapted to the specific recovery needs of each patient. These videos can demonstrate exercises with precise movements and provide visual feedback to patients as they perform them, adjusting the guidance in real time based on the patient's progress and performance.

10.6 Conclusions

The emergence of smart technologies and artificial intelligence in the healthcare field will have a drastic impact on patient care and healthcare diagnosis and management. Personalized medicine, widespread through smart environments, wearable technologies, and based on artificial intelligence will greatly improve patient outcomes and healthcare quality of life. However, advances in these areas have not been without challenges, such as achieving interoperability, ensuring data security, and gaining user acceptance.

Problems identified a decade ago remain relevant today, and solutions are still actively sought in research. Particularly in Smart Homes, where much progress has been made in the integration of technologies and services, demonstrating a sustained effort to make technology less visible but still functional at home. Artificial intelligence is emerging as an essential tool to propose viable solutions. In the healthcare sector, the integration of healthcare devices has advanced significantly, but the most important issues remain data security and privacy. Here, artificial intelligence is providing innovative answers at the research level, developing solutions that promise to achieve significant advances toward a digital health system capable of addressing the main social and economic challenges it must deal with.

The future of healthcare will be defined by AI-based predictive analytics and personalized health assistants enabling more accurate diagnoses, personalized treatments, and proactive health management. The move toward open systems architectures, integrated health informatics, and the Internet of Medical Things will further improve healthcare delivery, ensuring that patient-centered care continues to evolve alongside technological advances. In our opinion, the big impact will be carried out by emerging innovations such as generative AI, federated learning, and explainability to take solutions to an extraordinary level, transforming the way problems are approached. To fully realize the potential of these technologies in healthcare, the industry must prioritize patient privacy and data security, ensuring that the adoption of AI and smart systems aligns with regulatory standards and ethical considerations.

References

1. Omotunde H, Mouhamed MR (2023).The modern impact of artificial intelligence systems in healthcare: a concise analysis, Mesopotamian J Artif Intell Healthc, 66–70
2. Azorin-Lopez J, Fuster-Guilló A, Saval-Calvo M, Bradley D (2016) Home technologies, smart systems and ehealth, Mechatron Futures Challenges Solut Mechatron Syst Their Des, 179–200

3. Plattform Industrie 4.0—Startseite. (n.d.) https://www.plattform-i40.de/IP/Navigation/EN/Home/home.html. Accessed May 2024
4. Hermann M, Pentek T, Otto B (2016) Design principles for industrie 4.0 scenarios, Proc Annual Hawaii Intl Conf Syst Sci, 3928–3937
5. Li J, Carayon P (2021) Health care 4.0: a vision for smart and connected health care. IISE Trans Healthc Syst Eng 11(3):171–180
6. Mbunge E, Muchemwa B, Jiyane S, Batani J (2021) Sensors and healthcare 5.0: transformative shift in virtual care through emerging digital health technologies. Glob Health J 5(4):169–177
7. Chen C, Ding S, Wang J (2023) Digital health for aging populations. Nat Med 29(7):1623–1630. https://doi.org/10.1038/s41591-023-02391-8
8. Antic M, Papp I (2021) Smart home integration with external IoT device platforms and services. IEEE Consum Electron Mag 10(6):69–75
9. Zegeye W, Jemal A, Kornegay K (2023) Connected smart home over matter protocol, IEEE Intl Conf Consum Electron, 1–7
10. Orfanos VA, Kaminaris SD, Papageorgas P, Piromalis D, Kandris D (2023) A comprehensive review of IoT networking technologies for smart home automation applications. J Sensor Actuator Netw 12(2):30
11. Morita PP, Sahu KS, Oetomo A (2023) Health monitoring using smart home technologies: scoping review. JMIR mHealth uHealth 11:e37347
12. Lindsay D, Gill SS, Smirnova D, Garraghan P (2021) The evolution of distributed computing systems: from fundamental to new frontiers. Computing 103(8):1859–1878
13. Benayache A, Bilami A, Barkat S, Lorenz P, Taleb H (2019) MsM: a microservice middleware for smart WSN-based IoT application. J Netw Comput Appl 144:138–154
14. Gómez HD, Garcia-Rodriguez J, Azorin-Lopez J, Tomás D, Fuster-Guilló A, Mora-Mora H (2021) IA-CPS: intelligent architecture for cyber-physical systems management. J Comput Sci 53:101409. https://doi.org/10.1016/J.JOCS.2021.101409
15. Kraus M, Wagner N, Callejas Z, Minker W (2021) The role of trust in proactive conversational assistants. IEEE Access 9:112821–112836
16. Tettey F, Parupelli SK, Desai S (2023) A review of biomedical devices: classification regulatory guidelines, human factors, software as a medical device, and cybersecurity. Biomed Mater Dev 2(1):316–341
17. Deniz-Garcia A, Fabelo H, Rodriguez-Almeida AJ, Zamora-Zamorano G, Castro-Fernandez M, Alberiche Ruano MD, Solvoll T, Granja C, Schopf TR, Callico GM, Soguero-Ruiz C (2023) Quality, usability, and effectiveness of mHealth apps and the role of artificial intelligence: current scenario and challenges. J Med Internet Res 4(25):e44030. https://doi.org/10.2196/44030
18. Sleurs K, Seys SF, Bousquet J, Fokkens WJ, Gorris S, Pugin B, Hellings PW (2019) Mobile health tools for the management of chronic respiratory diseases. Allergy 74(7):1292–1306
19. Moreno-Ligero M, Moral-Munoz JA, Salazar A, Failde I (2023) MHealth intervention for improving pain quality of life, and functional disability in patients with chronic pain: systematic review. JMIR mHealth uHealth 11:40844. https://doi.org/10.2196/40844
20. Dawes J, Lin AY, Varghese C, Russell MM, Lin AY (2021) Mobile health technology for remote home monitoring after surgery: a meta-analysis. British J Surg 108(11):1304–1314
21. Bonnechère B, Kossi O, Mapinduzi J, Panda J, Rintala A, Guidetti S, Spooren A, Feys P (2023) Mobile health solutions: an opportunity for rehabilitation in low- and middle income countries? Front Public Health 10:1072322. https://doi.org/10.3389/fpubh.2022.1072322/full
22. Safdar Z, Farid S, Qadir M, Asghar K, Iqbal J, Hamdani FK (2020) A novel architecture for internet of things based E-health systems. J Med Imaging Health Inform 10(10):2378–2388
23. Rubí JN, Gondim PR (2020) Interoperable internet of medical things platform for e-health applications. Int J Distrib Sensor Netw 16(1):1550147719889591. https://doi.org/10.1177/1550147719889591
24. Razdan S, Sharma S (2022) Internet of medical things (IoMT): overview, emerging technologies, and case studies. IETE Tech Rev 39(4):775–788

25. Bosanac D, Stevanovic A (2022) Trust in E-health system and willingness to share personal health data. Stud Health Technol Inform 289:256–259
26. Antolis K, Jaksetic D (2023) Patients' perception of data security in healthcare, IEEE Intl Mediterr Con Commun Netw, 24–28
27. Graham M (2023) Data for sale: trust, confidence and sharing health data with commercial companies. J Med Ethics 49(7):515–522
28. Psarra E, Patiniotakis I, Verginadis Y, Apostolou D, Mentzas G (2020) Securing access to healthcare data with context-aware policies, 11th Intl Conf Inform Intell Syst Appl, 1–6
29. Yogesh MJ, Karthikeyan J (2022) Health informatics: engaging modern healthcare units: a brief overview. Front Public Health 10:854688. https://doi.org/10.3389/fpubh.2022.854688/full
30. Rekha PM, Dakshayini M (2020) Health informatics as a service, Innov Commun Comput, 183–198
31. Schork NJ (2019) Artificial intelligence and personalized medicine, Precision Med Cancer Therapy, Springer, pp 265–283
32. Ali O, Abdelbaki W, Shrestha A, Elbasi E, Alryalat MAA, Dwivedi YK (2023) A systematic literature review of artificial intelligence in the healthcare sector: benefits, challenges, methodologies, and functionalities. J Innov Knowl 8(1):100333. https://doi.org/10.1016/j.jik.2023.100333
33. An R, Chang G, Fan Y, Ji L, Wang X, Hong S (2021) Machine learning-based patient classification system for adult patients in intensive care units: a cross-sectional study. J Nurs Manag 29(6):1752–1762
34. Brom H, Brooks Carthon JM, Ikeaba U, Chittams J (2020) Leveraging electronic health records and machine learning to tailor nursing care for patients at high risk for readmissions. J Nurs Care Quality 35(1):27–33
35. Salgado CM, Vieira SM (2020) Machine learning for patient stratification and classification part 1: data preparation and analysis. Springer, In Leveraging Data Science for Global Health, pp 129–150
36. Badawy M, Ramadan N, Hefny HA (2023) Healthcare predictive analytics using machine learning and deep learning techniques: a survey. J Elect Syst Inform Technol 10(1):40. https://doi.org/10.1186/s43067-023-00108-y
37. Kibria HB, Nahiduzzaman M, Goni MdOF, Ahsan M, Haider J (2022) An ensemble approach for the prediction of diabetes mellitus using a soft voting classifier with an explainable AI. Sensors 22(19):7268. https://doi.org/10.3390/s22197268
38. Rastogi R, Bansal M (2023) Diabetes prediction model using data mining techniques. Measur Sens 25:100605. https://doi.org/10.1016/j.measen.2022.100605
39. Tariq MU, Ismail SB (2024) Deep learning in public health: comparative predictive models for COVID-19 case forecasting. PLoS ONE 19(3):e0294289. https://doi.org/10.1371/journal.pone.0294289
40. da Silva DB, Schmidt D, da Costa CA, da Rosa RR, Eskofier B (2021) DeepSigns: a predictive model based on deep learning for the early detection of patient health deterioration. Expert Syst Appl 165:113905. https://doi.org/10.1016/j.eswa.2020.113905
41. Sathyanarayana A, Joty S, Fernandez-Luque L, Ofli F, Srivastava J, Elmagarmid A, Arora T, Taheri S (2016) Sleep quality prediction from wearable data using deep learning. JMIR Mhealth Uhealth 4(4):e125
42. Wickramasinghe N, Bodendorf F, Kraus M (2023) Dimensions of intelligent analytics for smart digital health solutions, Chapman and Hall/CRC
43. Feng J, Phillips RV, Malenica I, Bishar A, Hubbard AE, Celi LA, Pirracchio R (2022) Clinical artificial intelligence quality improvement: towards continual monitoring and updating of AI algorithms in healthcare. Npj Digital Med 5(1):66. https://doi.org/10.1038/s41746-022-00611-y
44. Dubey A, Tiwari A (2023) Artificial intelligence and remote patient monitoring in US healthcare market: a literature review. J Market Access Health Policy 11(1):2205618. https://doi.org/10.1080/20016689.2023.2205618

45. Jin X, Cai A, Xu T, Zhang X (2023) Artificial intelligence biosensors for continuous glucose monitoring. Interdisc Mater 2(2):290–307
46. Bickmore TW, Trinh H, Olafsso S, O'Leary TK, Asad R, Rickles NM, Cruz R (2018) Patient and consumer safety risks when using conversational assistants for medical information: an observational study of Siri, Alexa, and Google assistant. J Med Internet Res 20(9):e11510. https://doi.org/10.2196/11510
47. Dave T, Athaluri SA, Singh S (2023) ChatGPT in medicine: an overview of its applications, advantages, limitations, future prospects, and ethical considerations. Front Artif Intell 6:1169595. https://doi.org/10.3389/frai.2023.1169595
48. Brown T, Mann B, Ryder N, Subbiah M, Kaplan JD, Dhariwal P, Neelakantan A, Shyam P, Sastry G, Askell A, Agarwal S, Herbert-Voss A, Dario Krueger G (2020). Language models are few-shot learners, Advances in Neural Information Processing Systems, Larochelle H, Ranzato M, Hadsell R, Balcan MF & Lin H (eds), pp 1877–1901
49. Norori N, Hu Q, Aellen FM, Faraci FD, Tzovara A (2021) Addressing bias in big data and AI for health care: a call for open science. Patterns 2(10):100347. https://doi.org/10.1016/j.patter.2021.100347
50. Dong C, Li Y, Gong H, Chen M, Li J, Shen Y, Yang M (2023) A survey of natural language generation. ACM Comput Surv 55(8):1–38
51. Tian YJ, Felber NA, Pageau F, Schwab DR, Wangmo T (2024) Benefits and barriers associated with the use of smart home health technologies in the care of older persons: a systematic review. BMC Geriatr 24(1):152. https://doi.org/10.1186/s12877-024-04702-1
52. Murdoch B (2021) Privacy and artificial intelligence: challenges for protecting health information in a new era. BMC Med Ethics 22(1):122. https://doi.org/10.1186/s12910-021-00687-3

Chapter 11
Mechatronics Design Methodologies: New Frontiers in Design and Technology

Martin Törngren, Georgios Andrikopoulos, Fredrik Asplund, Dejiu Chen, Lei Feng, and Martin Edin Grimheden

Abstract In this chapter, we explore how new technologies and requirements affect current design methodologies for mechatronics. We investigate gaps and directions needed for the methodologies of tomorrow in view of trends affecting mechatronics and current state of the art. To fully reap the opportunities of mechatronics with advances in materials, sensors, additive manufacturing, AI, computing and communication, but also to handle new requirements and regulations, there is a need for new methodologies and architectures. We introduce the concept of "MechaOps" and related considerations that promise to assist in enhancing scalability, smartness, performance and sustainability for extended mechatronic products that collaborate with a smart infrastructure, humans and other mechatronic systems. MechaOps refers to the integration of the concepts of Mechatronics and DevOps. As opposed to DevOps in software engineering, MechaOps encompasses data gathering, upgrades/downgrades as well as reconfigurations considering both mechanics and/or software in a mechatronic product. With the life-cycle view implied by the MechaOps concept, it becomes essential to design for upgrading, downgrading, maintenance, reuse and refurbishment. The development of new methodologies requires overcoming disciplinary gaps, with specific considerations of novel architectures including digital twins, interactions with humans, other systems and a smart infrastructure, the role of AI in mechatronics, and in assuring trustworthiness and sustainability. We believe that new methodologies and architectures will initially be especially relevant for high-end systems, supporting the creation of adaptable and flexible mechatronics products and services with improved performance and reduced environmental footprint.

M. Törngren (✉) · G. Andrikopoulos · F. Asplund · D. Chen · L. Feng · M. E. Grimheden
Mechatronics Unit, KTH Royal Institute of Technology, Stockholm, Sweden
e-mail: martint@kth.se

P. Hehenberger and D. Bradley (eds.), *Mechatronic Futures*,
https://doi.org/10.1007/978-3-031-83571-1_11

11.1 Introduction

Mechatronics has been an established field for several decades as manifested by academic and industrial institutions with educational curricula, research efforts and product development [1–3]. Mechatronics has traditionally been associated with the flexible generation of controlled motions in products and is more generally viewed as a synergistic combination of mechanics, electronics, software and controls. These views reflect both a product and development perspective.

The opportunities associated with mechatronics are expanding due to advancements in new technologies, e.g., in materials, computing, and manufacturing, enabling novel designs and enhanced support tools for automation and artificial intelligence (AI) assistance. Opportunities for innovation are seen in many fields where human–machine collaboration is essential, e.g. when enhanced physical capabilities are needed and where automation drives cost-efficient operations. Example applications include exoskeletons (robots designed to provide augmented support for human movement and strength, e.g., in rehabilitation and manufacturing), collaborative robots where robots work with humans in shared workspaces, assistive robotics in terms of robotic prostheses to assist individuals performing daily activities, medical robotics such as surgical robots, and agricultural robotics for automating tasks in farming such as planting, harvesting, and monitoring crops.

In this chapter, we are interested in the design methodologies of mechatronics, and to what extent they are prepared for the opportunities and challenges associated with the mechatronic products of tomorrow. We adopt Estafan's definition of methodology, as a "collection of related processes, methods, and tools" [4].

While new and emerging technologies offer opportunities, they also introduce new challenges such as new, possibly complex behaviors and potentially new failure modes that must be understood and managed. Further constraints are today also imposed on mechatronic products due to sustainability requirements (e.g., the digital product passports initiative[1]), concerns related to the use of AI and data (e.g., as addressed by standardization developments in several domains and by the recent European AI and Data acts), cybersecurity threats, and sovereignty in production and maintenance.

The corresponding opportunities result in a larger design space that combined with new requirements add to the inherent challenge in mechatronics; dealing with multiple (heterogeneous) technologies and aspects, such as geometry and kinematics, dynamics, materials, sensors, actuators and information processing and communication, with the purpose to satisfy multiple properties such as performance, robustness, reliability, safety and cost, [4, 5].

Overall, the ongoing changes lead to even more dependencies and trade-offs, a wider cone of uncertainty, and a corresponding need for enhanced information

---

[1] European Commission (2023). Communication from The Commission to The European Parliament, The European Council, The Council, ...—A Green Deal Industrial Plan for the Net-Zero Age. Available online: https://commission.europa.eu/document/41514677-9598-4d89-a572-abe21cb037f4_en [Accessed 20 April 2024].

management including lifecycle traceability, see e.g. [6, 7]. New technologies and tools, at the same time, promise to provide new ways of handling some of this complexity, for example in terms of enabling novel designs and enhanced support tools for automation and AI assistance. This leads us in this chapter to delve into the changing landscape of mechatronics design methodologies.

We address three questions: **(1) What are current mechatronics' design practices? (2) Which trends are affecting mechatronics and (3) What are the corresponding gaps and directions required for tomorrow's methodologies?**

Interdisciplinary education, training, skills and organizational development are essential for realizing design methodologies for mechatronic products. In this chapter we however focus on the design methodologies as such and refer to other related work for these aspects, see e.g. [3, 7, 8].

Our chapter begins by revisiting current design methodologies, in terms of state of the art and industrial experiences. We then turn to trends affecting mechatronics and case studies to further elicit gaps regarding existing methodologies. This leads us to discuss and propose directions found to be essential for renewing current mechatronics design methodologies.

11.2 Current Mechatronics Design Practices and State of the Art

In this section, we briefly review state of the art in terms of existing design methodologies and when relevant treat identified limitations of these.

The literature in mechatronics and related fields embodies vast amounts of knowledge in design methodologies. For example, the reference work by Pahl and Beitz describes a systematic approach to design including principles and guidelines for design, [5], with a starting point from mechanical engineering. The German engineering association VDI, has introduced guidelines for the development of mechatronic and cyber-physical systems, VDI2206 that describes the desired shift from sequential to concurrent engineering in developing mechatronic systems, [1].

In software engineering, agile development and continuous (integration and deployment) engineering, emphasize rapid iterations, integration and deployment of features, allowing for evolving requirements and frequent end-user/stakeholder interactions, see e.g. [9]. However, establishing concurrent and agile engineering for mechatronics faces a number of difficulties. As design progresses to concurrent design, complications of handling the integration arise. These include handling connections/dependencies between different design models and data in relation to early evaluation, testing and verification [10–13]. The issues are closely related to the need for establishing inter- and intra-communication for multidisciplinary teams of engineers [14–16]. The dependencies of mechatronic products and their tight integration further poses challenges when it comes to robust design [11].

Mechatronics design has a critical dependency on decisions regarding how functionality is partitioned between components and implemented in terms of technology. Such decisions shape the corresponding transfer functions, data flows, and energy transfer of the system. As these decisions are often made early in the design process, it is difficult to take the resulting characteristics of the implementation into account. Early stage function and behavior modeling, and closing the gap to more disciplinary/component models, represents a known key challenge in providing methodologies to enable appropriate modeling, analysis, synthesis, simulation, and prototyping, [2, 3]. This challenge is also reflected by the Function–Behavior–Structure (F-B-S) model of design, highlighting the presence of emerging behaviors in design, [17]. The idea of front-loaded development embraces this difficulty, with best practices emphasizing the use of multidisciplinary teams and the use of system simulations to support early evaluation and integration, as illustrated by the so-called W-model [18].

Composability, the ability to integrate mechatronic components without negative side effects, represents both a research vision in several disciplines as well as an overarching limitation of existing methodologies, [6]. It should be noted that the inherent complexity of the capabilities of a mechatronic product and its environment are reflected in the implementation of the system, as well as in the corresponding virtual environments (models, simulation environments) and their associated risks [19].

Spurred by the advances in electronic design automation (EDA) and software, a vision of the "Mechatronics compiler" has been put forward [20]. The vision stems from a desire to automate the design process of mechatronic systems, much like how software is compiled from high-level code to executable binaries. A mechatronics compiler tool would take the desired behavior and specifications of a mechatronic system as inputs, and then automatically generate the corresponding hardware and software components necessary to realize that system. As of today, such tools have been demonstrated, but with rather limited scope in terms of the types of components that can be handled, see e.g. [21]. One likely reason for the difficulty in fulfilling the vision is that a mechatronics compiler has to deal with multiple possible realization technologies, as well as multiple aspects and properties as described above [22]. This causes a large design space which makes analysis and synthesis much more difficult, requiring a diverse set of models and tools for assessing performance, reliability, cost and other metrics. This is aggravated by the typical fragmentation of the corresponding tools with different formalisms, methods and data, complicating automation, [4, 23].

Tool providers have worked to develop increasingly integrated capabilities across different stages of the mechatronics design process. Current state of the art tool suites incorporate capabilities for mechatronics design, with different starting points (e.g., from mechanical design, controls or electronics). Such tool suites may also be integrated with product data management databases, providing collaboration and version control for multidisciplinary teams. In some cases, such tools also address the connection to manufacturing, for virtual commissioning, robotics simulation, and process optimization. It is, however, very difficult to capture all facets of mechatronics development within one integrated tool suite. This becomes even more difficult due to the fast technological development pace, as further elaborated in the next section. Furthermore, integrated tool suites tend to have lock-in effects, making it difficult to integrate the best tools from different providers. This has resulted in many efforts toward open exchange standards and federated environments, see e.g. [24]. Finally, while current tool suites provide relevant capabilities, their appropriate use is not straightforward and frequently requires adequate methods and processes that do not yet exist, see e.g. [2, 25].

From industrial practices, studies have identified issues in accomplishing explicit communication of design changes across disciplines, in allocating design requirements to specific systems/subsystems/components, achieving the corresponding traceability, and in simulating integrated mechanical, electrical, and software components [26]. Integration and interoperability problems are further highlighted, arising since each discipline in mechatronics tends to use different methodologies, as elaborated above. A common industrial approach for design is that of assembling mechatronic components. Such an approach provides building blocks to facilitate design, but also constrains and limits optimization and synergistic integration. It has therefore been proposed that mechatronics research should also focus on interdisciplinary interactions [3].

In [13], the impact of connectivity and systems collaboration is discussed, concluding that many new considerations are needed during systems design. This is in line with the Cynefin guideline from systems engineering which says that there are needs for new methodologies when developing systems for which there are many unknowns and where there is limited understanding of causal relationships, [12]. Similar findings have been obtained in related fields such as cyber-physical systems (CPS) [6]. The Vision 2035, by the International Council on Systems Engineering (INCOSE) [27], recognizes the need for systems engineering to evolve and adapt in an increasingly complex, dynamic, and technologically driven world. The core aspects of this vision include the need to handle increasingly complex systems with more stringent requirements on sustainability and the incorporation of technologies such as AI including machine learning (ML), and digital twins.

11.3 Trends and Case Studies

Over the years, mechatronic products have maintained basically the same basic composition, featuring physical components/processes that generate motions, transfer energy, forces, or torques, along with sensors, actuators, software, and hardware components that regulate these operations. Regardless, the richness of functionality and level of performance have evolved with advancements in science and technology. Several technological trends have thus had, and are having, significant impacts on modern mechatronic products. In this section, we first summarize some of these trends—illustrated in Fig. 11.1 and then introduce two case studies related to robotics to highlight challenges and needed directions.

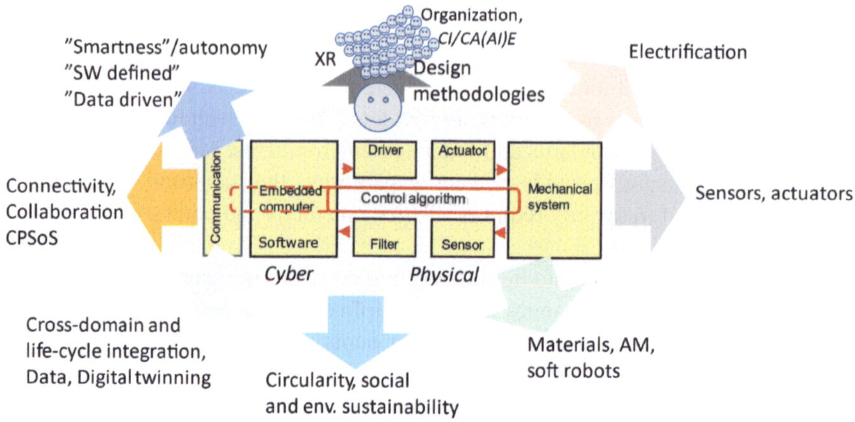

Fig. 11.1 Trends affecting Mechatronics through innovations in several dimensions, societal drivers and regulations, e.g., related to sustainability. XR stands for eXtended reality, CI for Collective Intelligence, and CAE for Computer-Aided Engineering, with AI support indicated

Advancements in materials and production technologies are reshaping the field of mechatronics, enabling the creation of lighter, smaller, more durable, smarter, and adaptable mechatronic products. Key to this trend are innovations in materials such as composites and smart materials that adjust their properties based on environmental stimuli. Additive manufacturing (AM) techniques have revolutionized production processes, paving the way for intricate and finely tuned interactions with the physical world and people. The trends toward customization and miniaturization supported by the advancements of printed electronics and electromechanical solutions (e.g., MEMS), are driving the creation of smaller, more effective components for interacting with the physical world and people. Advancements in sensory technology have opened new possibilities in areas like medical devices and wearable tech, showcasing the evolution of mechatronics role in integrating and optimizing for advanced features across diverse domains.

Modern mechatronic products are evolving toward higher levels of automation, effectiveness and efficiency. This trend can be seen in efforts to have humans and robots, and robots with robots, to collaborate on the factory floor to enable more efficient, effective and flexible manufacturing. This trend goes hand in hand with taking robots and machines "out of their cages", to operate in more open environments. A prime example of this is automated driving. In general, automation efforts tend to go beyond the traditional robotics application paradigm of "dirty, dull and dangerous", to add a "dear" (expensive) dimension, [28].

Closely related to automation, increased efforts are spent on adaptive systems and autonomic communication. The former is often motivated by the need for systems that can function dependably even when operational requirements change [29]. The discussion of the latter typically focuses on the four self-x properties (i.e., the ability to self-configure, self-monitor, self-adapt, and self-heal) [30]. Nevertheless, adaptive systems hide the complexity of the underlying technology to users, and autonomic communication decreases the complexity of network management by moving the associated handling to high-level policies.

The "smartness" of mechatronic products has traditionally been enhanced by integrating advanced "programmed" functions to implement, for example algorithms for sensor fusion, motion control and decision-making. Such functions have typically been realized through first-principles based modeling and through theoretical advances, for example in terms of robust control to deal with bounded modeling errors or external disturbances and model predictive control—to optimize the control actions based on system dynamic models..

A major change in emphasis has come through recent advances in machine learning and probabilistic reasoning, e.g., with stochastic process models and decision-making algorithms, like Bayesian Networks (BN) and Reinforcement Learning (RL). Bayesian Networks contribute by offering a probabilistic framework for modeling complex operational processes with uncertainty, enabling informed decisions based on likelihoods and evidence. RL algorithms enable mechatronic systems to learn optimal actions through explorative interactions with their environment, refining their operational strategies for maximum efficiency and effectiveness over time. Deep neural networks or deep learning systems (DL) extend these capabilities further by enabling a data-driven approach to advanced perception and decisions by learning from the collection of operations data. DL's can encode intricate relationships within data and make predictions or decisions, significantly enhancing the capabilities of mechatronic products. While DL's excel in performance they are also brittle, they are said to hallucinate, thus posing challenges in adopting them in critical applications [31–34].

The advancement of software and information technologies is driving mechatronic products into a new dimension of innovation, underscored by distributed functionalities, cooperative behaviors, modular design, adaptive performance, enhanced quality assurance, and effective lifecycle management. For example, the adoptions of real-time Linux and 5G networks showcase a broader trend where distributed control functionalities are becoming increasingly prevalent. Middleware technologies, such as ROS (Robot Operating System) and SOME/IP (Scalable service-Oriented MiddlewarE over IP), have added to the traditional (real-time) operating systems, significantly enhancing interoperability but on the other hand also increasing complexity and typically with gaps when it comes to predictable real-time performance. Advancements in computing, algorithms and connectivity are leading to a shift where more and more functionality is influenced, or largely defined, by software and data-driven (machine-learning-based) functionalities. This shift is also closely related to connectivity and capability augmentation, where additional services are directly provided by edge- or cloud computing systems [35]. This paves the way for data gathering and analysis at different levels (e.g., for pro-active maintenance or real-time anomaly detection), as well as continuous software integration and deployment, forming so-called DevOps loops that connect developments and operations [9].

Production and operation of mechatronic products and services add to the global environmental footprint. This challenge requires to develop sustainable systems that meet the needs of today without compromising the abilities for future generations. This is reflected in several initiatives such as Industry 5.0 [36] and the digital product passport initiative (DPP), addressing resource efficiency and life cycle management. The DPP aims to enhance the sustainability of products by maintaining their environmental data, and is expected to be required to be satisfied in future by suppliers to bring their products to the market. Additionally, The European Green Deal[2] focuses on achieving climate neutrality by 2050 through decarbonization, sustainable energy promotion, and biodiversity protection. The Circular Economy Action Plan[3] aims to transform the EU economy into a more sustainable and circular model, emphasizing resource reduction, recycling and reuse, and minimizing waste.

There is a corresponding growing demand and opportunity for mechatronic products that are energy-efficient, resource-efficient, and with minimized environmental impact throughout the entire product lifecycle. This implies a growing need for manufacturers to consider sustainability across entire supply chains. This in turn requires explicit support for traceability and improved sustainability assessment over product lifecycles [37].

As mechatronics products become more interconnected and data-driven, cybersecurity has emerged as a critical concern, and has become the norm for mechatronic systems, while proliferating in public spaces. This can be seen in the extension of the ordinary Confidentiality, Integrity and Availability (CIA) triad with the Possession/Control, Authenticity and Utility aspects [38]. Possession/Control refers to how highly distributed systems that, in part, occupy public spaces, are difficult to physically secure. Authenticity highlights challenges, due to the sheer volume of traffic and the scale of the involved hardware, for large mechatronic systems in verifying the authenticity of messages and physical artefacts. Simultaneously, utility underlines that a system can be secure, but not useful. For certain large-scale use cases it is difficult to find scalable security mitigations that do not impose an unreasonable computational burden. Also here, complying with regulations related to data privacy has become essential in many domains, e.g., with cybersecurity becoming part of type approval for modern vehicles.

Naturally, all these new and changing requirements regarding system properties are accompanied by an associated increase in demands on system engineering methods in the early design phases as well as an increase in the number of engineering standards that support the development of mechatronic systems. The need to adhere to standardized processes for developing mechatronic systems has become more intense, with associated implications for engineering organizations, engineering methodology and inter-organizational knowledge sharing [39–41]. Nonetheless, this has not led to a stronger imperative to make use of standardized system components. This could be due to the move toward goal-based standards, which

[2] European Commission (2023). Communication from The Commission to The European Parliament, The European Council, The Council, ...—A Green Deal Industrial Plan for the Net-Zero Age. Available online: https://commission.europa.eu/document/41514677-9598-4d89-a572-abe21cb037f4_en [Accessed 20 April 2024].

[3] https://environment.ec.europa.eu/strategy/circular-economy-action-plan_en

rather than prescribing the use of particular methods or technology require manufacturers to establish goals for the product at hand and provide an associated argument that prove compliance for a particular system property [42]. This allows flexibility in choosing different means to comply with standards, and thus does not hinder innovative solutions needlessly.

Combining these developments provides a rich and vast space for new innovations, as exemplified through the following case studies.

11.3.1 Case Study (1): Design and Control of Smart Soft Robots

Soft robots are advantageous over rigid robots for safe interactions with people and fragile objects, thanks to their flexible movements and compliant deformation in irregular and uncertain environments, and even force distribution for grasping and touching [43]. Nevertheless, soft materials are afflicted by nonlinearity, hysteresis, time-variant dynamics, large number of degrees-of-freedom (DOF), etc. Accurate sensing and precise control on the shape and/or force of soft robots are challenging [44, 45]. Consequently, some soft robots use open-loop control [43, 46], or the control methods are simple and do not require detailed physical models, such as manually tuned PID [47].

These soft robots are inadequate for more complex applications where the shapes/positions of multiple actuators must be precisely coordinated in time or the force between the robot and the object must be regulated. To this end, popular research topics for soft robotics include accurate mobile shape and force sensors tailored for the soft robots, flexible and compliant actuators, and feedback control methods addressing the difficulties of soft materials.

Mechatronic solutions to these challenges include the following.

(1) The movement and force of soft robots shall be measured by particularly designed soft sensors using responsive and flexible materials.
(2) Flexible and energy-efficient actuators suitable for soft robots shall be enabled by programmable smart materials [48].
(3) The feedback control of the soft robots must integrate effective methods from both model-based control and deep learning fields.

The study on soft robots promotes interdisciplinary research on smart materials and mechatronic systems design; see Fig. 11.2 for an actual soft robot developed at KTH that explores the mentioned solutions.

Soft materials with responsive physical and electrical properties are good candidates for soft sensors and actuators [49, 50]. The functional materials can change shape under various stimuli, such as four-dimensional printing technologies (4DP) [51], with the ability to generate highly complex geometries.

The unpredictable properties of soft materials prevent us from deriving concise mathematical models based on physical laws [52]. One approach is then to learn the

Fig. 11.2 A soft quadruped robot made by 3D printing

neural network model of the dynamics or the control strategy of the soft robot using deep learning (DL); however, the errors between the learned models or controllers and the ideal ones can be neither eliminated nor predicted. Hence the DL results cannot guarantee the desired properties, including safety critical ones, and may fail unexpectedly. Rigorous model-based control methods can be applied online to improve the DL-based control strategy. One potential approach to improve the control performance is with control theory [53].

A common challenge of applying DL methods to the real-time control of robots comes from the sharp contrast between the high complexity of the DL methods and the limited computing resources of the on-board microprocessors in the robots. During the learning process, the robot must collect a lot of measurement data, but the on-board microprocessor of the robot may not have enough memory space to store all the data and the deep learning algorithm requires extraordinary CPU resources. During the online control process, the learned model may be too large for the memory space of the robots and the computation time for combining DL inference and online predictive control may fail the real-time constraints leveraging robot onboard only. A potential solution to this challenge is to offload the training and online control tasks to a nearby edge server. The high computing power of the edge server and the high communication bandwidth between the robot and the nearby edge server ensure small end-to-end latencies for the offloaded tasks [54]. An edge server-based solution also facilitates updates of the DL algorithm to deal with needed changes (new tasks, unforeseen scenarios, performance improvements etc.).

In summary, case study 1 highlights the following gaps with respect to current mechatronics design methodologies:

- Incorporating DL (AI) together with traditional algorithms, to obtain increased performance and deal with uncertainties of DL.
- Design support and architectures for such mechatronic soft-robots, referring to the integration of sensors and actuators into the mechanical structure as well as supporting the new computations required through edge off-loading.

11.3.2 Case Study (2): Design, Development and Control of a Tabletop Robot for Child Interaction

An integral part for completing the master's degree in Mechatronics at the KTH Royal Institute of Technology is undertaking a capstone course, where student teams are required to solve a real-world problem by applying the mechatronics VDI 2206 design guidelines. The project proposals are provided to the groups by industrial partners as well as ongoing research projects of the mechatronics unit. The case study task described here [55] was provided by the Robot Design Lab as part of an active collaboration with Honda Research Institute of Japan on compliant robotics for social interaction [56], and revolved around the design, development and control of a tabletop robot that engages with children. Specifically, the project aimed to create a robot that embodied the principles of safety, transparency, reliability, and explainability as outlined in UNICEF's policy guidelines on AI for children [57]. These guidelines not only directed the design process that the students had to follow but were also integral in setting the foundational criteria for the robot's development throughout the project's 9-month duration.

Following a state-of-the-art analysis, the student team produced conceptual designs with the potential of addressing the extracted requirements. This stage's output sought to strike a balance between human-inspired traits that make the robot friendly and expressive, and those that could potentially scare children, prevent meaningful interaction, or create over-expectations [58]. This resulted in an egg-shaped design for the robot, selected for its inherent safety and simplicity, minimizing the risk of injury with its smooth edges and robust structure. The shape also provided a clear visual indication of the robot's state—closed when inactive, supporting transparency and safety by showing when the robot is not operational.

In addition, the robot's capability to mimic the six basic emotions (happiness, sadness, surprise, fear, anger, and disgust [59, 60]) was set as an important requirement to the robot's design. This feature was developed to enhance engagement with children, providing a relatable and interactive experience without overstimulation [61]. The requirement to achieve this task through non-verbal communication led the team to employ the use of the robot's motion sequences through its 11 degrees-of-freedom, the incorporation of robotic eyes, as well as color-based communication through an illuminated head cap. To ensure reliability and explainability, the robot included an interface to communicate its actions clearly to users—explaining, for instance, when it is recording or storing data. The implemented prototype is shown in Fig. 11.3.

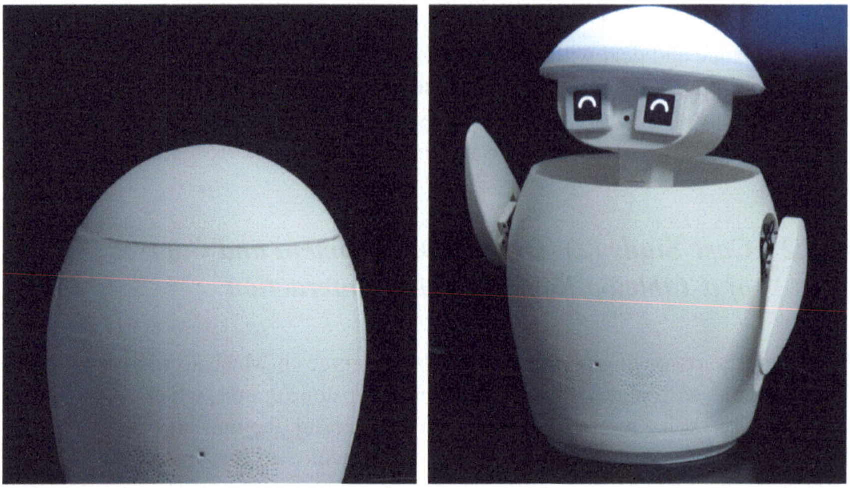

Fig. 11.3 The AKI robot prototype in its idle (left) and activated (right) state, as developed by the mechatronics student team [55]

The implementation attempted to address the requirements from UNICEF's policy guidelines on safety, transparency, reliability and explainability, specifically:

- Safety: The robot employs compliant actuation and uses soft materials to prevent injuries during interaction. Mechanical parts, such as the neck mechanism and arm extensions, are designed and programmed to halt upon detecting obstructions, preventing injuries. Temperature and current sensors are also used to always monitor the robot's states to further increase its operational safety.
- Transparency: The inclusion of features such as LED indicators were implemented to inform users about the robot's states and activities clearly and understandably. For example, different colors indicate different modes of operation.
- Reliability: The incorporation of robust electromechanical components in the robot's selected architecture increased the robot's reliability over long-hour operations, which is essential for maintaining trust and ensuring safety during interactions with children.
- Explainability: The robot was programmed to perform actions that are predictable and understandable, with displays and sounds that explain its actions in real-time, thus making its operations as transparent as possible for children.

The case study illustrates the integration of policy-driven design in mechatronics education through project-based learning, providing an effective way for practical application of theoretical knowledge in mechatronics to solve modern real-world challenges. It also highlights gaps in mechatronics design methodologies and areas where existing practices require additional focus:

- Policy-driven mechatronic design integration: This approach emphasizes the importance of aligning mechatronic designs with established ethical and safety standards, a practice that is often overlooked in more technically focused design methodologies.

- Durability and reliability in mechatronics project-based education: This work aimed to address a common gap in mechatronics-related academic projects which may not always emphasize long-term performance and maintenance as critical design requirements.
- Comprehensive safety features in mechatronics products: The need for integration of safety protocols throughout all phases of the design and development process, which creates an opportunity for a more structured approach to safety that is consistently applied across all mechatronics projects.
- Transparency and explainability in user-centered robot design: The need for AI-enabled robots to be transparent and explainable in their operations, which is crucial for trust and user-friendliness especially when vulnerable demographics (in this example, children) are the target user group.

11.4 Gaps and Directions for Future Design Methodologies

Mechatronics design is facing new opportunities and requirements. While existing design principles and methods are still relevant, challenges remain for dealing with the inherent complexity of mechatronic products of today; we are far from the "mechatronics compiler". Smaller-scale and innovative mechatronic products can for sure be developed using current methodologies as exemplified by the case studies. However, the limits of these methodologies become apparent when considering how to exploit advances in materials, sensors, additive manufacturing, AI, computing and communication, and in handling new requirements.

It appears evident that future mechatronic systems, and indeed the entire life-cycle, will be permeated by AI in different roles. New capabilities, the increasing level of automation and operation in more open environments, drive needs for new ways for connecting development and operations—drawing inspiration from software DevOps. These developments emphasize the need for novel architectures and platforms that go beyond the standalone mechatronic system, and more explicitly take interactions with humans into account. New requirements on sustainability as well as trustworthy AI further provide new constraints that need explicit consideration in design. Finally, the scope of the multitude of competences and skills required in design is increasing, requiring reconciling several (disciplinary) practices, processes, methods and tools—leveraging on the cross-disciplinary work practices already emerging in contemporary development [62, 63]. There is thus a need for new methodologies and architectures to address requirements on sustainability, agility and scalability, and operation in more open environments.

We introduce the concept of "MechaOps" and related considerations that promise to assist in enhancing scalability, smartness, performance and sustainability for extended mechatronic products that collaborate with a smart infrastructure, humans and other mechatronic systems. *MechaOps* refers to the integration of the concepts

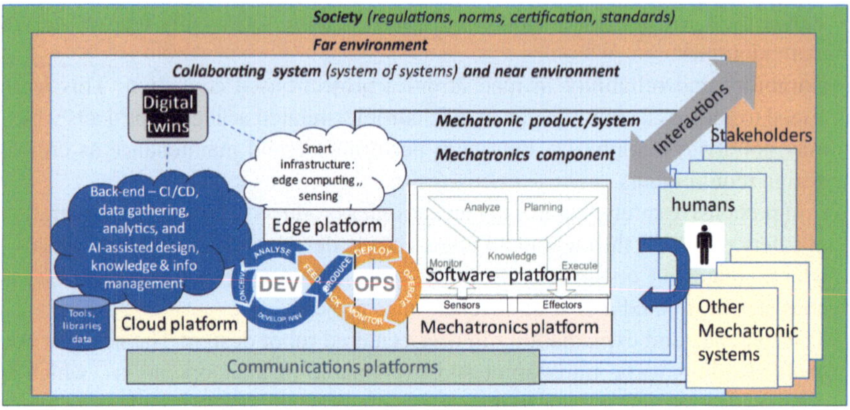

Fig. 11.4 A vision of MechaOps referring to the integration between Development and Operations, enabling more frequent reconfiguration, upgrades and data gathering relating to parts of a mechatronic product. The core of a mechatronics product is likely to evolve to incorporate higher level software—adding capabilities for situational awareness and planning based on a software platform. The additional computing needs will be supported by edge and cloud computing. The MechaOps concept is further characterized by using digital twins and explicit considerations of interactions and trustworthiness

of *Mechatronics* and *DevOps*. As opposed to DevOps in software engineering, MechaOps encompasses data gathering, upgrade/downgrades as well as reconfigurations considering both mechanics and/or software in a mechatronic product. Figure 11.4 depicts our view of the "expanded" nature of future mechatronic products, elaborated in the following.

11.4.1 MechaOps with Digital Twins

As depicted in Fig. 11.4, we draw upon the "DevOps" paradigm from software engineering but extend it to a mechatronics context. One implication of this is that the concept of what constitutes the "mechatronics product/system", becomes extended to incorporate interactions with several supporting systems.

Key to the concept of MechaOps is the ability to provide agility (adaptability), scaling of intelligence and collaboration through the concrete integration between Development and Operations. The approach brings abilities to gather data from a mechatronic product (or a range of products) over the lifecycle, to analyze their behavior, and to take decisions accordingly. Such decisions could take the form of pro-active maintenance and/or the deployment of entirely new functionalities to a product. This deployment would take place by changing/upgrading software and/or hardware parts, and/or its supporting systems including digital twins and other functionalities within the edge- or cloud platforms, recall Fig. 11.4. A key aspect for the data gathering is to understand what data to collect, corresponding to central performance indicators. Software upgrades could be deployed using over-the-air techniques, while hardware updates would require physical intervention. In both cases, modular architectures (in SW and HW) are a key requirement.

Multiple digital twins will likely exist in the envisioned kind of system, with some representing "live-models" executed on product—say a robot or on a nearby—real-time capable edge server computing system connected to the robot. These computing systems and the digital twin will support real-time use-cases for the robot, compare for example with case study 1. Another type of digital twin will exist in the cloud to support less time-critical and more detailed analytics, utilizing data from the field, and for developing and evaluating new functions before robot deployment.

We believe that the concept of digital twins will be essential for an effective integration of developments and operations for MechaOps, supporting virtual verification and validation before upgrades, and on the other hand to feed collected data back to the digital twin(s) for evaluation and validation. Without a digital twin between the Dev and Ops loops, the corresponding cycles could become very lengthy and far from agile practices. The digital twin approach on the other hand places requirements on dedicated efforts for digital twin development, validation and maintenance.

11.4.2 Novel Architectures and Platforms

Consider the example of the soft robot in case study 1, where Fig. 11.4 highlights four types of platforms. The robot itself is composed of a core mechatronics platform, through an integrated structure that includes load-bearing elements, sensors, actuators, embedded control systems and power. The trend to form smarter and automated robots will require adding an "upper" software platform that also incorporates machine learning functionalities and with connections to further external sensors such as cameras and lidars. The external sensors may either be connected to the mechatronics or the software platform.

Similarly, software and ML-based applications may reside in the mechatronics platform and/or on the software platform. The illustration in Fig. 11.4 should be interpreted such that a software platform supports software and ML-based applications (including their execution, supervision and upgrading). In general, the software platform supports advanced functionalities that rely on the mechatronics platform, for example for the robot to move around, illustrated by the MAPE-K pattern (Monitor, Analyze, Plan, Execute and Knowledge) [64]. The mechatronics and software platforms will be physically present on the robot—"on-board", whereas the cloud platform will reside off-board. Edge computing may take both on-board and off-board forms.

Interactions between the on-board and off-board systems, as well as with other external systems and human stakeholders will be supported through multiple communication systems, e.g., in the form of cellular communications and/or short-range communications.

A key requirement for the corresponding architectures is to support modularity, transparency and, as far as practically possible, independence and graceful degradation. Transparency will be important given the increasing complexity and to promote explainability, recall case study 2. Transparency will require emphasis on systematic monitoring for on-line as well as off-line diagnosis. Graceful degradation is important to enhance availability and in applications where there are no clear fail-safe states. For example, partial failures may occur in one or more of the systems depicted in Fig. 11.4, implying the importance of taking such failures into account and ensuring controlled degradation.

11.4.3 Interactions with Humans, Other Systems and a Smart Infrastructure

A key aspect of the operation in more open environments is the need to place much more emphasis on the near environment and interactions with humans and other systems/machines. This trend is clearly seen in the area of automated driving with intense ongoing efforts toward modeling, ontologies and standards to describe so-called operational design domains (ODD's), see e.g. [65, 66]. A key challenge is to handle uncertainty, and to reason about the completeness and correctness of such environmental descriptions, going well beyond the traditional "plant modeling" to support traditional control design in mechatronics. Operating in more open environments also requires contextual understanding by an automated system, an understanding of the "super-system". This is challenging for advanced automated systems, see e.g. [67].

Human–machine interaction requires dedicated efforts as technical systems engage in our lives, thus becoming more and more important. Problems with human machine interaction are often due to a lack of understanding of human user needs. Engineers tend to design systems based on what they understand and assume users might need, while users interact with systems based on their own experiences and expectations. This gap appears to be very difficult to overcome. Closing this gap requires a stronger focus on human-centered design, involving users throughout the design process, simplifying interfaces, improving communication, and continuously testing and refining based on user feedback [68]. As case study 2 showed, we believe it is essential to incorporate human-centered design into engineering curricula to bridge these gaps.

In automated systems, human-centered design becomes even more important. As Lisanne Bainbridge emphasized already in 1983, the higher the automation, the more difficult it will be for a human operator to step in when the automation fails—sometimes referred to as the automation paradox. A tragic illustration of this is represented by the Boeing 737 Max aircraft accidents, where the pilots (who were not trained for a new aircraft capability) had to fight the automation, and in two cases lost this fight [69]. More research is needed to reconcile the gap between human-centered design and automation. Promising directions in this regard include extending the concept of MAPE-K loops to involve human interactions [70] and attempts to reconcile design and systems thinking [71].

11.4.4 AI-Based Mechatronics

AI is likely to permeate most aspects in society including mechatronics. With AI we refer to the use of several techniques, encompassing traditional symbolic AI (e.g., symbolic reasoning and rule-based approaches), physics-based models and algorithms (e.g., from the field of automatic control and signal processing), as well as machine learning and deep-learning. The first two kinds of AI approaches are already prevalent in most mechatronic systems. During the past decade the topic of deep learning has had significant progress and has come to dominate the AI arena. We are now seeing a corresponding adoption in mechatronic products, for example for perception and anomaly detection. Referring to Fig. 11.4, such functions could be embedded into the mechatronic product itself or the cloud- or edge-based systems.

AI has a strong potential to support mechatronics development through design space exploration, for example regarding the potential use of new materials, alternative structures (e.g., combining load bearing, sensing, actuators and computing), and the corresponding trade-off analysis. The corresponding methods need to consider limitations and opportunities for manufacturing such as additive manufacturing and modularity. AI-assisted design bears the promise to boost engineering creativity and productivity, leveraging AI capabilities to capture state-of-the art—providing suggestions to designers, evaluating designs and in generating design alternatives as well as various system descriptions. The breakthroughs in large language models and foundational models have already provided indications of a potential for substantially increased productivity [72].

In an investigation from 2016, it was found that novice designers in the aerospace industry were aware of only 35% of the knowledge they needed to deal with a complex system. This was concluded to represent a competence barrier [73]. Leveraging generative AI in this context, as designer companions, would have the potential to fill in the missing competence gaps and deal with scalability issues in handling the enlarged design space and dependencies. However, one would also need to deal with the failure modes (hallucination) of DL systems. The strength of experienced humans in this context is that of "a common sense", and experiences that go beyond what (current) AI-based systems can provide. This implies the need to find ways where experienced designers oversee and check results obtained by less experienced engineers collaborating with an AI.

Another interesting aspect is the role of creativity in AI-assisted design. Pahl and Beitz [5], describes important developments in CAE and say that these will drastically facilitate the ease with which information can be managed and retrieved, … as well as design guidelines and processes. ", believing that many design tasks will become automated but that "These systems" (referring to CAE) "will also aid the analysis, optimization and combination of solutions, but they will not replace designers". This is in line with our discussion, since although current generative AI-systems can deliver designs, these cannot be fully trusted and thus we foresee a future with AI-assisted design where humans work closely with AI-assistants.

As illustrated in case study 1, there are also needs to combine DL-based algorithms with traditional algorithms/rule-based software, partly to deal with DL failure modes but also partly to provide explainable solutions when this is possible.

11.4.5 Assuring Trustworthiness and Sustainability

While the MechaOps concept, digital twins and AI promise to provide greatly enhanced capabilities, there is however no free lunch. These new capabilities come at a cost, including a corresponding increase in complexity. While this complexity is needed, and desired in the sense of providing new opportunities, it must also be managed.

Moreover, for mechatronic products developed according to the Mecha approach, there are open questions for how to assure their trustworthiness. In other words, how can we rely on and trust these more complex systems, and how can we assure that they are ready—fit for purpose? This is especially difficult considering the more open environments with needs to characterize them and establish sufficient testing scenarios, handling the brittleness of DL-technologies, and in providing trustworthy upgrades, accomplished in a way that eliminates the need to redo all verification and validation efforts.

Trustworthiness, e.g., in terms of the trustworthy AI guidelines, encompasses both traditional dependability attributes such as safety and availability, but also AI associated properties such as transparency and fairness [74]. On top of this there is the changing landscape of security threats. All these trustworthiness properties must to some extent be assured for mechatronic products, and especially for those that are safety, security and/or availability critical. Each property just mentioned represents a challenge on its own and is moreover represented by a fragmented and corresponding changing landscape of regulations and standards. For example, most safety–critical domains have ongoing (or recently completed) legislation and standardization work to address DL. The EU AI act takes another stance and introduces a horizontal regulation across domains [75]. This implies that the current situation could be seen to represent a "standardization and regulatory jungle", which has yet to converge and mature.

There are also trade-offs and conflicts between the properties that need considerations since the trustworthiness properties are not independent, compare for example with trade-offs related to safety-availability, and AI performance vs. privacy/confidentiality. Architecture frameworks for trustworthiness are needed to emphasize the corresponding trade-offs, [76].

In dealing with the trend toward software defined machines and increasing use of machine learning, it has also been recognized that there is a need to account for evolvability of systems, and thus that the assurance itself must be more adaptive. Current DevOps practices in software engineering face challenges when applied to heterogeneous systems as in mechatronics, and in assuring that critical properties get the attention they deserve. This has led to new research on concepts such as SafeOps and continuous safety assurance, see e.g. [77–79]. The purpose of such efforts is to reconcile and integrate software DevOps with safety critical system practices. In this context—and to support such integration—we described the role of digital twins (recall the section on "MechaOps with digital twins") and modular architectures (recall the section on "Novel architectures and platforms") as essential elements of the MechaOps approach.

A somewhat similar reasoning is relevant for sustainability, which traditionally has been catered for through design time life-cycle analysis (LCA), with less considerations of the remainder of the life cycle. New regulations such as the DPP are likely to change this, and thus sustainability naturally needs to consider the entire lifecycle and actual environmental impact. The concept of MechaOps, recall Fig. 11.4, supports such sustainability and circularity considerations by providing data gathering, (updated) LCA analysis, traceability (through the digital twins) and considerations for repairs, recycling or refurbishment of parts.

AI-assisted design promises to aid in investigating architectures to promote trustworthiness and sustainability, for example by considering relevant modularity in hardware and software, and in general the design for X (robustness, resilience, product reliability, etc.). This includes robust design optimization techniques, fault-tolerant control strategies, and redundancy management approaches, to minimize the risk of failures and accidents.

11.5 Conclusions

Mechatronics, as a subset of the field of CPS, is more relevant than ever as we are entering an era with expectations on sustainable and smart machines. Both opportunities and challenges arise as such machines act and interact in more open environments with humans and other systems.

To enable future mechatronic systems there is a need for a broader perspective on mechatronics that consequently, beyond the immediate product, closely involves the near environment and integration with a supporting infrastructure.

Key ingredients of our presented MechaOps approach encompass digital twins, with real-time and off-line analytics with the potential for off-loading to edge computing (adding further flexibility), and upgrading/downgrading, considering both software and hardware parts. Methodologies prepared for the MechaOps approach will need to take all of the aspects described into the previous section into account. The corresponding approaches also need to consider different settings of development, from new products to extensions of existing systems. With the life-cycle view of the MechaOps concept, it becomes essential to design for upgrading, downgrading, maintenance, and also reuse and refurbishment in the spirit of circular systems. We believe that new methodologies and architectures will initially be especially relevant for high-end systems, supporting the creation of adaptable and flexible mechatronics products and services with improved performance and reduced environmental footprint.

Paving the way for the new methodologies further requires reconciling several worldviews—“cultural gaps”, encompassing key stakeholders and disciplines. These include software engineering, machine learning, safety critical systems, core mechatronics, materials and mechanical engineering, sustainability engineering (e.g., as embodied by life-cycle analysis and circular economy) and human-centered design.

New methodologies will support the creation of adaptable and flexible mechatronics products and services with improved performance and reduced environmental footprint, leveraging novel sensors, actuators, and materials, such as those found in soft robots. A relevant topic for further work (beyond the scope of this paper) is how to prepare relevant stakeholders, such as engineers, managers, operators, and end-users with the right set of skills to deal with future mechatronic systems.

Acknowledgements We acknowledge inspirational collaboration with Professor Jan Wikander that paved the way for this book chapter. Jan contributed an initial version of Figure 1.

References

1. Graessler I, Hentze J (2020) The new V-model of VDI 2206 and its validation. Automatisierungstechnik 68(5):312–324. https://doi.org/10.1515/auto-2020-0015
2. Hehenberger P, Bradley D (2016) Mechatronics futures—challenges and solutions for mechatronic systems and their designers. Springer International Publishing, ISBN (Print) 9783319321547
3. Wikander J et al (2001) The science and education of mechatronics engineering. IEEE Robot Autom Mag 8(2):20–26. https://doi.org/10.1109/100.932753
4. Cabrera AA et al (2010) Towards automation of control software: a review of challenges in mechatronic design. Mechatronics 20(8):876–886. https://doi.org/10.1016/j.mechatronics.2010.05.003
5. Pahl G, Beitz W, Feldhusen J, Grote KH (2007) Engineering design: a systematic approach, 3rd ed. London, UK: Springer London Limited
6. Törngren M, Grogan PT (2018) How to deal with the complexity of future cyber-physical systems? Designs 2:40. https://doi.org/10.3390/designs2040040
7. Törngren M et al (2016) Strategies and considerations in shaping cyber-physical systems education. SIGBED Rev 14(1):53–60. https://doi.org/10.1145/3036686.3036693
8. Törngren M, Bensalem S, McDermid J, Passerone R, Sangiovanni-Vincentelli A, Schätz B (2015) Education and training challenges in the era of cyber-physical systems: beyond traditional engineering. In: Proceedings of the WESE’15: workshop on embedded and cyber-physical systems education (WESE’15). Association for Computing Machinery, New York, NY, USA, Article 8, 1–5. https://doi.org/10.1145/2832920.2832928
9. Leite L (2019), o.a. A survey of DevOps concepts and challenges. ACM Comput Surv (CSUR). Vol 52, 6
10. Bathelt J et al (2005) Applying the new VDI design guideline 2206 on mechatronic systems controlled by a PLC. In: International conference on engineering design (ICED 05), Melbourne, p 2601

11. Hehenberger P et al (2016) From mechatronic systems to cyber-physical systems: demands or a new design methodology?. In: Hehenberger, P., Bradley, D. (eds) Mechatronic Futures. Springer, Cham. https://doi.org/10.1007/978-3-319-32156-1_10
12. Snowden D, Boone MA (2007) A leader's framework for decision making. Harvard Bus Rev 85(11):68–76
13. Bricogne M et al (2016) Design processes of mechatronic systems. In: Hehenberger, P., Bradley, D. (eds) Mechatronic Futures. Springer, Cham. https://doi.org/10.1007/978-3-319-32156-1_10
14. Asplund F (2019) Exploratory testing: do contextual factors influence software fault identification? Inf Softw Technol 107:101–111
15. Edmondson AC, Harvey J-F (2018) Cross-boundary teaming for innovation: integrating research on teams and knowledge in organizations. Hum Resour Manag Rev 28(4):347–360
16. Santos V, Goldman A, De Souza CR (2015) Fostering effective inter-team knowledge sharing in agile software development. Empirical Softw Eng 20:1006–1051
17. Gero JS, Kannengiesser U (2014) The function-behaviour-structure ontology of design, In: A. Chakrabarti and L.T.M. Blessing (eds) An Anthology of Theories and Models of Design, Springer, pp 263–283.
18. Mcharek M, Hammadi M, Azib T, Larouci C, Choley J (2019) Collaborative design process and product knowledge methodology for mechatronic systems. Comput Ind 105:213–228. https://doi.org/10.1016/j.compind.2018.12.008
19. Asplund F (2014) Risks related to the use of software tools when developing cyber-physical systems: a critical perspective on the future of developing complex, safety-critical systems. Diss. KTH Royal Institute of Technology
20. Van Brussel H et al (2001) Towards a mechatronic compiler. IEEE/ASME Trans Mechatron 6(1):90–105. https://doi.org/10.1109/3516.914395
21. Rahimi F, Wikander J (2020) Concurrent and optimal structure, control and implementation design. 2020 IEEE 11th International conference on mechanical and intelligent manufacturing technologies (ICMIMT). IEEE
22. Törngren M (2021) Cyber-physical systems have far-reaching implications. In M. Duranton et al., editors. HiPEAC Vision 2021, p 12–19
23. Broman D et al (2012) Viewpoints, formalisms, languages, and tools for cyber-physical systems. Proc. 6th International workshop on multi-paradigm modeling (MPM '12). ACM, New York, USA, 49–54. https://doi.org/10.1145/2508443.2508452
24. Asplund F, Biehl M, Loiret F (2012) Towards the automated qualification of tool chain design. Computer safety, reliability, and security: SAFECOMP 2012 Workshops, Magdeburg, Germany, September 25-28, 2012. Proceedings 31. Springer Berlin Heidelberg
25. Asplund F (2015) The future of software tool chain safety qualification. Saf Sci 74:37–43
26. Aberdeen Group (2008), System design: new product development for mechatronics, Boston, MA, January. (www.aberdeen.com)
27. Incose Systems Engineering Vision (2035) https://www.incose.org/publications/se-vision-2035
28. McAfee A, Brynjolfsson E (2017) Machine, platform, crowd: Harnessing our digital future. WW Norton & Company
29. Macías-Escrivá FD et al (2013) Self-adaptive systems: a survey of current approaches, research challenges and applications. Expert Syst Appl 4018:7267–7279
30. Dobson S et al (2006) A survey of autonomic communications. ACM Trans Auton Adapt Syst (TAAS) 12:223–259
31. Preece A (2018) Asking "why" in AI: explainability of intelligent systems–perspectives and challenges. Intell Syst Account Finance Manag 25(2):63–72. https://doi.org/10.1002/isaf.1422
32. Tao X, Lu J, Chen D, Törngren M (2020) Probabilistic inference of fault condition of cyber-physical systems under uncertainty. IEEE Syst J 14(3):3256–3266. https://doi.org/10.1109/JSYST.2020.2965400
33. Hussain H, Tamizharasan PS, Rahul CS (2022) Design possibilities and challenges of DNN models: a review on the perspective of end devices. Artif Intell Rev 55(7):5109–5167. https://doi.org/10.1007/s10462-022-10138-z

34. Rychener Y, Kuhn D, Sutter T (2023) End-to-end learning for stochastic optimization: a Bayesian perspective. In: Proceedings of the 40th international conference on machine learning (ICML'23), Vol 202. JMLR.org, Article 1225, 29455–29472
35. Sánchez JM, Gaspar et al (2022) Edge computing for cyber-physical systems: a systematic mapping study emphasizing trustworthiness. ACM Trans Cyber-Phys Syst (TCPS) 63: 1–28
36. Renda D, Serger AS, Tataj S, Hidalgo D, Giovannini C, Huang E, Isaksson A (2022) Industry 5.0, a transformative vision for Europe. ESIR Policy Brief No.3. Luxembourg: Publications Office of the European Union
37. Teodosiu C, Fiore S, Hospido A (2022) Assessing progress towards sustainability: frameworks (eds), Tools and Case Studies. Elsevier
38. Parker DB (1998) Fighting computer crime: a new framework for protecting information. John Wiley & Sons, Inc.,
39. Asplund F, Holland G, Odeh S (2020) Conflict as software levels diversify: tactical elimination or strategic transformation of practice? Saf Sci 126:104682
40. Habli I (2017) Safety standards: chronic challenges and emerging principles, Handb Saf Prin, 732–746
41. Asplund F et al (2021) The genesis of public-private innovation ecosystems: bias and challenges. Technol Forecast Soc Change 162:120378
42. Penny J et al (2001) The practicalities of goal-based safety regulation. Aspects of safety management: proceedings of the ninth safety-critical systems symposium, Bristol, UK 2001. London: Springer London
43. Lee C et al (2017) Soft robot review. Int J Control Autom Syst 15:3–15
44. Chin K, Hellebrekers T, Majidi C (2020) Machine learning for soft robotic sensing and control. Adv Intell Syst 2(6):1900171
45. Kim D et al (2021) Review of machine learning methods in soft robotics. PLoS ONE 162:e0246102
46. Thuruthel TG, Falotico E, Manti M, Laschi C (2018) Stable open loop control of soft robotic manipulators. IEEE Rob Autom Lett 3(2):1292–1298
47. Chikhaoui MT, Cot A, Rabenorosoa K, Rougeot K, Andreff N (2016) Design and closed-loop control of a tri-layer polypyrrole based telescoptic soft robot, In: IEEE/RSJ International conference on intelligent robots and systems (IROS), Daejeon, South Korea
48. Li M et al (2022) Soft actuators for real-world applications. Nat Rev Mater 73:235–249
49. Leigh SJ, Bradley RJ, Purssell CP, Billson DR, Hutchins DA (2012) A simple, low-cost conductive composite material for 3D printing of electronic sensors. PLoS ONE 7(11):1–6
50. Zolfagharian A, Kouzani AZ, Khoo SY, Moghadam AAA, Gibson I, Kaynak A (2016) Evolution of 3D printed soft actuators. Sens Actuators A 250:258–272
51. Pei E (2014) 4D printing: dawn of an emerging technology cycle. Assem Autom 34(4):310–314
52. George Thuruthel T et al (2018) Control strategies for soft robotic manipulators: a survey. Soft Robot 52:149–163
53. Bertsekas D (2022) Lessons from AlphaZero for optimal, model predictive, and adaptive control. Athena Scientific
54. Tan K, Ji Q, Feng L, Törngren M (2024) Edge-enabled adaptive shape estimation of 3D printed soft actuators with Gaussian processes and unscented Kalman filters. IEEE Trans Ind Electron 71:3
55. Andreanidis C et al (2023) On the design and development of a tabletop robot for interaction with children, 2023 IEEE/ASME International conference on advanced intelligent mechatronics (AIM), Seattle, WA, USA, pp 1232–1237, https://doi.org/10.1109/AIM46323.2023.10196188
56. Andrikopoulos G et al (2024) On the design and development of a soft robotic neck for the social robot HARU, accepted for presentation In: IEEE/ASME International conference on advanced intelligent mechatronics (AIM), Boston, MA, USA, July 15–19
57. UNICEF (2021) Policy guidance on AI for children, 2021. Available online: https://www.unicef.org/globalinsight/reports/policy-guidance-ai-children
58. Mori M et al (2012) The uncanny valley [from the field]. IEEE Robot Autom Mag 19(2):98–100

59. Ekman P (1992) An argument for basic emotions. Cogn Emot 6:169–200
60. Martinez L et al (2016) Contributions of facial expressions and body language to the rapid perception of dynamic emotions. Tanfondline 30(5):939–952
61. Fisher AV et al (2014) Visual environment, attention allocation, and learning in young children: when too much of a good thing may be bad. Psychol Sci 25(7):1362–1370
62. Asplund F, Ulfvengren P (2021) Work functions shaping the ability to innovate: insights from the case of the safety engineer. Cogn Technol Work 23(1):143–159
63. Asplund F et al (2022) Which skills? A critical perspective on the skills facilitating the transfer of third-cycle students to knowledge-intensive SMEs. 2022 IEEE Frontiers in education conference (FIE). IEEE
64. Kephart JO, Chess DM (2003) The vision of autonomic computing. Computer 36(1):41–50
65. Lee CW et al (2020) Identifying the operational design domain for an automated driving system through assessed risk. 2020 IEEE Intelligent vehicles symposium (IV). IEEE
66. BSI (2020) PAS 1883:2020 operational design domain (ODD) taxonomy for an automated driving system (ADS) —Specification
67. Koopman P (2024) Anatomy of a Robotaxi crash: lessons from the cruise pedestrian dragging mishap. https://users.ece.cmu.edu/~koopman/pubs/Koopman2024_CruiseMishap_arXiv_Preprint.pdf (Preprint of paper accepted SafeComp 2024)
68. Bartneck C et al (2020) Human-robot interaction: an introduction. Cambridge University Press
69. Appicharla S (2022) Lessons learnt from Boeing 737 max 8 crashes as safety data: needles in the haystack, Int Syst Safety Conf
70. Cleland-Huang J et al (2022) Extending MAPE-K to support human-machine teaming. Proceedings of the 17th symposium on software engineering for adaptive and self-managing systems
71. Gürdür D, Törngren M (2018) Visual analytics for cyber-physical systems development: Blending design thinking and systems thinking. 15th Annual NordDesign Conference (NordDesign 2018)
72. Dell'Acqua F et al (2023) Navigating the jagged technological frontier: field experimental evidence of the effects of AI on knowledge worker productivity and quality. Harvard Business School Technology & Operations Mgt. Unit Working Paper. 24–013
73. Melville C et al (2016) TiV-Model—an attempt at breaching the industry adoption barrier for new complex system design methodologies. In: Hehenberger, P., Bradley, D. (eds) Mechatronic Futures. Springer, Cham. https://doi.org/10.1007/978-3-319-32156-1_10
74. AI HLEG (2019) High-level expert group on artificial intelligence. Ethics guidelines for trustworthy AI 6
75. Commission (2021) Proposal for a regulation of the European Parliament and of the Council laying down harmonised rules on Artificial Intelligence (Artifical Intelligence Act) and amending certain Union legislative acts" COM/2021/206 final
76. Ramli MR, Törngren M (2022) Towards an architectural framework and method for realizing trustworthy complex cyber-physical systems. 16th International conference on research challenges in information science (RCIS 2022), 17–20 May 2022, Barcelona, Spain. Vol. 3144. CEUR-WS
77. Fayollas C et al (2020) SafeOps: a concept of continuous safety. 2020 16th European dependable computing conference (EDCC), Munich, Germany, pp 65-68, https://doi.org/10.1109/EDCC51268.2020.00020
78. Johansson R, Koopman P (2022) Continuous learning approach to safety engineering. CARS—Critical Automotive applications: Robustness & Safety, Zaragoza, Spain
79. Munk P, Schweizer M (2022) DevOps and safety? SafeOps! towards ensuring safety in feature-driven development with frequent releases. Safecomp 2022. Lecture Notes in Computer Science, vol 13415

Part III
Digital Technologies

Chapter 12
Artificial Intelligence (Applied)

Abbas A. Dehghani and John Mitchell

Abstract The application of Artificial Intelligence has been growing rapidly, and is now seen in many aspects of our daily lives. Mechatronic systems are becoming more and more intelligent by using AI in different ways. This chapter looks at some of the ways in which AI is being applied in Mechatronics and provides a case study on the application of AI in the operation of lower-limb Assistive Robotic Exoskeletons. For these systems to function properly, user intent recognition is crucial. One way to achieve this is to accurately analyse captured gait data by using AI. The chapter focusses on how different AI algorithms can be used for this purpose.

12.1 Introduction

Applications of Artificial Intelligence (AI) have been growing rapidly in recent years, and we are now witnessing the impact of its presence in all aspects of human life. This impact is so huge and threatening that we frequently see publications and conferences on the '*dangers of AI*' and how to prevent them.

AI is nowadays applied to almost all aspects of human life: from weather forecasting to economics and business; engineering and automation; driverless cars and unmanned vehicles; aircraft control as well as medicine and robotic systems.

In this chapter, the focus will be on assistive robotic systems (as a case study) to support the mobility of an ageing population and those with muscle dysfunctions. The idea will be to see how AI is applied in this field and can impact on such robotic systems, and how these systems can be controlled to not only support the mobility functions but also provide additional intelligent functionalities such as fall prediction and prevention. It will be seen that the impact of this application of AI will enable user intent capture, along with fast and accurate gait data analysis.

A. A. Dehghani (✉) · J. Mitchell
University of Leeds, Leeds, UK
e-mail: a.a.dehghani-sanij@leeds.ac.uk

P. Hehenberger and D. Bradley (eds.), *Mechatronic Futures*,
https://doi.org/10.1007/978-3-031-83571-1_12

12.2 Some Examples of Applied AI Methods

A variety of statistical and AI-based methods are used in Mechatronics. These methods have different characteristics which affect accuracy and computer processing times. In this section, some of the most common methods are briefly reviewed, and their advantages and disadvantages highlighted.

12.2.1 Artificial Neural Networks (ANN)

The artificial neural network is a classification model comprised of layers of densely connected neurons. Each neuron is a simple form of classifier which receives input values with associated '*weight*' values, and uses a threshold value (or '*bias*') along with an activation function to classify the inputs. Large networks comprised of multiple layers of these simple neurons can be constructed to form a network capable of multiclass classifications. As the inputs are fixed for a given datapoint, the network performs supervised learning by adjusting the weights and biases of each neuron according to a known output value. ANNs can be created in various sizes depending on the given problem, enabling the creation of large, deep neural networks (DNNs) which enable high accuracies across many applications whilst keeping computational complexity low [1, 2].

ANNs can suffer from overfitting if training data are lacking or the number of hidden layers is too large, reducing their generalisation and heavily impacting their accuracy when handling data outside the training set [1, 2]. When utilising sufficient data and methods such as cross-validation to mitigate this problem, ANNs provide powerful classification capabilities with low computation times.

12.2.2 Support Vector Machine (SVM)

Support Vector Machines (SVMs) are a supervised classical machine learning approach which can be implemented as support vector classifiers (SVCs) to classify data by finding a hyperplane that separates classes whilst maximising the margins between datapoints and the hyperplane [3]. SVMs have the capacity to make use of kernel functions, such as polynomials and the radial basis function, to simulate an increase in data dimensionality and enable a hyperplane to more easily separate data with maximal margins [4].

Whilst generally yielding good performance, SVMs can suffer from long decision times when dealing with multiclass data due to their being binary classifiers by nature, which is overcome using computationally expensive methods [3, 5]. Despite this limitation, SVMs are an extremely popular and high-performing classical machine learning model across multiple fields and applications [4].

12.2.3 Convolutional Neural Networks (CNN)

Convolutional neural networks (CNNs) are networks designed to mimic the function of the human brain based on the work of Hubel and Wiesel in 1968 [6]. CNNs mimic this functionality by featuring a series of complex convolution neurons in the early layers of the network, which use kernel functions to apply filters and extract features from the input space [7, 8]. Among the convolution layers, '*pooling*' layers are positioned which use a simple function, such as the maximum value, to reduce the input space by extracting a single value from a sliding window, reducing the computational complexity of the algorithm [7, 8]. The process of repeatedly convolving and pooling data allows the algorithm to extract features from the raw data, at which point a series of dense, fully connected layers can perform classification on these features.

As a result of their architecture, CNNs excel at image processing, seeing success in many applications such as face and speech recognition due to their capacity to identify complex relationships and work with raw data [7, 8]. Whilst extremely versatile and accurate, CNNs can suffer from large decision times and computational complexity due to the number of operations and layers required when compared to many classical machine learning techniques [9].

12.2.4 K-Nearest Neighbours (KNN)

K-Nearest Neighbours (KNN) is a simple form of machine learning classifier which typically uses the Euclidian distance between a given, unclassified datapoint and the k-nearest classified datapoints (placed during training) to perform classification [10, 11]. The value for k in a KNN is often decided using a supervised learning technique where accuracies are generated for different values of k, which can then be obtained from the model with the highest accuracy [10].

Whilst KNN generally exhibits high accuracies, some major issues with KNN involve the inability to make accurate decisions for overlapping classes [11] and the algorithm's computational complexity, as large values of k with high dimensionality can have extremely large classification times [12, 13]. This can be mitigated using dimensionality reduction techniques such as Principle Component Analysis (PCA) [14, 15], feature selection, and the optimisation of the parameter 'k', although decision times are still generally higher than other classification algorithms [13].

12.2.5 Decision Trees (DT) and Random Forests (RF)

Decision trees are algorithmically generated from a set of training data to maximise classification accuracy through the use of a univariate splitting criteria [16]. The most common of these criteria select features and separation parameters by minimising the number of misclassifications (the '*impurity*') at each node, starting from the root. For the root node, the impurity for all individual features is calculated and the feature with the lowest impurity is selected. Next, a single branch of the root node is selected, and impurity is again calculated for the remaining features. A leaf node is placed when either all features have been exhausted, impurity is zero, or when the impurity of a node would be greater than the impurity of its preceding node [16, 17].

Whilst simple, decision trees are limited by their lack of capacity for generalisation [17, 18]. One solution for this is a method of ensemble learning named '*Random Forest*' (RF) which combines many decision trees to increase accuracy whilst maintaining computational efficiency for large datasets [19, 20]. Random Forests generate '*bootstrap*' datasets, which consist of randomly selected datapoints from the original dataset. A subset of features are then randomly selected from the bootstrap dataset and a decision tree is created from this limited data [19, 20]. This process is then repeated until many decision trees have been created, each with different datapoints and feature subsets. For classification, each decision tree in the random forest performs classification on the given datapoint and the class with the most '*votes*' from the individual decision tree classifications is selected as the definitive classification of that datapoint [19, 20].

Random Forest classifiers have many benefits such as computational efficiency, high accuracy and generalisability, applicability in a wide range of areas and few tuning parameters [20]. For these reasons, Random Forest is one of the most popular classical ML methods in modern literature due to their competitive accuracies and decision times to other classical machine learning methods [19–21].

12.2.6 Recurrent Neural Networks (RNN) and Long Short-Term Memory (LSTM) Models

Recurrent neural networks (RNNs) are a form of neural network in which the outputs of certain neurons in the network are connected to their inputs to form loops which allow the network to use elements of the previous classification in the current classification [22]. This property makes RNNs extremely useful for sequential applications such as predictive text, where the words in a sentence are used as a sequence to predict what the next word will be. However, RNNs are greatly affected by the vanishing/exploding gradient problem which strictly limits the length of possible sequences and therefore the number of possible applications [23].

Long Short-Term Memory (LSTM) models are a specific implementation of gated recurrent neural networks which mitigate the vanishing/exploding gradient problem by providing the network with a '*cell state*' variable that gives the model control over what to remember from a given time step to the next [23]. This method allows LSTMs to process much longer sequences, enabling them to find applications in many fields, particularly relating to speech recognition and vocal processing [24, 25].

12.3 Case Study—Assistive Robotic Systems and AI Applications

Assistive technologies in the form of biomechatronic systems can support patients in two distinctive ways. (i) For those patients who have lost a limb due to various reasons such as diseases and traumatic accidents, a biomechatronic system can be provided to replace the lost limb. For this system to fully function in coordination with other body parts there should be an appropriate interaction between the biological system (human body) and the biomechatronic system (e.g. a lower limb prosthetic device); (ii) For the older people and those with weak muscles, where the limbs are intact, wearable assistive robotic exoskeletons can be used to help with mobility and some additional functionalities.

In order for these systems to function properly in coordination with the body segments and to provide optimum support, user intent is a key parameter. There are various methods available to capture user intent by using mechanical sensors and/or biological signals. A typical non-invasive method would be based on using Inertial Measurement Units (IMUs) as an example for the mechanical sensors and surface ElectroMyoGraphy (sEMG) as an example for biopotential sensors.

12.3.1 Human Gait Analysis Using AI

One of the very common sensors used in many applications is Inertial Measurement Units (IMUs). This is a device that typically consists of gyroscopes for measuring angular rate, accelerometers for measuring specific force, and magnetometers to determine orientation.

In the gait analysis application, IMUs can be used for assessing spatio-temporal and kinematic parameters and to identify and estimate different locomotion modes and activities. This can help in detecting and identification of human intent and provide guidelines for controlling smart prostheses and exoskeletons.

IMUs can be attached to different segments of the body, and a huge amount of the steady-state locomotion data from participants in a trial can be collected to be analysed. The question then would be how the data can be analysed to link them to user intent recognition. The answer is that advanced AI-based classification methods can be used to obtain the required results. It then becomes evident that the collected data can be used for training and testing the AI-based intelligent classifiers.

12.3.2 *The Application of AI for Gait Analysis*

Participants and data collection

As the focus of this chapter is the application of AI in this particular field, the process of gait data collection is only briefly covered. Healthy volunteers took part in data collection using seven wireless portable IMUs. As mentioned earlier, each IMU comprises a 3-axis accelerometer, gyroscope, and magnetometer. They were attached at the body centre of mass (CoM), thigh, shank, and foot segments of subjects as shown in Fig. 12.1. The participants were asked to perform certain Activities of Daily Living (ADLs) such as level ground walking (LGW) with self-selected pace, ascending/descending a ramp, sitting down, and standing up. Each of these 5 activities was captured from 4 participants, and 3 repeats of each trial were captured, resulting in 12 data files for each activity.

Data cleaning and pre-processing

The raw data captured by the IMUs contains some non-numerical (NaN) values and additional, erroneously-captured data channels which must be addressed before data

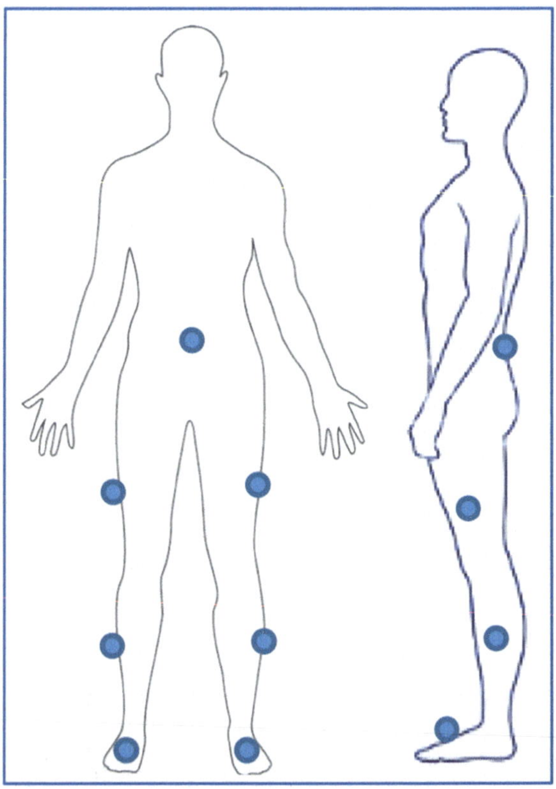

Fig. 12.1 IMU locations

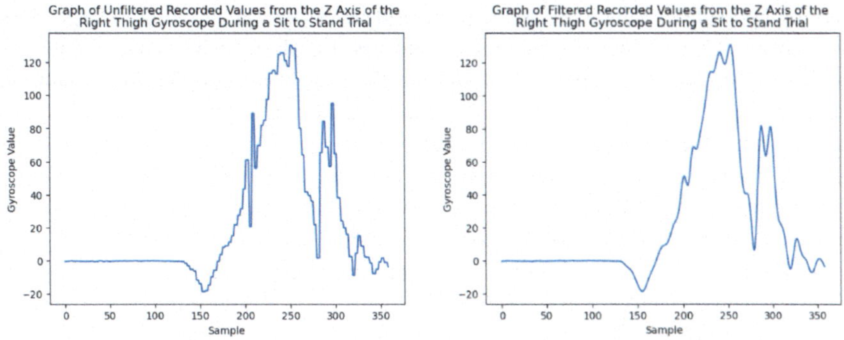

(a): Unfiltered (b): Filtered

Fig. 12.2 Unfiltered **a** and filtered **b** values from the *Z* axis of the gyroscope positioned on the right thigh

can be prepared for machine learning. As this is time-series data, and the number of NaN values is small, simple linear interpolation can be used to restore a suitable approximation of the lost values. For larger or more complex signals, spline, quadratic, or higher-degree polynomial interpolation can be applied to restore the lost data more accurately. Another data cleaning process involves removing irrelevant data channels, such as the '*sync timestamp*'. Once these channels are removed, 9 data channels remain for each of the IMUs (3-axis accelerometer, 3-axis gyroscope, and 3-axis magnetometer), totalling 63 data channels, along with one-time channel for each activity. The time channel can be used to determine that the data was captured at a frequency of 100 Hz.

Once the data have been cleaned, a 10 Hz, 4th order lowpass filter is applied to remove high-frequency noise which can be introduced electrically, or through the imperfect mounting of the IMUs to the subject, causing the sensor to vibrate during the trials. Graphs of raw and filtered data for the sit to stand activity can be found in Fig. 12.2.

Semi non-overlapping window feature engineering

As we are dealing with time-series data, it is insufficient to provide the machine learning models with the samples at this stage, which each represent just a single moment in time. The Semi Non-Overlapping Window (SNOW) approach to feature engineering is a method of generating samples which represent a larger '*window*' of time-series data. This approach introduces two variables: the size of the window, and the time increment at which the next window is taken, which can be made smaller than the window size to introduce an overlap between windows. From these windows, statistical properties can be extracted from each data channel to represent the larger window, condensing a time-series sequence into individual values dubbed '*features*'.

This approach has the disadvantage of reducing the number of samples by compressing them into features, which can severely affect results for machine learning approaches which are prone to overfitting or poor at adapting to large class imbalances. To mitigate this problem, windows are allowed to overlap with one another, bolstering the number of samples at the cost of introducing bias when the train and test set are separated. A balance must be struck between the window size and step size (the number of raw data samples in a window, and the increment at which the next window is taken, respectively) to create features from large enough windows to contain sufficient information about the time-series data, whilst retaining enough samples to train a model capable of accurate activity classification.

For this example, the maximum value, minimum value, mean, standard deviation, root mean square, maximum gradient, and number of zero crossings are extracted from each data channel, resulting in 441 features in total after calculating these for each of the 63 data channels. Regarding windows, this example will explore the results for several different window sizes and step sizes to demonstrate the effect these variables have on the classification accuracy.

It should be noted that features are not always useful, however, and extracting too many features can result in overfitting or introduce noise to the model if the provided feature does not correlate with any of the classes. To prevent this, feature selection techniques can be applied to the extracted features to remove noisy features using statistical analysis. In this example, mutual information gain is used to determine the statistical dependence between each feature and class using entropy, enabling the identification of the top 15 features which are expected to produce a high accuracy model. These 15 features are:

(1) Pelvis magnetometer *X* max;
(2) Pelvis magnetometer *X* min;
(3) Foot *L* magnetometer *Y* max;
(4) Foot *R* magnetometer *Y* max;
(5) Shank *L* magnetometer *X* max;
(6) Foot *L* magnetometer *Y* min;
(7) Pelvis magnetometer *X* mean;
(8) Foot *R* magnetometer *X* min;
(9) Foot *L* magnetometer *Z* max;
(10) Thigh *R* magnetometer *Z* max;
(11) Foot *R* magnetometer *Z* min;
(12) Thigh *R* magnetometer *Z* min;
(13) Pelvis magnetometer *Y* min;
(14) Foot *R* magnetometer *Z* max;
(15) Pelvis magnetometer *Z* max.

This process is repeated each time the window size is changed. The rest of the 441 features are discarded, and these 15 features alone are used to train the machine learning models.

Each activity now contains a different number of samples, with LGW containing the most at around 1200 samples, and stand-to-sit containing the least at around 350 samples when using a window size of 500 ms and a step size of 150 ms. Large class imbalances can result in reduced accuracies, as models learn the base probability of a sample belonging to a larger class is higher than a smaller class. In this example, models will bias towards classifying samples as LGW which, given the similar actions required for LGW and ascending/descending a ramp, can increase the number of misclassifications between these activities.

Classification with classical machine-learning approaches

The classical machine learning approaches ANN, KNN, RF, SVM, and DT are trained on the extracted features for a range of window sizes and step sizes. For each model, hyperparameter tuning is applied to optimise the model parameters using tenfold cross validation. This process involves the iterative training of a model whilst changing the parameters used to construct it. In the case of an ANN, the number of neurons in the hidden layer, and number of training iterations are two examples of hyperparameters which can be changed to affect model performance. To reduce the number of iterations required to identify the optimal hyperparameters, searching is performed using Bayes' theorem. This approach is a balance between the typical '*grid*' and '*randomised*' searches, which search through every possible combination of hyperparameters, and a random selection of hyperparameters respectively. Bayesian searching converges on a good set of hyperparameters in a relatively short time, unlike grid searching which finds the best set in an extremely long time, or randomised searching which is fast but not guaranteed to find a good set of hyperparameters. For this example, 25 iterations of the Bayesian search are used.

For testing the model, 15% of the total number of samples are removed from the training set and used to test the model. This randomised approach to a train-test split is more suitable for smaller datasets featuring fewer participants, such as the one in this example. For larger datasets, the data from one or more subjects can be used as the test set, which will give a better insight into whether the model has sufficiently learned the activities independent of the person performing them. As we have just 4 participants in this study, leaving out one person for the testing and another for validation when performing hyperparameter tuning would leave insufficient data for the models.

Confusion matrices for each model with a window size of 500 ms and a step size of 50 ms can be found in Fig. 12.3.

For large overlaps, KNN, RF, and SVM are the highest performing models. At these smaller values of step size, overfitting is more prevalent in limited datasets such as the one in this analysis. This can be seen through the results of the hyperparameter tuning, where both the variables 'k' for the KNN the box constraint 'C' for the SVM were found to provide the highest validation score at a value of 1. On the other hand, smaller overlaps resulted in less overfitting due to the reduction in shared values between the windows. At these values, the SVM outperforms all other models, demonstrating the model's generalisation capabilities and effectiveness with small datasets.

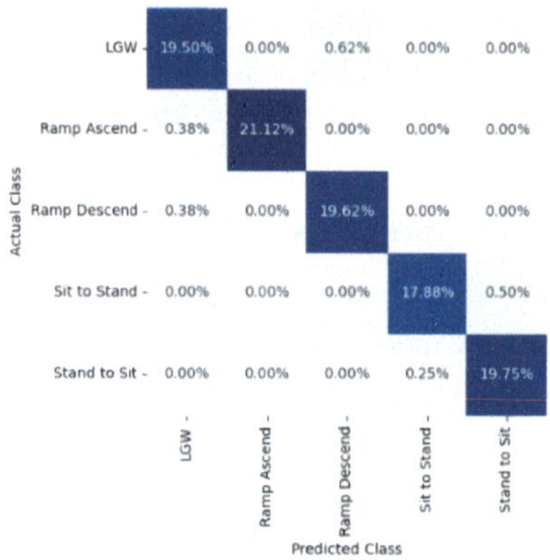

(a) Artificial Neural Network

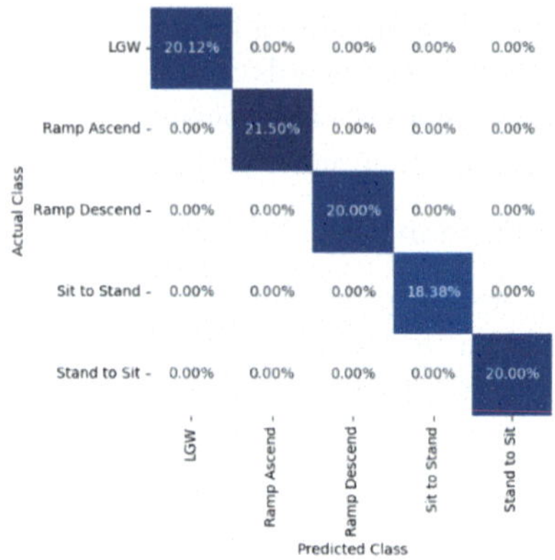

(b) K-Nearest Neighbours

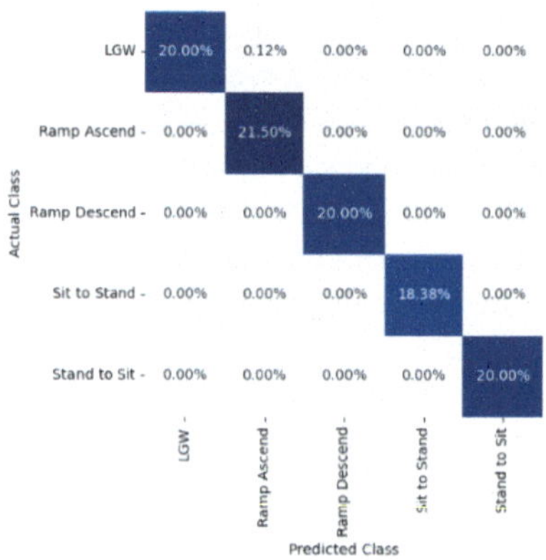

(c) Random Forest

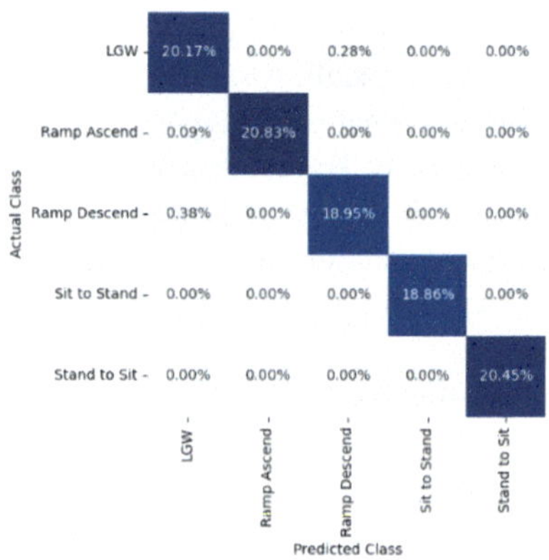

(d) Support Vector Machine

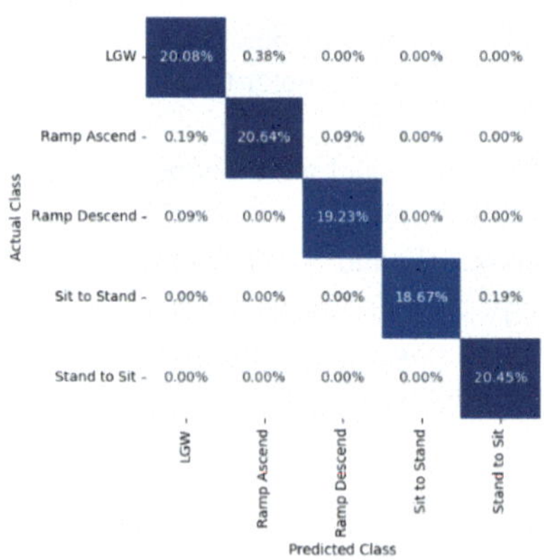

(e) Decision Tree

Fig. 12.3 Confusion matrices for the various classical machine learning models trained on samples generated from windows of size 500 ms with a step size of 50 ms

Classification with deep learning approaches

For the deep learning models: LSTM and CNN, the sliding window process is adapted due to the capacity for these models to automatically extract features from raw sequence/image data. As such, the window extraction approach is performed, but features are not extracted from each window. Instead, the windows themselves are stacked into a 3-dimensional matrix, where each 2D '*slice*' is a window, with the 3rd dimension containing the number of windows. Each widow is given in its entirety to the models as a sample, allowing them to perform feature extraction through the LSTM and convolutional filter neurons. These automatic features are not bound by human statistics, enabling them to be either more or less complex than the features engineered through SNOW, depending on what the model can learn from the data. Typically, this results in deep learning approaches being capable of high accuracies, at the cost of requiring much more data than classical machine learning approaches.

In addition, for the CNN, each window was reshaped to be squarer. This process was done to mitigate the impact of windows being long and thin which was found to result in reduced accuracies for the CNN in some cases. This process was carried out by selecting the middle, or middle 2, factors of the total number of samples in a window and reshaping the window to have a width and height equal to those factors.

As with the classical machine learning approaches, 15% of the samples were randomly left out as a test set. However, hyperparameter optimisation was not performed for the deep learning models. This was due to hardware limitations, as each deep learning model took much longer to train than the classical machine learning models, making hyperparameter optimisation infeasible. Another 15% of the data is used as a validation set with the deep learning models, which allowed the models to stop learning early if there was no reduction in validation loss over the course of 20 epochs.

The CNN consistently outperformed the LSTM in this example, although the architecture for these models was kept simple. More advanced architectures which make use of bidirectional LSTM layers, word embedding, or more complex combinations of LSTM and dense layers may further improve the classification accuracy. These architectural changes can be explored where the time and hardware are available to overcome the computational complexity involved in hyperparameter tuning deep learning models.

Confusion matrices for the deep learning approaches are found in Fig. 12.4.

12.4 Discussion

From the confusion matrices in Figs. 12.3 and 12.4, the classical machine learning models outperform the deep learning models in this example. To investigate this phenomenon further, several different window and step sizes are explored to identify potential contributing factors. Figure 12.5 shows the performance metrics for each model at a range of window and step sizes. From these charts, the effects of varying

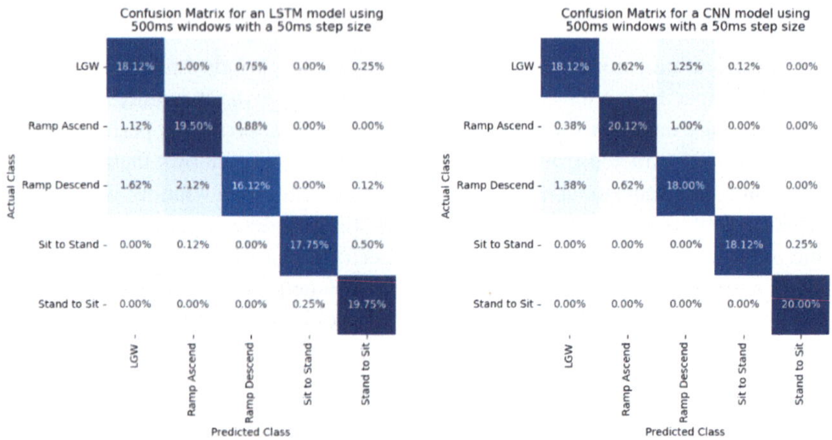

(a) Long Short-Term Memory (b) Convolutional Neural Network

Fig. 12.4 The confusion matrices for the deep learning models trained on windows of size 500 ms with a step size of 50 ms

the amount of available data on the deep learning models is pronounced, with larger window sizes increasing the performance of deep learning, and smaller window steps appearing to increasing the performance of all models due to overfitting. With sufficient data, windows can be made large enough that the performance of deep learning models increases greatly. These plots also show the effect that small amounts of overlap have on overfitting, with larger overlaps causing inflated accuracies due to this issue. One solution to this issue is to introduce '*leave-one-subject-out*' cross validation, though this requires a larger number of participants than were available in this dataset.

Figure 5e and f display the results in the case that the window size and window step are equal, resulting in fully non-overlapping windows. In this case, overfitting risk is at a minimum, as no between-window data bleeding occurs. These examples clearly highlight SVM as the most capable classifier where generalisation is concerned.

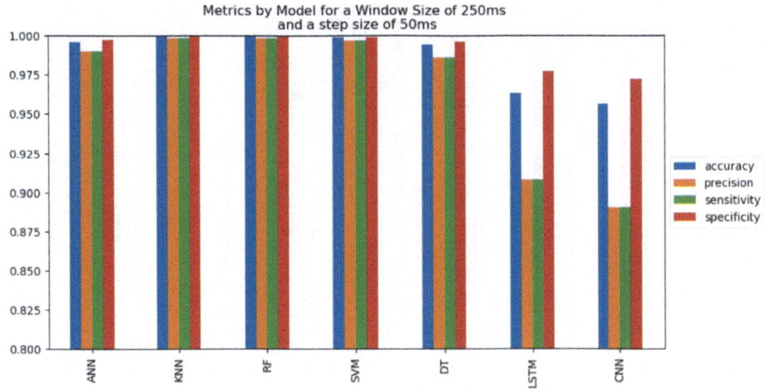

(a) Performance metrics for a window size of 250ms and a step size of 50ms.

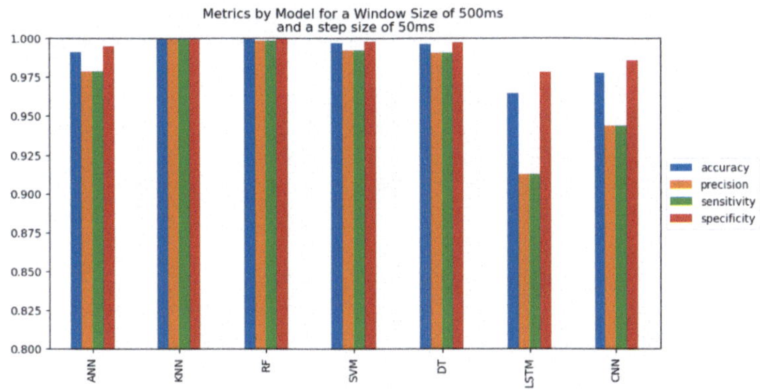

(b) Performance metrics for a window size of 500ms and a step size of 50ms.

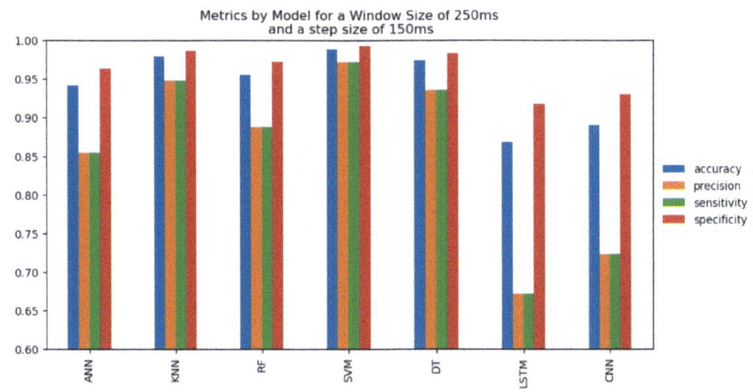

(c) Performance metrics for a window size of 250ms and a step size of 150ms.

Fig. 12.5 The performance of each model with a range of window sizes and step sizes

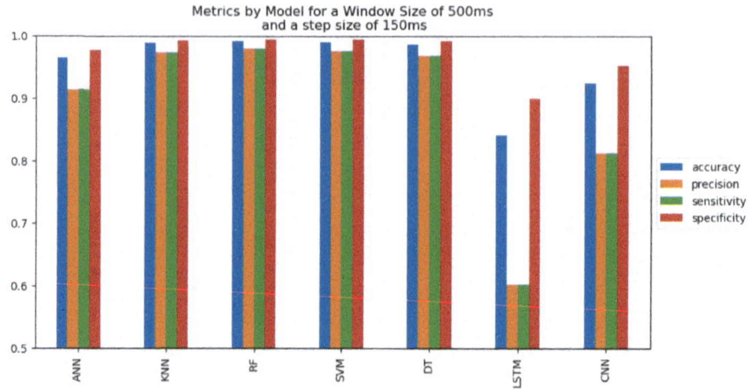

(d) Performance metrics for a window size of 500ms and a step size of 150ms.

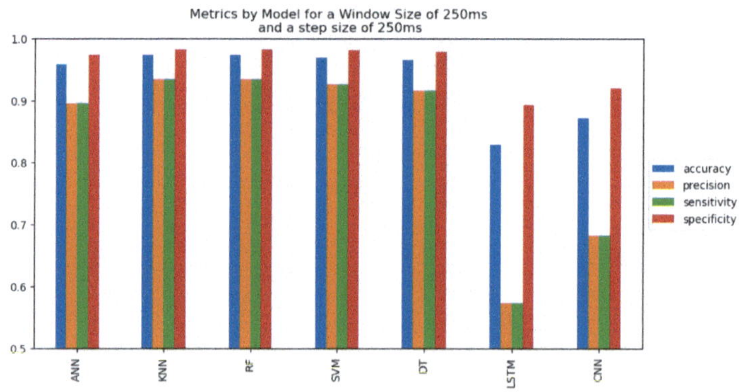

(e) Performance metrics for a window size of 250ms and a step size of 250ms.

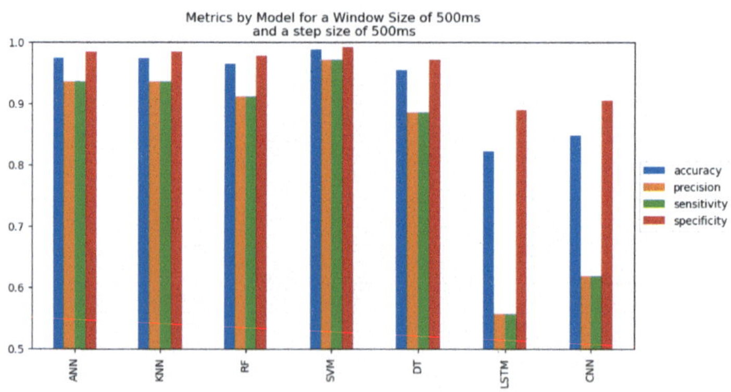

(f) Performance metrics for a window size of 500ms and a step size of 500ms

Fig. 12.5 (continued)

12.5 Conclusion

The chapter provides a case study as an example for the application of AI in an aspect of assistive robotic exoskeletons. In this example, an application of the use of AI in achieving high-accuracy activity recognition in gait analysis is explored, demonstrating the capabilities, advantages, and disadvantages of various models and how the performance of these models changes with the amount of available data, and the chosen pre-processing methods. For smaller values of step size, overfitting is a risk due to the high number of windows which share similar values for max, min, mean, etc. At these smaller values, all classical machine learning models perform well. When increasing the step size to be equal to the window size, SVMs are the highest performing classifier due to their capabilities to generalise well to small datasets through the optimal positioning of the separating hyperplane using support vectors. Deep learning models demonstrate reduced accuracies in this example due to the small amount of available data but show a clear trend towards higher accuracies when sufficient data is made available to them. This case study demonstrates how AI can be used for adaptive control and fall prediction and prevention in assistive robotic systems. AI are also increasingly used in many other mechatronics and robotic systems as we witness advanced technologies all around us nowadays and in the future.

References

1. Jain AK, Jianchang M, Mohiuddin KM (1996) Artificial neural networks: a tutorial. Computer 29(3):31–44
2. Hemalatha K, Rani KU (2017) Advancements in multi-layer perceptron training to improve classification accuracy. Intl J Recent Innov Trends Comput Commun 5(6):353–357
3. Attal F, Mohammed S, Dedabrishvili M, Chamroukhi F, Oukhellou L, Amirat Y (2015) Physical human activity recognition using wearable sensors. Sensors 15(12):31314–31338
4. Noble WS (2006) What is a support vector machine? Nat Biotechnol 24(12):1565–1567
5. Meyer D, Leisch F, Hornik K (2003) The support vector machine under test. Neurocomputing 55(1):169–186
6. Hubel DH, Wiesel TN (1968) Receptive fields and functional architecture of monkey striate cortex. J Physiol 195(1):215–243
7. O'Shea K, Nash R (2015) An introduction to convolutional neural networks, arXiv preprint arXiv:1511.08458 @ arxiv.org/abs/1511.08458. Accessed September 2024
8. Albawi S, Mohammed TA, Al-Zawi S (2017) Understanding of a convolutional neural network, 2017 Intl Conf Eng Technol (ICET), Vol 10
9. Yu X, Qiu H, Xiong S (2020) A novel hybrid deep neural network to predict pre-impact fall for older people based on wearable inertial sensors. Front Bioeng Biotechnol 8:63–63
10. Zhang Z (2016) Introduction to machine learning: k-nearest neighbors. Ann Trans Med 4(11):218–218
11. Hu L-Y, Huang MW, Ke SW, Tsai CF (2016) The distance function effect on k-nearest neighbor classification for medical datasets. Springerplus 5(1):1304–1304
12. Deng Z, Zhu X, Cheng D, Zong M, Zhang S (2016) Efficient kNN classification algorithm for big data. Neurocomputing 195:143–148

13. Halim A, Ibrahim AA, Awad MI, Atia MR (2020) Optimization of sensor number for lower limb prosthetics using genetic algorithm, 2020 IEEE International conference innovative trends in communication and computer engineering (ITCE), pp 210–215
14. Wold S, Esbensen K, Geladi P (1987) Principal component analysis. Chemom Intell Lab Syst 2(1):37–52
15. Abdi H, Williams LJ (2010) Principal component analysis. Wiley Interdisc Rev Comput Stat 2(4):433–459
16. Rokach L, Maimon O (2005) Decision trees, data mining and knowledge discovery handbook. Springer. pp 165–192
17. Song Y-Y, Lu Y (2015) Decision tree methods: applications for classification and prediction. Shanghai Arch Psychiatry 27(2):130–135
18. Zurbuchen N, Bruegger P, Wilde A (2020) A comparison of machine learning algorithms for fall detection using wearable sensors, 2020 IEEE International conference artificial intelligence in information and communication (ICAIIC), pp 427–431
19. Oshiro TM, Perez PS, Baranauskas JA (2012) How many trees in a random forest? Proceedings 8th international conference machine learning & data mining in pattern recognition, pp 154–168
20. Biau G, Scornet E (2016) A random forest guided tour. TEST 25(2):197–227
21. Harris A, True H, Hu Z, Cho J, Fell N, Sartipi M (2016) Fall recognition using wearable technologies and machine learning algorithms, IEEE International conference on big data, pp 3974–3976
22. Medsker LR, Jain L (2001) Recurrent neural networks: design & applications (eds). CRC Press international series on computational intelligence. (Vol 5)
23. Hochreiter S, Schmidhuber J (1997) Long short-term memory. Neural Comput 9(8):1735–1780
24. Sak H, Senior A, Beaufays FJ (2014) Long short-term memory based recurrent neural network architectures for large vocabulary speech recognition, arXiv preprint arXiv:1402.1128 @ arxiv.org/abs/1402.1128. Accessed September 2024
25. Sun L, Kang S, Li K, Meng H (2015) Voice conversion using deep bidirectional long short-term memory based recurrent neural networks, Proceedings IEEE international conference acoustics, speech & signal processing, pp 4869–4873

Chapter 13
Impact of Future Digital Twins for Mechatronic Systems

Roland Rosen and Stefan Boschert

Abstract The chapter provides a brief overview of the historical development of the digital twin and explains related terms. The first focus is the consideration of digital twins as simulation-based, executable models. These can be used in the development of mechatronic systems and in the form of software solutions for various tasks in operation and service. Based on expected future requirements for mechatronic systems, a digital twin methodology is presented to support those involved in developing mechatronic systems and applications for their use. To this end, an extended concept is presented on how the digital twin should be developed over the life cycle of the mechatronic system. The use of digital twins in linking different value chains is also described.

13.1 Introduction

The topic of digital twins has been subject to a significant change in recent years due to the beginning of the establishment of new technologies, such as artificial intelligence, and its increased use, e.g., in smart grids [1, 2] and industrial applications [3]. The use of digital twins has become more complex and diverse. Examples include using digital twins to exchange information along the supply chain in the automotive industry [4] and implementing a Digital Product Passport via digital twin based on the Asset Administration Shell [5]. This is also reflected in the definition of digital twins. While writing the paper [6], the authors created a collection of digital twin definitions [7], which in August 2024 had 112 entries. This shows that a unified definition is not possible. However, two aspects can be found in most definitions. Firstly, it is a virtual ("digital") representation of a (planned or already existing) physical counterpart ("twin") with which, secondly, the digital twin is in a specific

R. Rosen (✉) · S. Boschert
Siemens AG, Garching, Germany
e-mail: roland.rosen@siemens.com

S. Boschert
e-mail: stefan.boschert@siemens.com

P. Hehenberger and D. Bradley (eds.), *Mechatronic Futures*,
https://doi.org/10.1007/978-3-031-83571-1_13

exchange. In this chapter we will introduce the digital twin as a concept pragmatically applied in all life cycles of a mechatronic system and very different scenarios. One perspective here is the expected further development of mechatronic systems and the resulting requirements for the further concept development of the digital twin and its technology.

As with all products and solutions, the planning, development, production and use of mechatronic systems are also about the three familiar goals of cost, quality and time. Due to COVID-19 and geopolitical changes, these are significantly supplemented by the three additional goals of resilience, sustainability and innovation. Implementing the digital twin concept can and will be essential for the improved achievement of objectives. We want to demonstrate this in this chapter.

The technology and concept of the digital twin can be applied to (almost) any type of component, product, system, factory, infrastructure and process. A comprehensive view on general basic research is given in [8]. The focus in this chapter on mechatronic systems makes it possible to take a more concrete look at interdisciplinarity from the areas of mechanics, electronics and computer science. Already today and increasingly with a view to the future development of mechatronic systems, some significant expansions are important here. Instead of just mechanics, you have to consider (at least) other areas of physics, otherwise a fully automatic coffee machine, for example, would not be a mechatronic system. Furthermore, the area of computer science has expanded significantly. The realization of mechatronic systems no longer only involves software-based control systems, but also extensive applications that include complex simulation models and artificial intelligence (AI) methods. Further points include these systems' connectivity, internet capability, and their increasingly modular structure (system-of-systems). In short, we expand the view to cyber-physical systems, because a strict restriction to a classic definition of mechatronics only makes limited sense from a practice- and industry-oriented perspective.

In the following sections, we discuss the different viewpoints on digital twins and related terms and look at the characteristics of (future) digital twins and their impact on (future) mechatronic systems. We propose a digital twin methodology that intends to support stakeholders in the creating and using mechatronic systems in their various tasks. Furthermore, we discuss the effects in the individual life cycle phases and the possibilities across life cycles and value chains.

13.2 Specific Viewpoints on Digital Twins

13.2.1 Digital Twin History and Related Terms

The concept of digital twins originates back to the works of Michael Grieves [9] in 2002. It is based on the idea that digital information about a physical system could be treated as an entity of its own. This digital information gives relevant information about the physical system and is linked to it through the entire system lifecycle. In

2010, this was first mentioned as digital twin in [10]. Over time, several Digital-X were introduced to highlight specific aspects. Popular Digital-X's are digital models which are characterized by the fact that the dataflow between the model and physical object is purely manual. The digital shadow is characterized by automatic data flow from the physical object, whereas the digital twin has bidirectional automated data exchange. In light of the original idea of digital twins this distinction is not helpful, as a digital twin requires a continuous exchange of information along the whole lifecycle.

Grieves introduced a more helpful distinction oriented on the lifecycle of the digital twins in 2017 [11], where he proposed digital twin prototype (DTP) and digital twin instance (DTI). The DTP contains all information necessary to describe and produce a physical version of the DTP. This can include requirements, 3D CAD, bill of materials, bill of processes, simulation models, etc.

The DTI describes a specific instance of the product throughout its lifecycle. Depending on the actual usage of the product different information is relevant. However, most of the general information of the DTP is now concretized to the individual asset. For example, the CAD is enhanced with the actual dimensioning and tolerances, the service record will describe all maintenance operations and replaced parts, operational states are captured from actual sensor data, and future states can be predicted via adopted simulation models. So, the DTP can be seen as a description of the object "as planned", whereas the DTI is the description "as built".

Another concept related to the digital twin is the digital thread: The digital thread is a digitalized and automated sequence of specific business workflows within the product or production lifecycle. As such it enables tracing all relevant information about an object. Taking an electric motor in a pumping station for example:

The control system of the pumping station issued a warning about the reduced performance of the pump. During the maintenance it was discovered that a wrong bearing was used in the motor, which caused the malfunction. Tracing down the business workflows the supplier of the motor and from there, the bearing supplier could be identified. Going the other direction along the business workflows, all instances with gears of the same batch can be identified for a potential check. The digital thread thus requires a common and shared understanding (semantic integration) of data related to objects or processes, which are part of the digital twin.

13.2.2 Executable Models as Core Pillar of Digital Twin

An essential pillar of the digital twin is an executable model that describes the behavior of the physical twin. There are several ways to derive these models. The simplest way is to use existing data and knowledge and derive a heuristic. A more sophisticated way is to use methods from data analytics to derive the model. Both methods require a sufficient amount of data or previous experience. As these kinds of models are mainly based on interpolating between datapoints, the extrapolation quality beyond the range of existing data is limited. In particular, when several dimensions of input data are present, it becomes difficult to tell if a new datapoint is inside or already outside of known data and also in the valid range of the model. Further, it is difficult to give a good estimation of the model error, as no stringent error estimation is available. However, data-based models are used successfully for suitable applications due to their simplicity. A continuous update and recalibration of the models based on additional data collected during the operation phase can continuously improve the model quality further, at least for the "usual" operating conditions, whereas the description of failure states can become difficult if they are sparse.

The numerical simulation of the physical equations that describe the physical object offers an alternative approach. This approach has been well developed over the last decades. Most sophisticated products are developed using the power of advanced numerical simulation like FEM for structural mechanics or FVM for CFD problems to mention a few. The models used during the development can be adapted to become the "calculation engine" of the digital twin (see Fig. 13.1).

The models used during the development can in most cases not directly be used without modifications, which can be identified as the step from DTP to DTI. During the design phase, the simulation aims to improve the design and validate the requirements. In particular, general scenarios (like most probable operating conditions or worst cases) are usually assumed as input for the simulations for a general-purpose object like an electric drive. In the operation phase, the purpose of the simulation changes. The objective of the digital twin mainly defines it. This determines which physical effects must be considered to a given accuracy and in a predetermined time. This could lead to the necessity to combine the results of several different models. Further, the specific environment, operation mode and real-time sensor information also need to be considered. Consequently, the same type of object can have different digital twin models or a combination thereof, with different capabilities depending

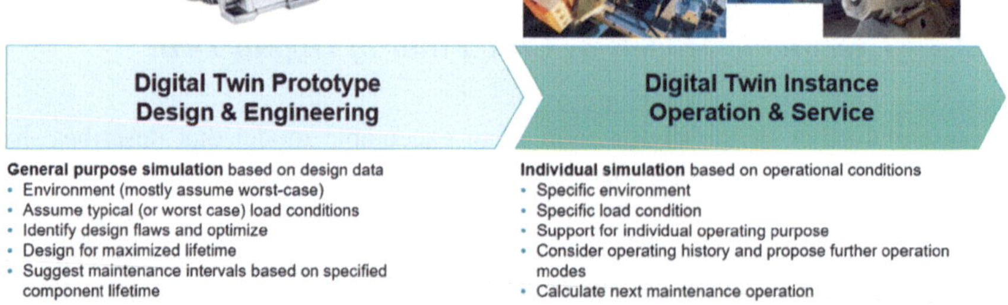

Fig. 13.1 The scope of simulation model changes from DTP to DTI

on the final application. As models always simplify reality, they focus on specific aspects and neglect irrelevant parts for the purpose for the current focus. Therefore, the results of a simulation are always specific to the given problem and usually cannot answer questions beyond its scope. So, it always has to be questioned if the chosen combination of models is appropriate to the specific purpose of the digital twin. This will again be illustrated by an electric motor, that is used in different applications:

In a pumping station the motor drives water pumps. A digital twin of this motor is used to evaluate sensor signals of a vibration sensor to identify bearing damages before they lead to a severe failure of the pumping system. For this application a Finite Element Model from the design phase can be used to create artificial vibration signals for the current operation condition, that can be compared with the signals from the sensor. A severe deviation of the signals is an indicator of an emerging failure.

In a different scenario, the same drive is used in a conveying system to deliver raw materials to a blast furnace. Similar to the other scenario, the real sensor values are compared with those of artificially created ones. But here, the energy consumption of the motor is monitored. This requires that the digital twin comprises the motor and the whole conveyor system, including its load. A derivation of the measured and the simulated power can indicate, e.g., wet raw material (due to higher weight the power consumption for the conveyor increases) and issue a warning for the operator.

In an even larger scenario of a parcel distribution system, the electric drives are still the main components of the physical realization. Still, the goal of the digital twin shifts toward the description of the system behavior. For the system model only generic parameters like power consumption, torque or hours of operation are relevant.

13.3 Future of Digital Twins for Mechatronic Systems

13.3.1 Requirement from Tomorrow's Mechatronic Systems to Digital Twins

The reason for the expanded requirements for digital twins lies in three trends that have already begun. Firstly, in many cases, mechatronic systems will hardly differ from cyber-physical systems. Secondly, sustainability and resilience goals give rise to new requirements. Thirdly, executable digital twins are emerging, i.e., digital twin applications that represent product components or even products in their own right:

- In future, mechatronic systems will increasingly incorporate features of cyber-physical systems. This means that, in addition to mechanics, at least other areas of physics will have to be simulated in the digital twin. In computer science, the mechatronic system will have more powerful applications beyond software-based control and automation systems. These will use methods from simulation, data analysis and AI. Mechatronic systems will also increasingly have connectivity to other systems and the internet (e.g., downloading recipes for the food processor or updating a device in the machine tool).

Figure 13.2 shows on the left side a typical representation of a mechatronic system consisting of the physical part, the actuators and sensors and the information processing with the usual representation for material, energy and information flows.

In the middle, this representation is expanded to include the digital twin of this mechatronic system, which can be exchanged with other digital twins or IT/SW-systems. In the figure on the right, the representation of the mechatronic system is compressed, the actuators and sensors are added to the physical system, the change of term from information processing to SW-based processing emphasizes the software part and ultimately a representation is created that corresponds to that of cyber-physical systems. A digital twin—shown in turquoise—can exist for this CPS. Since this digital twin, implemented as software, is now in exchange with the software part of the CPS, from a practical perspective, these two software parts can also be combined. Mechatronic systems for which digital twins exist are migrating to CPS.

- Combined with these developments, mechatronic systems will have a modular, partially hierarchical structure. The individual subsystems, which can be mechatronic systems, will communicate with each other. However, these subsystems can also implement external communication. An example could be the energy analysis and optimization of a drive.

In Fig. 13.3, the compressed representation from Fig. 13.2 is used and the basic representation from [12] is reused to represent a networked or hierarchical system that consists of several mechatronic systems (or CPS). The distinction between material and energy flows is dissolved in the selected abstraction. Examples of such networked/hierarchical systems are machine tools, consisting of various drives,

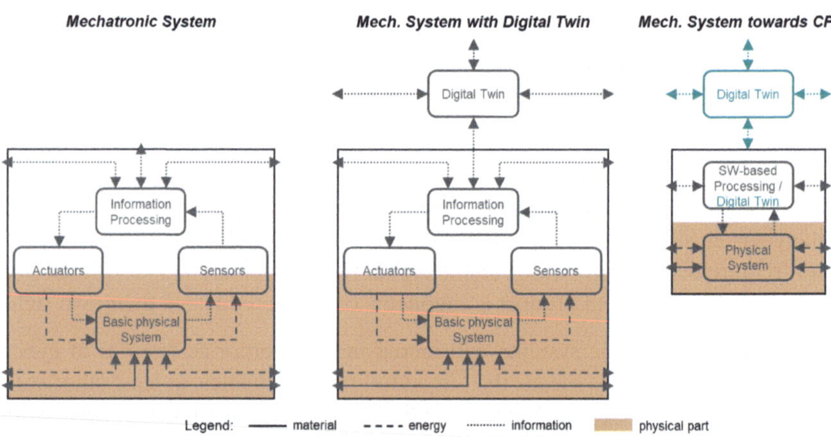

Fig. 13.2 From mechatronic to cyber-physical systems and the role of digital twins

subsystems and parts, or modular production systems (e.g., additive manufacturing), which comprise many (autonomous) production units. Here too, digital twins can exist for such networked/hierarchical systems. With the migration from mechatronic systems to CPS outlined in the point above and the approach of integrating the digital twin into the software part of the CPS, networked digital twins are created. From a practical perspective, the unity of networked digital twins can be understood as a digital twin of the entire system. This distributed digital twin shows an alternative implementation option, so there should be no theoretical or terminological conflict because there are two digital twins for a system (the explicitly shown digital twin and the distributed digital twin).

- Concerning the different definitions of the digital twin and the understandings underlying these definitions, the digital twin can also be a software application. The digital twin application can be an integrated part of the mechatronic system (and, for example, take on extended control functions) or represent a separate application that enables additional functions (for example to increase the flexibility and adaptability of the mechatronic system). In both cases, the digital twin becomes a product that must meet quality criteria. The user or customer will also expect an explicit description of the functionality.
- The topic of sustainability also affects mechatronic systems. In many cases, the ideas and implementations of the circular economy are based on digital twins.

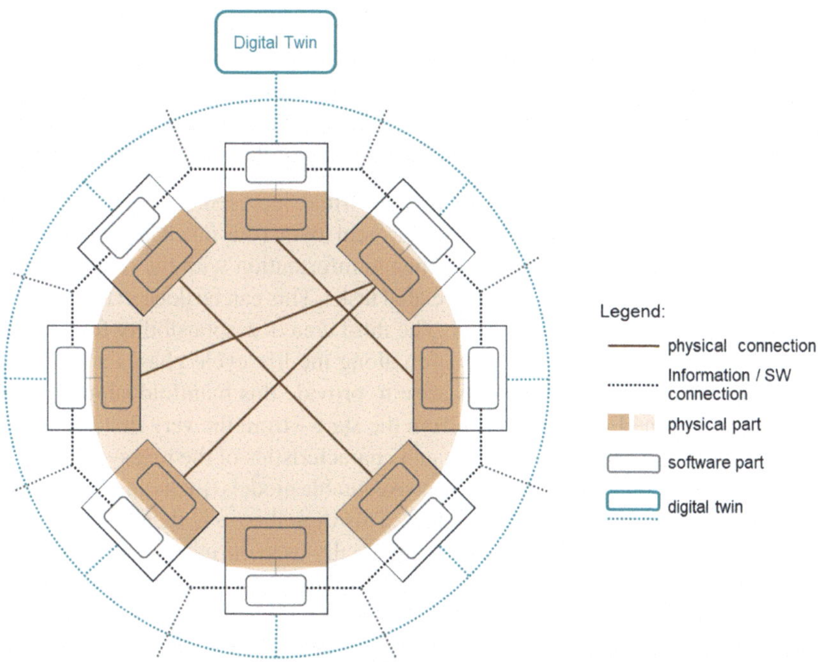

Fig. 13.3 A networked / hierarchical mechatronic system and its digital twin

These are product passports but also solutions for the six "Rs" (Reduce, Reuse, Repair, Remanufacture, Recycle, and Recover)

- The aspect of cross-company collaboration is deferred in many definitions of the digital twin. In practice, however, this is an existing requirement. This means that digital twins or key artifacts of a digital twin must be exchanged. Or to put it more precisely, a (standardized) digital twin becomes an exchange object in the supply chain.

13.3.2 Proposal for a Digital-Twin-Methodology for Mechatronic Systems

The Digital-Twin-Methodology proposed here intends to support the different objectives in the application area of mechatronic systems. This support relates to the participants in the assigned value chains and the various tasks involved in creating and using mechatronic systems. The view of the digital twin and thus the understanding of what a digital twin is or should be is strongly influenced by the area of responsibility. We consciously differentiate between the following task areas in a rough classification:

1. Creation support: conception, design and engineering tasks in the development of a mechatronic system (e.g., a validation of a planned function),
2. Improved usage: creating solutions for the use of the mechatronic system (e.g., a software solution for energy-optimized operation)
3. Effective collaboration: solutions for challenges in the cooperation of different partners in one or more value chains (e.g., the exchange of complex, extensive information between two partners in the life cycle of the mechatronic system)

In the first task area, the digital twin is a virtual representation of a planned mechatronic system; in the second area, the real twin (the mechatronic system) exists. Therefore, the digital twin can exchange information with the real twin and thereby carry out operational or service-related tasks. This can be done in the form of user support or in an automated manner. The third area of responsibility focuses on the (cross-partner) exchange of information along the life-cycle phases and across the value chains. Various requirements arise to provide this manifold support. The digital twin is designed and developed from the start—from the very first life-cycle phase—to capture all relevant properties and characteristics of the real system. This digital twin usually consists of descriptive or executable models (including simulation models) of the planned mechatronic system. This digital twin is then continuously updated during the manufacturing process and finally used during the use of the real system. Here, it is mainly an executable digital twin, i.e., software that knows the status of the real twin by exchanging data with it and can provide precise (automated) support. Artificial intelligence methods are increasingly being used in this area of application, as the real twin has sufficient data available for learning methods.

Another important aspect of the digital twin is the exchange of information between the various partners along the life cycle of the system and between different value chains. The digital twin can offer a standardized way to do this. The Asset Administration Shell designed by the IDTA (Industrial Digital Twin Association e. V.) uses the digital twin in this way [13]. The digital twin enables seamless communication and collaboration between the various stakeholders, regardless of whether they are involved in the development, production, use or maintenance of the system. Using standardized interfaces and data formats, information can be exchanged efficiently to improve effectiveness and efficiency in the various value chains.

This leads to the proposal of our Digital-Twin-Methodology for mechatronic systems. The digital twin refers to a description of a mechatronic system[1] by a set of well-aligned, descriptive and executable models. It is a semantically linked collection of all relevant digital artifacts, including design and engineering data, operational data, and behavioral descriptions.

The digital twin has several characteristic features:

- It evolves along the lifecycle—use and content should be pre-planned but are open for additional, un-planned artifacts.
- It integrates the currently available and commonly required information and knowledge.
- It addresses (proto)type and instance aspects.
- It is synchronized with the physical world—the type of synchronization changes over the different lifecycle phases.
- It offers an efficient way to generate new services and applications—beside extended solutions in development, new offerings in operation and service phases promise new business opportunities.
- It closes the feedback loop—on the one hand information from the field can be used to design and develop system variants or new mechatronic systems, and, on the other hand, information is available for end-of-life purposes such as recycling and reuse.
- It is a tool for cross-partner exchange of information along the value chains and across different value chains.

[1] It is the intention of the authors to generalize mechatronic systems to components, products, systems, infrastructures, or processes. This is not discussed further in this paper.

13.4 Characteristics of Future Digital Twins for Mechatronic Systems

13.4.1 The Digital Twin Concept for Life-Cycle Aspects

A characteristic of the proposed Digital-Twin-Methodology is the structuring as a set of well-aligned, semantically linked artifacts. We will use the term artifact as a synonym for all kind of information from a single datapoint to a complex behavior model like a simulation or an AI application. The digital twin contains the artifacts, that are widely required for the digital twin's user in the current phase, and it is kept for the possible use in the subsequent phases, see Fig. 13.4. This figure is an evolution of previous work, started from 2016 in [14]. Figures with comparable statements can also be found in other publications, e.g., by C. Wagner from 2017 [15]. The digital twin is not a replacement for a PLM / PDM database or a SCADA data management system. For example, it does not store all design alternatives that engineers have developed, evaluated, and discarded during the development of a system. Only an essential selection of these artifacts is included and semantically linked in the digital twin. Which artifacts are stored depends on the kind of mechatronic system, the application domain, and the intended solutions which should be developed in later phases. Regarding the implementation of digital twins, they can store a copy of the artifacts or provide links to the original IT system. In anticipation of the section on the digital twin in the value networks, it should be noted that this information can also be stored as Asset Administration Shells with standardized sub-models.

The figure shows that applications (apps) can also be created from or with the artifacts, which can be used for a specific task in the life cycle phase. Because the term digital twin has no fixed definition, these applications are often equated with

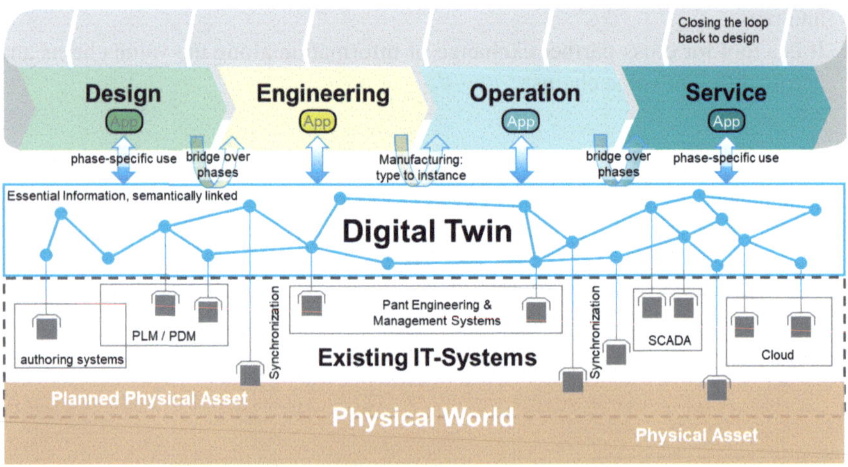

Fig. 13.4 Concept of digital twin along the life-cycle phases

the digital twin. This particularly applies to the operating and service phase if the application (or digital twin) solves a specific task in synchronization with the real twin. The illustration also shows that the digital twin helps to bridge the life cycle phases through its continuous further development. The transition in the production of the mechatronic system from type to instance is indicated.

On the one hand, the digital twin is an open concept that allows the addition of artifacts at any time. On the other hand, it requires pre-planning and management of the content, see Fig. 13.5. This principal illustration shows the evolution concept of a digital twin along the life cycles. Starting from a mixture of connected planed and existing artefacts (left) the complexity of the digital twin increases through further artefacts and novel mutual relations (middle). This covers all application areas (product, production and performance). The preplanned artifacts are replaced gradually by existing artifacts (right). An example from the manufacturing industry can illustrate this. If, for example, a new machine tool is to be commissioned and a virtual commissioning solution must be carried out for this purpose, then it is easy to anticipate that corresponding simulation models of all system elements and the associated equipment (e.g., tools, workpiece holder) must be available as well. Appropriate pre-planning makes sense so that no artifacts are forgotten. Suppose specific components are exchanged in the later course of commissioning, then the twin must also be updated to ensure, for example, that the appropriate models are available and can be reused in a training simulator or for operation support (e.g., production planning) in a whole production scenario. The mix between pre-planning, update of artifacts, and new ad hoc included content forms the evolving character of the digital twin.

This example also suggests that information (digital twins) about the mechatronic system as a product can be used in its production and in its usage phase as well (e.g., to improve performance).

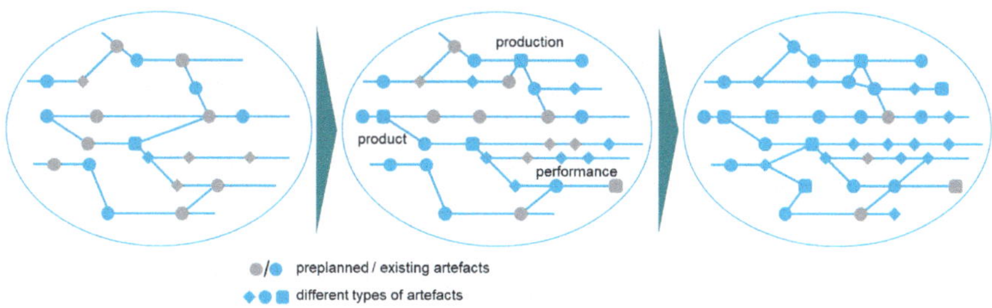

Fig. 13.5 Concept for evolution of a digital twin along the life cycle (principal illustration)

13.4.2 Manifestations of Digital Twins Along its Life Path

Figure 13.4 shows the digital twin along the general life cycle with the design, engineering, operation and service phases. This diagram illustrates the classified task areas (see Sect. 13.3.2), such as “creation support” and “improved usage” for mechatronic systems. The “effective collaboration” task area is only shown indirectly in the diagram. Therefore, in this section, we look at the manifestations of the digital twin. The following section will discuss this in the context of value chains.

The digital twin initially provides all relevant information about the planned mechatronic system in the development phases. These can be requirements regarding planned functions, as well as properties, concept designs, or models. Regarding mechatronic systems, this includes artifacts at the system level and for the individual disciplines involved (mechanics, electronics, IT). As already described, this information should be up to date, i.e., representing the current valid status of the planned mechatronic system. In a figurative sense, this digital twin—the set of artifacts—is synchronized with the imaginary real twin. The digital twin thus provides the basis for using the artifacts required for a specific task from the stored artifacts. This can be an animated 3D model for carrying out motion studies or a simulation model for calculating system behavior. From a pragmatic and industrial user perspective, both examples can be understood as digital twins of the same mechatronic system. The more scientific-methodological conflict with the uniqueness of the twin is not considered here. In operation and service, the main task of a digital twin is to improve the use of the mechatronic system. Here, the digital twin is executed as a software application. Through synchronization (data exchange) with the real mechatronic system, the digital twin can use all processes that can be implemented by software. In particular, this includes data analyses and AI processes that can operate on the data of the real twin. The appearance of these digital twins can vary:

- As a software application, the digital twin offers functionality that improves the operation of the mechatronic system or supports its service. The digital twin can be an independent product with a corresponding business model.
- The digital twin is required to operate the mechatronic system, which transforms the mechatronic system into a CPS (see Fig. 13.3) or at least it is a mechatronic-system-integrated-digital-twin.
- If there are many instances of the mechatronic system, e.g., machine tools, printers, fully automatic coffee machines, then digital twins can be implemented as fleet solutions that offer the users of the individual system additional functions.

Many partners are already involved in the development process of the mechatronic system. This is where the digital twin offers a reasonable basis for increasing digital value creation. By understanding the DT methodology presented, artifacts—in the complete range between simple data points and independent applications—can be exchanged between the partners. The corresponding standardization of these artifacts increases their direct usability by the user. The Asset Administration Shell is particularly suitable for this, as it standardizes both the content (standardized submodels) and the path to exchange. The digital product passport and the general use of information for circular manufacturing are linked to the digital twin and its ability to provide information across the life cycle. The data from the product passport, which in principle is a specific digital twin, can be used for improved reuse or the best possible recycling, for example.

13.4.3 Building Blocks for Simulation-Based Digital Twin Applications

The examples given in 13.2.2 show the range of different configurations that are possible for individual simulation-based digital twin applications. In Fig. 13.6 a non-exhaustive collection of building blocks for digital twin applications is shown. This collection is divided into a section describing the virtual representation of the physical asset and a section for digital twin integration services. Each section is divided into 3 different groups, starting from bottom with common infrastructure elements like databases, versioning and updating infrastructure to mention a few. The next group are basic models that describe the general phenomena relevant for the application. On a next level, these models are tuned toward the specific application needs. For example, if the basic models to describe fluid flow are the Navier–Stokes equations then the specific models are the addition of suitable turbulence models to describe the physics in the appropriate approximation. The digital twin integration services start with a group to configure the model to the concrete situations—a digital twin model usually has to cover a range of different operating conditions—and provided with interfaces to the actual environment, e.g., sensor interfaces. The last two groups contain the elements to configure, execute and access the digital twin for the concrete digital twin application. It is worth to mention that for a typical digital twin application not all building blocks from all groups are used. Depending on the actual use-case, the necessary elements are chosen to.

13.4.4 Linkage of Value Chains by Digital Twins

Figure 13.4 basically represents the Product Lifecyle Management (PLM) which is the systematic process of managing the product — in our case, a mechatronic system—over its whole lifecycle. This includes ideation, design, and engineering tasks as well as its manufacturing and service up to the disposal of the mechatronic system. The produced mechatronic system is the output of the production system. The Production System Lifecycle Management (PSLM) describes the development of the production system that manufactures parts, components and ultimately the mechatronic system. This includes using the means of production and the entire system

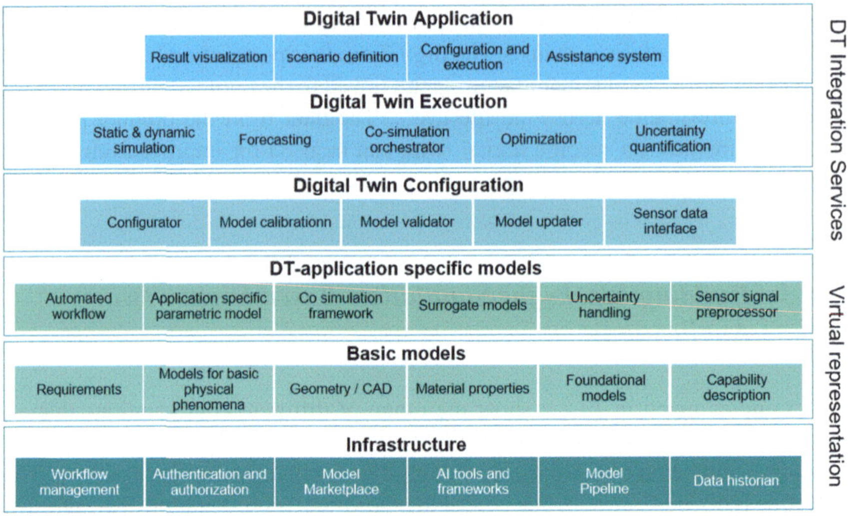

Fig. 13.6 Non-exhaustive collection of building blocks for simulation-based digital twin applications

itself with its environment, such as the building. In many cases, production systems are rebuilt for operational reasons (e.g., to produce new mechatronic systems), so that a re-design accompanies the engineering phase in the operation of systems. A further value chain is the Supply Chain Management (SCM), which describes the strategic coordination of goods and information flows. Important aspects are the logistics of raw materials and supply parts, warehouse aspects, and the delivery of manufactured mechatronic systems. Figure 13.7 shows the value chains with a crossing in the manufacturing of the mechatronic systems which is also the operation of the production system.

In each of these value chains several partners are involved. These are, for example, development partners, suppliers of parts and components, the factory and in the end the customers who use these mechatronic systems or, in turn, integrate them into their system-of-systems and sell them to their customers. This illustrates that the value chains are connected, and precisely tailored information exchange is required for process efficiency. The digital twin approach will significantly increase the effectiveness of information exchange, as more complex artifacts can be exchanged, e.g., complex or executable models. Figure 13.7 illustrates five examples of the exchange of digital twins:

1. A digital twin of the planned mechatronic system is made available for production preparation to be able to plan and validate conversions at an early stage.
2. Parts, components or subsystems required for the production of the mechatronic system are made available to the suppliers as digital twins at an early stage in order to ensure the timely start of production.

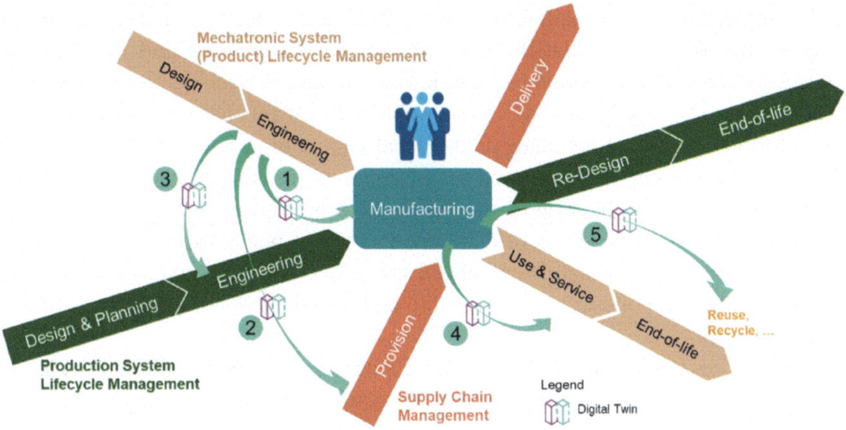

Fig. 13.7 Digital twins across value chains

3. The planned mechatronic system (e.g., a machine tool) can be the equipment of a planned production system. The digital twin of this production unit can be used in the engineering of the production system.
4. Information on the produced mechatronic system is used for service purposes.
5. Information on the manufacture of the mechatronic system, e.g., on the materials used, on dismantling, or information for recycling, is provided.

The above examples show the many possible uses of digital twins. The economic benefit increases if the digital twin is standardized regardless of its scope, modular structure, size, and complexity and is up to date, i.e., correctly depicts the real twin.

13.5 Summary and Outlook

This chapter presents a Digital-Twin-Methodology intended to support the stakeholders involved in creating and using mechatronic systems. It considers potential future needs for digital twins arising from emerging trends in mechatronic systems: the convergence of mechatronic and cyber-physical systems, the goals of sustainability and resilience, and the emergence of executable digital twins.

The proposed Digital-Twin-Methodology aims to support the different objectives and tasks in applying mechatronic systems, including design, engineering, use, and cooperation among partners in different value chains. The digital twin is described as a collection of well-aligned models that evolve throughout the system's lifecycle, integrating relevant information and offering efficient ways to generate new services and applications. It also facilitates the exchange of information along and across value chains.

The importance of sustainability will continue to increase. For mechatronic systems, this also means that the materials used in production and the footprint in terms of CO_2 and water must be recorded and (as an overarching social objective) reduced. This also applies to the operation and use of mechatronic systems. This means further increasing demands on the development of systems and the provision of digital information. The increased use of digital twins can provide support here.

As one possible application of the digital twin is to trace the whole life of the product, the re-use of products or components thereof becomes easier. Looking at the information on the usage it becomes easy to decide at the end of the nominal lifetime, which components are still good to be re-used maybe in a different product or application.

The advances in AI technologies will have a significant impact on digital twins. On one side data driven models, that represent the actual behavior of the physical system will become more powerful. The generation of high-level system models will be simplified as well, as the digital twin is a semantically linked collection of all artifacts and therefore is the perfect ground for AI methods. And also, the creation of physics-based models will greatly profit from AI technologies. In particular, generative AI will ease the creation of those models, by simplifying the cumbersome set-up of the simulations, e.g., by suggestion of similar existing models for a particular problem. However, it will be important at least in the near future that the experienced simulation expert will check the applicability and validity of those models to guarantee a valid result.

One of the most important aspects for future digital twins is that digital twins will become independent products as stand-alone solutions or applications. This offers the possibility that corresponding business models for digital twins will emerge. Digital twins must be carefully planned and meet the requirements applicable to a product.

References

1. Djebali S, Guerard G, Taleb I (2024) Survey and insights on digital twins design and smart grid's applications Fut. Gener Comp Syst 153
2. Jafari M, Kavousi-Fard A, Chen T, Karimi M (2023) A review on digital twin technology in smart grid, transportation system and smart city: challenges and future. IEEE Access. pp 1–1. https://doi.org/10.1109/ACCESS.2023.3241588
3. Boschert S, Heinrich C, Malik V, Rosen R, Söllner U (2024) On the importance of simulation and digital twin for industrial applications, In: Simulation foundations, methods and applications (ed) Grieves, M.; Hua, E. Springer, ISBN 978–3–031–69106–5
4. Catena-X (2024) Semantic Layer/digital twins https://catena-x.net/en/offers-standards/semantic-layer. Last access August 03
5. Industrial digital twin association, DPP4.0—the digital product passport for industry 4.0, https://dpp40.eu/. Last access August 03
6. Dalibor M, Jansen N, Rumpe B, Schmalzing D, Wachtmeister L, Wimmer M, Wortmann A (2022) A cross-domain systematic mapping study on software engineering for digital twins. J Syst Softw 1(193):111361. https://doi.org/10.1016/j.jss.2022.111361

7. Wortmann A (2024) Digital twin definitions, https://awortmann.github.io/research/digital_twin_definitions/. Last access August 03
8. Foundational Research Gaps and Future Directions for Digital Twins, ISBN 978–0–309–70042–9, National Academies of Sciences
9. Grieves M (2002) Completing the cycle: using PLM information in the sales and service functions [Slides]. SME Management forum, Troy, MI
10. Piascik R, Vickers J, Lowry D, Scotti S, Stewart J Calomino A (2010). Technology Area 12: materials, structures, mechanical systems, and manufacturing road map, NASA Office Chief Technol
11. Grieves M, Vickers J (2017) Digital twin: mitigating unpredictable, undesirable emergent behavior in complex systems, In: Kahlen FJ et al.: Transdisciplinary perspectives on complex systems, Springer
12. Hehenberger P, Bricogne M, Le Duigou J, Zheng C, Eynard B (2016) Using meta-models to manage information change in the design process of systems of systems. Int J Product Lifecycle Manag 9(4):285–304
13. Industrial Digital Twin Association (2024) Digital twin—the heart of industry 4.0 https://industrialdigitaltwin.org/en/technology. Last access August 03
14. Boschert S, Rosen R (2016) Digital Twin—The Simulation Aspect. In: Hehenberger P, Bradley D (eds) Mechatronic futures. Springer International Publishing, Cham, pp 59–74
15. Wagner C, Grothoff J, Epple U, Drath R, Malakuti S, Grüner S, Hoffmeister M, Zimmermann P (2017) The role of the industry 4.0 asset administration shell and the digital twin during the life cycle of a plant. In: Proceedings of the 2017 IEEE conference on emerging technologies factory automation (ETFA 2017), Limassol, Cyprus

Chapter 14
How Money Makes Lifecycle Management Go Round

John Ryan

Abstract Product lifecycle management is often presented as a seamless, closed-loop process—but in reality, it is a fragmented and complex series of independent actions, driven by profit at every stage. Each stakeholder, from manufacturers to end-users, operates with their own interests in mind, prioritising what benefits them financially. When maintaining or repairing a product becomes too costly, it's often skipped in favour of disposal and replacement, with taxpayers left to bear the burden of managing waste. Traditionally, manufacturers have designed products to meet the needs of their initial client, focussing on price and functionality. However, evolving demands now emphasise repairability, usability, cybersecurity, and the looming challenge of planned obsolescence. Each stage of a product's lifecycle, from extraction to end-of-life, offers an opportunity for profit—but also potential conflict, as different parties' interests' clash. This chapter delves into the forces driving each stakeholder across a product's life, uncovering the compromises made at every step. It also highlights the tension between maximising profitability and delivering sustainable, user-friendly products, revealing that, ultimately, the idea of a '*perfect product*' remains elusive.

14.1 What is Lifecycle Analysis

The standard methodology for assessing a product's lifecycle typically focuses on its environmental impact, the raw materials required, and the energy needed for its creation. Life Cycle Assessment (LCA), also referred to as Life Cycle Analysis or Cradle-to-Grave Analysis, is a standardised methodology designed to assess the environmental impact of a product or service at all stages of its lifecycle. This lifecycle starts from raw material extraction and continues through manufacture, distribution, use, and finally, disposal or recycling at the end of its life [1].

The LCA study covers the environmental impact of a product by undertaking a thorough review of the energy and materials used across the supply chain of a

J. Ryan (✉)
SymTerra, London, UK
e-mail: John@symterra.com

P. Hehenberger and D. Bradley (eds.), *Mechatronic Futures*,
https://doi.org/10.1007/978-3-031-83571-1_14

product at each stage in the life cycle, documenting its findings. Part of the output is the cumulative environmental impact score in terms of CO2-equivalent for different activities or stages. This can help identify areas of improvement that can be made to better reduce its environmental impact.

A standard example is the high-level lifecycle for a bicycle shown in Fig. 14.1:

This assessment example revealed that the manufacturing phase accounted for the majority of the environmental impact, with the wheels, fork, and frame being the components with the highest impacts. The study concluded that using carbon fibre instead of steel or aluminium could reduce the overall environmental footprint [3].

There are different procedures for conducting LCAs with the accepted approach outlined in the 14,000 series of environmental management standards of the International Organisation for Standardisation (ISO). However, key issues with the LCA remain, particularly regarding the consistency of the methodology, the difficulty and costs associated with assessments, and the latitude given to auditors' views or the economic and political incentives of the sponsoring entity. An LCA conducted by 10 different parties could yield 10 varying results. While the standards aim to create a level playing field, the guidelines are not overly restrictive and are open to the assessor's interpretation.

For instance, in the Nestlé Nespresso Lifecycle Assessment comparing their Nespresso system to other coffee-making methods, the results appeared biased to favour Nespresso. The assessment suggested that a Moka pot could have a lower impact than the Nespresso Original coffee system only under specific conditions. This raises questions about the reliability of LCA results when a ~ 10–15 component pot with minimal moving parts and no electronics is considered to have 86% of the CO2-equivalent of a small coffee machine with more than 150 components shown in Fig. 14.2 [4]. Such discrepancies challenge the standards' goal of providing a level playing field for comparing products and lifecycle stages.

LCA focuses solely on the environmental impact of a product, not the financial, market, or social aspects. These aspects are covered more thoroughly in Product Line Analysis (PLA) and similar methods such as Life Cycle Inventory (LCI) analysis, which creates an inventory quantifying raw materials, energy, atmospheric/land/water emissions, resource uses, and other releases over the product's lifecycle.

While these approaches are becoming standardised, they do not address the key drivers and influences that different parties have on various stages of product design, development, or upgrade/redesign. They do not address the fundamentals of good product design and the balance required between conflicting needs and requirements (e.g., cheap but robust, advanced but easy to use). This chapter does not review the techniques involved in LCA, PLA, or related assessments but aims to understand the reasons behind sometimes-conflicting design decisions made during product development and their impact throughout a product's lifecycle.

The perfect product would be cheap to manufacture using standard off-the-shelf raw materials, have a high-profit margin, be inexpensive to buy but simple to use, and desirable to consumers. It would have high uptime (availability/reliability), be easy to maintain with readily available, affordable spare parts, simple to disassemble and repair, and have a long lifespan. At the end of life, it would be easy to recycle or

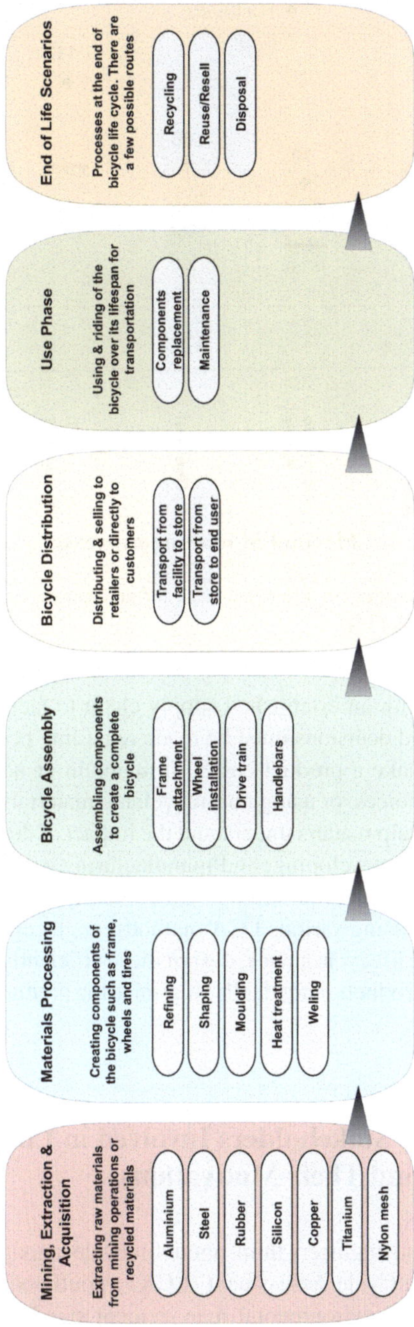

Fig. 14.1 Lifecycle for a bicycle (after [2])

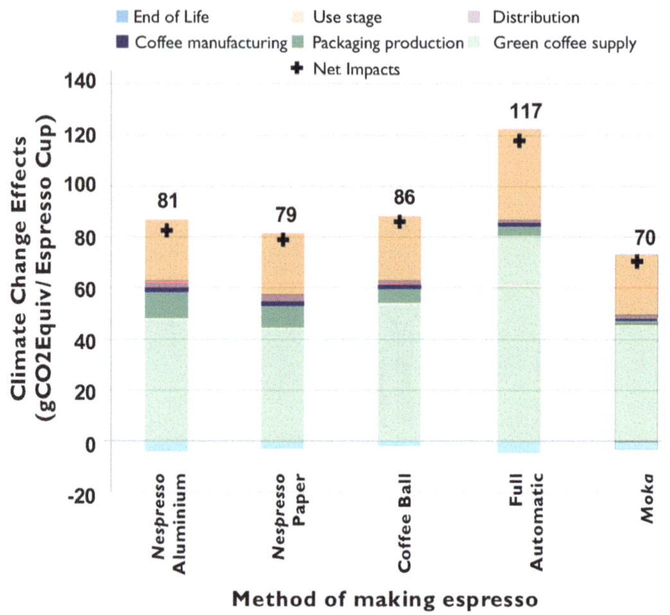

Fig. 14.2 Greenhouse gas emissions per life cycle stage for an espresso cup for selected coffee systems on the Swiss market (after [5])

reuse most components, with an established supply chain to facilitate this process. However, compromises and decisions must be made regarding priorities.

Deciding whether to make a product easier to maintain or not maintainable at all, simplifying material choices, or using non-recyclable materials all involve trade-offs. This chapter aims to help readers understand the impact of design decisions and the challenges in designing, developing, and manufacturing a competitively priced product that is market-comparable in terms of functionality while being optimised for the entire lifecycle. By using various LCA methods, engineers can better understand the needs at different lifecycle stages, design more efficiently, consider the full lifecycle when procuring products, and ideally minimise the product's environmental footprint.

14.2 Section 1—The Stakeholders Involved in Lifecycle Management and Their Motivations

When product designers and engineers think about lifecycle management, they often consider it in terms of Life Cycle Assessment (LCA) as outlined in standards such as the ISO 14000 series of environmental management standards. This approach considers the full lifecycle, from extraction of materials to decommissioning, disassembly, and recovery/recycling and disposal of the remaining materials. Throughout the various stages of a product's life, various stakeholders are involved in its design, creation, and use, ranging from plastic widgets to medical devices and aircraft airframes. This chapter focuses on the people involved in the various lifecycle stages, their specific interests, drivers, and needs, both conflicting and complementary. Other parties covering regulations, market trends, and taxation (on materials, goods, import/export duties, etc.) also significantly impact a product's design and lifespan but will not be considered here.

In this section, we will consider the drivers for each stakeholder at various lifecycle stages, how these drivers shape the overall product, and how they impact the lifecycle of the product as it moves through the various stages.

In an ideal world, the lifecycle is simple and linear, with minimal friction between stages. In reality, for the vast majority of products, it is broken up into a significant number of fragmented and disconnected steps involving unrelated parties whose only shared interest is the product itself. Each stakeholder has their own drivers and motivations, which do not always align with a more simplistic model. In short, lifecycle management is far from simple or straightforward.

14.2.1 Extractors

Extractors create the raw materials that go on to make the products. These stakeholders can be at the start, through the raw extraction or creation of materials from virgin sources, or at the end, using partial or fully recycled materials as the base to form new standardised compositions and sizes for the manufacturing stage to use.

The key drivers at this stage are the costs of obtaining the raw materials, production costs, and the demands and prices in the market for certain compositions, quality, sizes, and, in some cases, '*green*' attributes expressed as a percentage of recycled materials. As LCA and other standards become more popular, extractors are increasingly seen as a significant source of a product's CO2 emissions and are under pressure to reduce their overall energy consumption and environmental impact without corresponding price increases to offset the rising cost of manufacturing.

Using steel as an example, according to the World Steel Association, the steel industry accounts for 7–9% of global carbon dioxide emissions [6]. This covers scope 1 (direct emissions during production) and scope 2 (e.g. offsite energy suppliers used for production) emissions only and rises to 28% of global emissions when accounting for scope 3 emissions (e.g. supply chain and customer emissions not covered in scope 1 or 2). The primary method of creating steel is by converting iron ore into iron and then into steel inside blast furnaces using coking coal as a fuel, which emits about 2 tonnes of CO_2E for every tonne of steel produced.

One emerging solution involves electric arc furnaces (EAF), which are much smaller and more flexible than blast furnaces. They are also much cheaper to build. Because they use electricity to melt scrap, making steel using an EAF can reduce emissions by up to 85% compared with a blast furnace [7]. Today, electric arc furnaces account for about 29% of global production. However, challenges still exist, such as limited scrap supplies and unreliable quality, meaning the quality of steel won't meet requirements for certain industries, such as automotive. Electric costs remain an issue for the industry, and while there is a move towards green power and green steel, there is still no strong economic case for the industry to move more quickly towards this [8].

14.2.2 Manufacture / Assembly

Manufacturers are primarily interested in the creation (either manufacture or assembly), selling, and distribution of the product. They also further develop products to ensure they are desirable, competitive, and profit-generating. The product should meet relevant regulatory and legal requirements at a minimum. Once the core requirements are achieved, the focus switches to production costs and minimising these as much as possible. This can be to the detriment of future stages in the product lifecycle. For example, achieving specifications such as water resistance in Bluetooth earbuds is often done not by complex gaskets but by partially filling the unit with glue to ensure waterproofing, protect sensitive components and reduce vibration. This eliminates expensive QA processes, maximises performance, and keeps production costs down. However, it makes the product non-maintainable and difficult to recycle at the end of its life due to the inability to cost effectively disassemble and recover materials [9].

Once sold, for most products, the manufacturer is no longer involved. However, in some cases, such as John Deere with farm equipment, Pratt & Whitney with jet engines, and HP for printers, the manufacturer's interests lie in value-added services through selling consumables, providing regular servicing directly or via approved dealerships, and offering data/analytics services for the product's lifecycle.

14.2.3 Owner

However, once sold and for most products, typically the manufacturer is no longer involved. The owner is price-sensitive and will likely have several different manufacturers to choose from to achieve their required end goal. Whether that is a dishwasher or an assembly line robot, there will be choices to consider, weighing features, quality, availability (uptime), and operational costs for the product's active life stages. Within this category, there are two types of owners to consider:

Owner-Operator: The company purchasing the product uses it to deliver its own services or product development (e.g., purchasing vehicles or equipment).

Leasing Companies: These entities buy the asset and lease it to a third party to operate in their own business. The owner's driver here is price and the ability to profit from leasing the product with minimal financial risk (e.g. vehicle leasing companies).

Within each type of owner, a requirements specification and a cost–benefit analysis will be undertaken. When there is a change in the product being used, such as switching from Boeing to Airbus, additional considerations must be made. These changes can introduce friction in the client's transition, so the alternative's cost–benefit must strongly outweigh the decision to stick with the current product range. Key considerations include the work required to integrate the new product into the current working environment, decommissioning previous products, roll-out including training and support, and business impacts such as operational costs, changes in service level uptime, availability of parts/consumables, warranties, and changes in external systems.

If the Owners specification is met, and the current product is still in production and not cost-prohibitive, there will not be a strong drive by owners to adopt new products even if it has benefits for others in the product's lifecycle due to risk introduced and minimal commercial incentive to change.

14.2.4 Owner/Operator / Operator / Passive End User

For any operator, the core interests are simplicity of use, safety, reliability, and low operating costs. The easier it is to operate, the higher the value generated from the product. The easier it is to train users, the lower the staffing costs and the faster new staff can be on boarded.

Where the operator isn't within the owner's organisation, interests can diverge quickly. Cheaper, difficult-to-use products will impact the operator's output to the extent that the owner will be forced to address/replace the product due to potential breaches in contract for uptime of the unit. This is typically where the operator is a labour supply subcontractor whose interests are driven by margins on labour supply rather than work quality or productivity rates.

A passive end user is someone benefiting from the product but not directly engaged in operating it. Their key interest is in it functioning as designed. For example, airline passengers or hotel guests. The seamless operation of the mechatronics world around them is something most are oblivious to until something goes wrong. Their voice can damage the reputation of the manufacturer and impact the owner's profit margins, providing a negative feedback loop for manufacturers and potentially handing market share to competitors (e.g. for an airline company and aircraft manufacturer after a plane crash).

14.2.5 Maintainer

Maintainers are becoming a critical part of driving better lifecycle management. For products requiring maintenance during their life, there will be a maintenance component to consider along with any consumables required to operate. The simpler and easier it is to service and maintain a product with good access to spare parts, the longer the life of that product and lower maintenance costs.

Whether servicing a car, recalibrating instruments, or replacing high-wear components, the needs of the maintainer should be considered by the manufacturer in the design and disassembly methods. The balance here is the reduced operational costs for the Owner once purchased versus the time to design and the potential additional cost to manufacture which may be something owners are not willing to pay for. Owners may not fully understand the needs of the many assets they own and might not provide easy access to products such as air conditioning systems, assembly line motors, or suitable working areas for maintainers. The faster and easier the maintainer can access, assess, and repair/service a product, the less downtime the owner/operator will experience.

Some maintainers will be specialised in certain manufacturers' products and certified to service them, from cars to lifts. In some cases, manufacturers specify that their products can only be serviced by approved companies and parts via authorised suppliers/dealers, or it will invalidate any warranties in place. For some clients with less complex assets, it may simply be a matter of the lowest price or nearest company winning the work to maintain their assets. Maintainers in this case price based on experience with that product, known issues, accessibility to the assets, availability of spares, and the complexity of the work involved.

There has been growing pushback from owners against manufacturers specifying that they can only use authorised dealerships and OEM components. For example, Deere & Co., one of the world's largest makers of agricultural equipment, has faced legal action from farmers for the right to repair their own equipment. John Deere had for years refused access to software, designs, and electrical schematics, citing IP and trade secret issues with making such information public.

In terms of consumers, the UK introduced right-to-repair rules that legally require manufacturers to make spare parts available for electrical appliances. In 2022, Apple started a '*self-service repair*' programme so '*customers who are comfortable*' can fix their own devices. This covers replacing the batteries, screens, and cameras in their phones. Other companies like Fairphone are going further and creating modular devices that can be repaired simply by the phone's owner. For consumers looking to buy devices with longer life and more environmentally friendly credentials, there are a growing number of review websites that also include environmental impact assessments.

Maintainers are key to extending the life of products and ensuring they remain functional and efficient throughout their lifecycle. Their role and requirements must be integrated into the design and manufacturing process to achieve a truly sustainable lifecycle management strategy.

14.2.6 End-of-Life Management

This final stakeholder consideration covers refurbishment, reuse, recycling, and waste management. Their primary goal is to maximise the value recovered from discarded items while minimising the waste that must be sent for final disposal. To achieve this, end-of-life managers salvage usable components for reuse, extract recyclable materials (metals, plastics, glass, etc.), and ensure the safe disposal of any hazardous or non-recyclable residues in line with environmental regulations. This role is crucial for reducing the overall environmental footprint of products by reclaiming parts and materials that would otherwise become waste. Their profit margin is dependent on the ability to recover materials or components cheaply for maximum resale value and minimising disposal costs of non-recyclable materials.

The profit margin is strongly influenced by decisions made during a product's design and manufacturing phases. The key cost is generally that of labour due to the levels of manual handling needed to dismantle and process end-of-life products. This has led to much of the Western World's recycling being offshored to poorer countries, sometimes with more lax environmental regulations for waste disposal. For example, in December 2005 the French aircraft carrier Le Clemenceau was infamously intended to be sent to Alang in India for disposal while containing large amounts of asbestos and other hazardous materials. This resulted in the ship being returned to France to eventually be broken up in Hartlepool where they were equipped to manage the toxic materials.

Additionally, design choices, which prioritise reduced manufacturing costs, or product performance over end-of-life considerations, can also significantly reduce the portion of materials that can be reclaimed. In these cases, more material may end up as waste, underscoring how upstream design decisions reverberate through to the final stage.

As sustainability and circular economy goals gain prominence, the role of end-of-life managers is increasingly recognised as an integral part of the Extractors stage. Their position at the conclusion of the product life cycle gives them a unique perspective on how products fare when dismantled and processed for disposal or recovery. In fact, involving end-of-life specialists during the product development process can provide a feedback loop to manufacturers and designers to help create products that retain a higher residual value at the end of life which benefits all stakeholders.

14.2.7 Section 1 Comments

Thus it is apparent that each stakeholder involved in lifecycle management operates with distinct motivators shaped by their specific roles and responsibilities within the product lifecycle. Extractors are driven by the cost and market demand for raw materials, while striving to reduce their environmental impact without compromising on price. Manufacturers focus on efficient production, regulatory compliance, and profitability, often balancing immediate cost savings against long-term maintenance and end-of-life considerations. Owners prioritise cost—benefit analysis, operational efficiency, and seamless integration, whether they are direct users or leasing companies. Operators and passive end-users emphasise ease of use, safety, and reliability, with their satisfaction influencing both the owner's profitability and the manufacturer's reputation. Maintainers seek accessibility and ease of service to minimise downtime and operational costs, often pushing back against restrictive manufacturer policies. Finally, end-of-life managers aim to maximise the value recovered from products, whether through recycling, refurbishing, or disposal, influenced by the design choices made earlier in the lifecycle. Each stakeholder's unique drivers highlight the complexity of lifecycle management and underscore the need for a holistic approach to minimise environmental impact while balancing economic and operational realities.

14.3 Section 2—Using the R Framework to Design Better Products for a Longer Lifespan

Downstream Scope 3 emissions encompass the use/operation, maintenance, and end-of-life processes for a product. A core component of extending a product's lifespan is simplifying its maintenance process. When considering maintenance, we refer to the nine R-strategies developed by PBL Netherlands Environmental Assessment Agency and Utrecht University's study on circular economy transition as set out in Fig. 14.3 [10]. These strategies focus on smarter product use and manufacture, extending product life, and end-of-life recycling and material recovery.

14.3.1 Circular Strategies Within the Production Chain, in Order of Priority:

Using this framework, we will examine the challenges of maintaining a product to ensure optimal uptime, availability, and efficiency in a cost-effective manner. Designing a product involves considering the needs of clients and end-users to ensure the system is fit for purpose and competitive in terms of initial price and ongoing running and maintenance costs. Often, what the product designer envisions for how or where it will be operated and maintained doesn't always align with reality, introducing additional unknowns that can't always be accounted for.

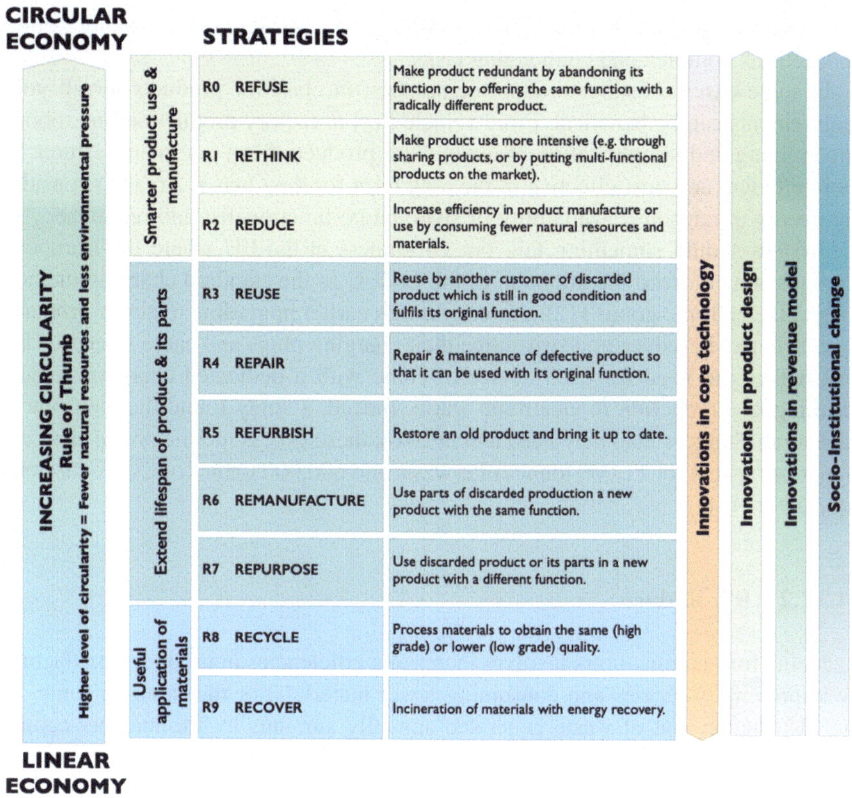

Fig. 14.3 From a linear economy to a circular economy (after [10])

14.3.1.1 R0 & R1: Refuse and Rethink

The first R-strategies are to Refuse and Rethink. Typically, this is for the owner to consider, but it can also apply to the manufacturer when streamlining product design and optimising components and sub-assemblies. Owners can assess their current assets to determine which ones are no longer needed, have redundant functions, or could potentially consolidate multiple functions across fewer products. Process optimisation is key to reducing operational costs, from energy usage and training simplification to reducing factory floor area requirements and streamlining maintenance.

Rethinking can involve extending the life of the product to extract more value from it. For example, in the aviation industry, determining the optimal age to withdraw aircraft from service involves considering fuel pricing, maintenance costs for older aircraft, airworthiness directives, flight range expansion plans, customer satisfaction, and changing market conditions [11]. However, this approach presents challenges, such as lack of spare capacity to expand or cover when other products break down.

Owners need products that are both specialised and multifunctional, making Refuse and Rethink complex and challenging tasks.

In some cases refusing to adopt or use best in class for products for all your requirements can be benefical. From vehicle fleet to factory production line robots, streamlining and simplification of assets with products from one manufacturer be more efficient and cost effective in the long term for the Owner. Areas to consider are energy efficiencies, availability of spare parts, functionalist and redundancy in the system should something fail. For consumers in the EU where the European Commission enforced the adoption of the USB-C as the standard charging method for small electronic goods [12], this encourages earlier upgrading to newer products by consumers to reduce and streamline their charging plugs and cable's needs. This will reduce the need for each device to come with a dedicated charging device, resulting in a reduction in electronic waste. Indeed, a study found that consumers have more chargers than they need and in 2018, these excess mobile phone chargers were responsible for 11,000 tonnes of e-waste and 600,000 tonnes of CO2 emissions equivalent [13].

14.3.1.2 R2: Reduce

Reducing for manufacturers involves increasing efficiencies in product manufacture by improving processes and consuming fewer materials for the same outcome. In the EU, the amount of waste generated annually amounts to 5 tonnes per capita, prompting a focus on more stringent waste laws aligned with the Circular Economy Action Plan [14].

Breaking down a product into its key parts—components, sub-assemblies, assemblies, and the final product—helps in understanding what to consider when designing (for manufacturers) or assessing products for purchase (for owners). Components should ideally avoid specialist manufacturing where possible to minimise cost and Quality Assurance issues. Sub-assemblies can be streamlined to be lighter, simpler, and more efficient, using standardised components to simplify supply chains and maintenance. For other stages within the lifecycle Reduce can simply be using one product instead of multiple specialised ones, however this isn't always possible or advisable in specialised jobs or industries.

14.3.1.3 R3: Reuse

Reusing covers repurposing a product for another purpose once it can no longer fulfil its original function. Examples include converting passenger aircraft to cargo planes [11], repurposing old train carriages for coffee shops, book shops, or holiday homes [15], and creatively reusing old furniture.

14.3.1.4 R4 & R5: Repair and Refurbish

For products designed for repair (R4) or refurbishment (R5), these typically have a long life and are designed with servicing in mind from the outset. Products with designed consumable or replaceable components, such as filtration meshes and batteries, require considerations for ease of access, simplicity of replacement, and availability of parts for the product's full lifespan.

14.3.1.5 R6—R9: Remanufacture, Repurpose, Recycle, and Recover

Disassembly spans the final four strategies. Remanufacture (R6) and repurpose (R7) involve recovering and reusing components, while recycling (R8) and energy recovery (R9) focus on reclaiming materials and energy from products at the end of their life. The ease and cost-effectiveness of disassembly play a crucial role in determining the feasibility of these strategies. Complex disassembly can make repair, remanufacture, or repurpose economically unviable, leaving recycling or energy recovery as the remaining options. In a paper that considered the environmental impact of demolishing a steel structure for disassembly it was found that there was a 6 × reduction in C02E in reusing the steel components versus remitting & new component production [16].

Outside of a manufacturer's downstream Scope 3 emissions, disassembly is critical for two main parties: maintainers performing complex system overhauls (typically for systems with decades of design life) and end-of-life management. For these parties, disassembly offers several benefits: lower time and cost for disassembly, better control of the outcome (e.g., fewer mistakes and breakages if disassembly is difficult), and ultimately, a more profitable undertaking compared to simply discarding items and generating unnecessary waste. However, some products, like computers with soldered on components or Bluetooth earphones with glue covering key components, are not designed for maintainability, disassembly or upgradability. This design choice makes it impossible or unnecessarily complex to replace individual units, leading operators and end-users to discard them, contributing to the growing problem of e-waste. This issue is exacerbated with small electronics like mobile phones and tablets, which contribute significant amounts of e-waste annually due to their design for easy manufacture but impossible recyclability.

A critical emerging issue impacting the goal of achieving a circular economy is software obsolescence. Products that fail due to outdated software need to be better understood in terms of risk and lifespan. For instance, if a product requires a cloud connection that is no longer available, or if the manufacturing company goes out of business and stops providing necessary software updates, the product can become unusable. This presents a significant challenge for owners planning integrated systems of products without knowing the reliability of manufacturers to stay in business and keep software updated.

This risk was highlighted in 2024 when the Fisker car company filed for Chap. 11 bankruptcy, leaving customers uncertain about the safety and maintenance of their vehicles due to the reliance on software updates [17]. The dependence on manufacturers to maintain connected devices' functionality and software for the duration of their lifespan is a focal point of the UK's cybersecurity laws for consumer goods [18]. Moreover, as some manufacturers also become software providers, this introduces additional risks. For example, AT&T in the United States discontinued its 3G service in 2022 which left Volkswagen car owners unable to add their Car-Net service [19]. The average phone has a lifespan of 2.8 years, but cars have an average age of 12.6 years. Sonos experienced reputational damage when a software update for their consumer audio products left users unable to control the volume, effectively rendering the devices unusable [20].

When manufacturers design products with R6 to R9 in mind, considering disassembly and the CO2 impact of end-of-life processes is crucial. A study on aircraft disassembly found that sustainable decommissioning is challenging due to current assembly methods (e.g., use of rivets) [21]. Semi-destructive disassembly methods may increase material recovery compared to traditional destructive methods. Another study on the economic viability of manual disassembly lines for component recovery at the end of a product's life suggested that even as a design and manufacturing analysis exercise, it would benefit manufacturers to align their products with emerging circular economy requirements in the EU and elsewhere [22].

Finally, materials and any treatments applied to them during manufacturing, such as chemical coatings and assembly methods (e.g., glues, rivets), significantly impact their recyclability. Normal usage and designed lifespan—considering wear and tear, environmental factors (indoor/outdoor, temperature range, exposure to water/saltwater, corrosive environments), and expected vibration and impact—are also key considerations for manufacturers. It is worth noting that it is not always possible to develop a more sustainable product due to product function requirements or where it is being used such as biomedical devices where the excessive consumables used is a product of the requirement to reduce spread of disease or cross contamination of test samples.

To design better products for a longer lifespan, it is essential to consider the various R-strategies and their impact on product design, manufacture, and end-of-life management. By focussing on reducing material use, simplifying maintenance, and planning for reuse, repair, and refurbishment, manufacturers can create products that are more sustainable and economically viable over their entire lifecycle.

14.4 Section 3—The Future and How It Will Impact Key Players

14.4.1 Policy and Regulatory Support

There is a belief among some that technology will save us. McKinsey discusses the 4th Industrial Revolution and the use of technology to '*unlock data and digital tools to provide analytical insights for data-informed actions*' [23]. However, digital tools are arguably a distraction that helps sell consultancy services rather than driving change or eco-efficiency in the industry. True change can only come from what matters most—profitability.

Only when extractors, manufacturers, owners, and maintainers have a single viable route to profitability, driven by regulations and taxes that prioritise lifecycle management, CO2 reduction, and environmental compliance, will we see a significant shift in the adoption of lifecycle management. Until then, companies at the forefront of LCM and carbon footprint reduction will struggle to compete with cheaper competitors offering similar functionality but with higher carbon footprints.

Without taxation or grants on a product's footprint across all stages, a LCA will become a tick box exercise for companies to show some green credentials rather than drive organisational decision-making and improved product design. Compromises will be necessary if we are to achieve net zero, the circular economy, and other environmental goals.

The Kyoto Protocol and the Paris Agreement of 2015 established international CO2 emissions goals and national emissions targets as routes to address climate change through carbon emissions reduction [24]. Carbon taxation, which emerged in the 1970s as a type of pollution tax, is an excise duty on fossil fuel use. Research has shown that carbon taxes could significantly reduce energy usage [25]. For countries, it can be a way to raise additional taxes and encourage investment in renewable energy technologies [26]. A 2020 proposal by the UK government forecasted a planned UK-wide carbon tax to generate £17 billion by 2030 [27]. However, challenges from lobby groups, complexities around existing taxes and subsidies, and the impact on disadvantaged groups remain significant hurdles [28].

Companies throughout the lifecycle can reduce their overall carbon tax by minimising their carbon footprint through renewable energy adoption, energy-efficient production techniques, and material reduction. Some will be easy quick wins, however, others require significant capital investment which may make the business uncompetitive due to increased prices.

On the flip side, carbon credits can be generated by reducing emissions, removing pre-existing greenhouse gases, or avoiding emissions altogether [29]. Companies can purchase these credits to offset emissions that can't be further reduced or during the transition period to implementing lower-emission solutions without facing penalties. This can either offset emissions that cannot be reduced any further (e.g., for Extractors or Manufacturing) or during the transition period to implementing solutions for lower emissions without being penalised during this more capital intensive phase for a business. This will also help growth in other industries such as carbon capture and renewable energy.

However, there is a '*green-washing*' risk that as carbon tax continues to rise and with it the market for carbon credits, that companies misuse LCA methodologies to present a more eco-friendly image than reality [30].

14.4.2 Practical Steps for Stakeholders

14.4.2.1 Extractors

The pressure on extractors to provide raw materials will continue to increase due to several factors:

- The United Nations estimates a population growth of over 20% by 2100 [31].
- Per capita CO2-equivalent consumption continues to rise [32].
- Global municipal solid waste generation is expected to increase nearly 80% by 2050 [33].

With growing pressure on energy, water usage, labour supply, and raw material accessibility, the cost of obtaining and processing raw materials will rise. This will make it economically necessary for extractors to engage in R8 and R9, recovering materials or using them for energy recovery to provide electricity for new products.

14.4.2.2 Manufacturing

Pressure from extractors and those involved in recycling and recovery will likely require manufacturers to design products that are easier to disassemble and recycle. Will manufacturers invest in product and supply chain optimisation to become more energy-efficient and reduce water usage thus reducing Scope 1 & 2 emissions? Will more companies adopt design principles around R4-R7? These companies might offer more competitive pricing in the long run or face significant capital investment with little return on investment or simply focus on product lines with higher profit-margins.

The balance between product complexity, lifespan, cost, and other factors is crucial. For example, the cost and energy used to recycle computer chips often outweigh the gains in recovered materials. Designers should consider reducing component and subsystem complexity, such as EV car batteries or simplifying headlight assemblies to make them cost effective to repair.

The R4-R7 support infrastructure for most products is missing. Where it exists and for example for vehicles where there are dedicated garages, yearly vehicle testing and a spare parts supply chain; the infrastructure is only in place due to the need and the financial opportunity for business rather than the collective benefit on the environment. This brings into question the role of government to take a lead in addressing funding and incentivising small goods recycling. However, this is not a popular industry strategy.

14.4.2.3 Owners

The Owner's needs don't change but will be augmented with additional considerations on the right to repair, the usable life of products & as applicable ongoing software support and any cost associated with it. With many devices becoming smarter-data ownership and the ability to access and use it becomes key considerations. Will Owners accept monthly fees for updates/support for Products or start to push for open access to their data and be able to manage internally the products they have purchased?

14.4.2.4 Operator/End User—Consumer Awareness and Engagement

In direct conflict with the Owner's views, where the Operator isn't the Owner, the drive will be for more connectivity, less friction, and shorter training/adoption times.

As products become more complex and labour supply issues reduce the number of Maintainers in the workforce, manufacturers may directly support end-users in self-diagnosing issues, identifying solutions, and, where necessary, ordering parts and dispatching technicians for maintenance support to fix more difficult problems.

The growing voice and influence of the Operator and End User have become significant factors for Owners and Manufacturers to consider. Whether it is a business or consumer-grade product, there is probably a YouTube video on it. This impacts manufacturers, as Owners are influenced by the online views of their Employees, Operators, and End-Users [34].

While ease of use is a key attraction for Owners to ensure operators can use the product, it is a double-edged sword. When it comes to replacement, Owners don't want to undergo significant retraining on new product lines and will likely stick with the existing manufacturer. Equally, for manufacturers, the decision lies between upgrading an existing product or enduring the painful steps of introducing a new product with the associated retraining and retooling.

Boeing's experience with the 737 *Max* is a cautionary tale. Owners had minimal training requirements for pilots to fly this variant. Unfortunately, this cost-cutting measure went too far, resulting in dire consequences and significant reputational and commercial damage for both the Manufacturer and Owners [35].

14.4.2.5 Maintainer

The lower cost of products and higher incomes in first-world countries mean that replacement will continue to be preferred over repair unless government intervention makes repair significantly cheaper and easier. The cost and time of calling a technician to assess, order parts, and fit them for a washing machine are often comparable to buying a new one, discouraging repair.

Maintaining new products is challenging due to their increased complexity. In 2024, Hertz Global's shares slumped 24% due to losses and struggles with electric vehicles (EVs). The Financial Times reported that Hertz's EVs were being damaged at an unexpectedly high rate, proving more costly to repair than traditional cars, with lower-than-expected resale values.

The right to repair issue will continue to be a contentious issue across all industries.

14.4.2.6 Data Aggregators

Data aggregators are an emerging group feeding into lifecycle management. While they don't make, own, operate or use, they are now feeding into lifecycle management of a product—they are the data aggregators.

From health apps taking data from athlete's smart devices through to IoT sensors and Building Management Systems, this group are able to capture and analyse vast amounts of data across different product Manufacturers, Owners & End-Users/ Operators. This data can be used across the entire life of a product helping Manufacturers design more reliable products, understand use cases & market opportunities, and ultimately support Operators identify issues before they become problems and inform Owners/Maintainers which products are not operating efficiently or are in need of a service. However, data ownership and privacy are significant concerns. Most data aggregators sell aggregated data back to manufacturers or other interested parties and charge access fees to the data creators in this case owners and end-users.

14.5 Technological Innovations

Manufacturers with internet-connected devices face more stringent requirements. The UK's Product Security and Telecommunications Infrastructure Act 2022 (PSTIA) and the upcoming EU Cyber Resilience Act will enhance the security of internet-connected products. Will manufacturers reduce product lifespans or ensure products can continue to be used under grandfathered standards?

Manufacturers may simplify control systems and offer plug-and-play upgrades to extend product lifespans. For example, Rivian Automotive optimised their R1 vehicle's electrical and compute platforms, reducing the number of engine control units and cabling. This could pave the way for owners to upgrade vehicle systems and mitigate software issues [36].

Replacement parts manufacture and on-demand 3D printing are emerging markets for manufacturers and maintainers. Licencing 3D designs to approved maintainers for printing and installation can be a win–win, but quality control and setup costs remain challenges.

Manufacturers may generate additional revenue by offering subscription services for features like self-driving car updates and security camera footage storage. However, excessive fees may lead to pushback and regulation, as seen when BMW tried to charge a subscription for heated seats.

14.6 Challenges, Opportunities and Conclusions

Significant pressures exist on all stakeholders to increase product life, reduce CO2E footprints, keep up with technological advances, increase profit margins, and be greener. However no company will be able to financially survive implementing any of these improvements due to the uncomeptetive cost of the final product against cheaper competitors. A core mechanism for driving change is through hitting profit margins, from taxation and grants for compliant products such as carbon taxes and carbon credits, through to changes in product conformity standard requirements to force improvements.

Waste management and repair/recycling infrastructure need to be addressed. Higher landfill taxes for non-repairable products and government support for repair shops could promote a repair-first approach. Leading manufacturers looking to future growth opportunities and getting ahead of the competition should prioritise lifecycle management in their design development to get ahead of upcoming changes both in terms of regulation and consumer needs.

Only time will tell if we can achieve a society where aspirational goods align with lifecycle management goals. We will not get there with LCA alone and the role of subsidies and government will be significant in addition to consumer re-education.

References

1. Koffler C, Amor B, Carbajales-Dale M, Cascio J, Conroy A, Fava JA, Gaudreault C, Gloria T, Hensler C, Horvath A, Humbert S (2020) On the reporting and review requirements of ISO 14044. Intl Life Cycle Assess 25:478–482
2. www.bsigroup.com/en-US/insights-and-media/insights/blogs/lca-and-its-role-in-sustainability/. Accessed August 2024
3. Coelho MC, Almeida D (2015) Cycling mobility–a life cycle assessment based approach. Transp Res Proced 10:443–451
4. www.mavink.com/explore/Nespresso-Vertuo-Parts-Diagram. Accessed August 2024
5. Quantis (2022) Life cycle assessment (LCA) of an espresso cup of coffee made from a Nespresso original capsule compared with other coffee systems in Switzerland. www.nestle-nespresso.com/sites/site.prod.nestle-nespresso.com/files/Summary_CH_B2CComparativeLCAISOReport_2023_Quantis.pdf. Accessed August 2024
6. World Steel Association (2021) Climate change and the production of iron and steel, Policy paper, www.worldsteel.org/wp-content/uploads/Climate-policy-paper-2021-1.pdf. Accessed August 2024
7. Service Steel Warehouse (2022) Electric arc furnaces v traditional blast furnaces, www.servicesteel.org/resources/electric-arc-furnace-vs-blast-furnace#:~:text=Environmental%20impact%20-%20because%20of%20the,carbon%20dioxide%20than%20blast%20furnaces. Accessed August 2024

8. Financial Times (2024) Climate tech explained: low-emission steel plants, 17 July, www.ft.com/content/735d00a8-678f-4e0c-8dcc-9d82518299ed. Accessed August 2024
9. www.ifixit.com/Device/AirPods_2019. Accessed August 2024
10. Potting J, Hekkert MP, Worrell E, Hanemaaijer A (2017) Circular economy: measuring innovation in the product chain, Planbureau voor de Leefomgeving, Pub. No. 2544, www.pbl.nl/uploads/default/downloads/pbl-2016-circular-economy-measuring-innovation-in-product-chains-2544.pdf. Accessed August 2024
11. de Oliveira Junior FS, Fernandes E, Bahiense L, Grandi CM (2021) A practical approach to support end-of-life commercial aircraft parking, market relocation, retirement and decommissioning strategic decisions. Intl J Prod Res 59(17):5144–5163
12. www.europarl.europa.eu/topics/en/article/20220413STO27211/usb-type-c-to-become-eu-s-common-charger-by-end-of-2024. Accessed August 2024
13. European Commission (2021) Impact assessment study to assess unbundling of chargers, Final report, www.op.europa.eu/en/publication-detail/-/publication/90e9a07d-1054-11ec-9151-01aa75ed71a1. Accessed August 2024
14. European Commission (2020) A new circular economy action plan: for a cleaner and more competitive Europe, www.eur-lex.europa.eu/legal-content/EN/TXT/?uri=CELEX:52020DC0098. Accessed August 2024
15. Five unexpected uses for decommissioned trains, www.rail.nridigital.com/future_rail_jan23/five_unexpected_uses_for_decommissioned_trains. Accessed August 2024
16. Broniewicz E, Dec K (2022) Environmental impact of demolishing a steel structure design for disassembly. Energies 15(19):7358
17. Fisker Is Bankrupt. Owners are angry and scrambling for answers, www.insideevs.com/news/724522/fisker-bankruptcy-ocean-ev-owners/. Accessed August 2024
18. UK Government (2024) The UK product security and telecommunications infrastructure (product security) regime, www.gov.uk/government/publications/the-uk-product-security-and-telecommunications-infrastructure-product-security-regime. Accessed August 2024
19. Marshal A (2024) Cars are rolling computers now. So what happens when they stop getting updates? Wired, www.wired.com/story/cars-are-now-rolling-computers-so-how-long-will-they-get-updates-automakers-cant-say/. Accessed August 2024
20. So A (2024) Sonos admits its recent app update was a colossal mistake, Wired, 25 Jul, 25 July www.wired.com/story/sonos-admits-its-recent-app-update-was-a-colossal-mistake/. Accessed August 2024
21. Zahedi H, Mascle C, Baptiste P (2016) A quantitative evaluation model to measure the disassembly difficulty; application of the semi-destructive methods in aviation End-of-Life. Intl J Prod Res 54(12):3736–3748
22. Sergio M, Franciosi C, Iannone R (2022) An approach to evaluate the impact of the introduction of a disassembly line in traditional manufacturing systems. J Ind Eng Manag 15(2):215–232
23. Betti F, de Boer E, Giraud Y (2021) Lighthouses unlock sustainability through 4IR technologies, McKinsey & Co, www.mckinsey.com/capabilities/operations/our-insights/lighthouses-unlock-sustainability-through-4ir-technologies/. Accessed August 2024
24. United Nations (1998) Kyoto protocol to the United Nations framework convention on climate change, www.unfccc.int/resource/docs/convkp/kpeng.pdf. Accessed Autumn 2024
25. Chandler DL (2018) Carbon taxes could make significant dent in climate change, study finds, MIT News, www.news.mit.edu/2018/carbon-taxes-could-make-significant-dent-climate-change-0406. Accessed August 2024
26. World Economic Forum (2021) Addressing climate change through carbon taxes, www.weforum.org/agenda/2021/06/addressing-climate-change-through-carbon-taxes/. Accessed August 2024
27. Shackleton L (2020) Carbon taxes: trickier than you think, Institute of Economic Affairs, https://iea.org.uk/carbon-taxes-trickier-than-you-think/#. Accessed August 2024
28. UK Government (2022) The VAT treatment of passenger transport, www.gov.uk/guidance/the-vat-treatment-of-passenger-transport-notice-744a. Accessed August 2024

29. Wadhawan S (2022) Explainer: what are carbon credits and how do they work?, Earth Org, www.earth.org/what-are-carbon-credits/. Accessed August 2024
30. Gelmini S (2021) We're living in a golden age of greenwash, Greenpeace, www.greenpeace.org.uk/news/golden-age-of-greenwash/. Accessed August 2024
31. United Nations (2024) World population prospects 2024, Dept. of Economic & Social Affairs, www.population.un.org/wpp/Graphs/Probabilistic/POP/TOT/900. Accessed August 2024
32. United Nations (2024) Global Resources Outlook 2024, UN environment programme, www.unep.org/resources/Global-Resource-Outlook-2024. Accessed August 2024
33. United Nations (2024) Global waste management outlook 2024, UN Environment Programme, www.wedocs.unep.org/bitstream/handle/20.500.11822/44939/global_waste_management_outlook_2024.pdf?sequence=3. Accessed August 2024
34. Qiu K, Zhang L (2024) How online reviews affect purchase intention: a meta-analysis across contextual and cultural factors. Data Inform Manag 8(2):100058
35. Collings D, Corbet S, Hou YG, Hu Y, Larkin C, Oxley L (2022) The effects of negative reputational contagion on international airlines: the case of the Boeing 737-MAX disasters. Intl Rev Financial Anal 80:102048
36. Rivian Introduces Second Generation R1S, R1T, www.rivian.com/en-GB/newsroom/article/rivian-introduces-second-generation-r1s-r1t. Accessed August 2024

Chapter 15
The Internet of Things—Further Promise of a Better Connected World

George R. S. Weir

Abstract Rapid progress in communication technologies, in combination with the ubiquity of computers and the profusion of personal mobile computing devices have contributed to the development of many extended and interconnected systems. These facilities promise easier interaction, greater flexibility, remote monitoring and remote control in operation, as well as the prospect of enhanced integration with other systems, to establish the Internet of Things (IoT). This chapter outlines the nature of IoT, describes application areas where IoT has traction and plausible benefits. The adoption and direction of these technologies is also addressed. While the progress of IoT is tangible and continues apace, there are areas of concern, both technical and societal, that give cause for reflection on the credibility and promise heralded in many quarters for IoT. Despite facing issues, including reliability, security, privacy, integrity and locus of responsibility, the future of IoT may be assured, but the benefits may be limited by cost, availability and negative social impact.

15.1 Introduction

The rapid development of interpersonal communication technology over the past thirty years has seen progress in applications beyond the traditional inherently asynchronous communication channel of email, to synchronous (live) interaction with audio and video digital channels. Applications such as Skype, Zoom, MS-Teams, Facebook Messenger and Apple Facetime now support immediate multimedia interaction in a one-to-one or one-to-many setting. In parallel to the appealing immediacy of these technologies, users have also embraced developments in social networking as a basis for selective broadcast and group interaction.

These relatively new synchronous and asynchronous applications are already widely adopted internationally across sectors, and age groups, with a corresponding growth in personal digital devices, particularly smartphones and tablets. Supporting

G. R. S. Weir (✉)
Mellow Internet Research, 7 Woodend Road, Ayr KA7 4QP, Scotland
e-mail: george@mellow.co.uk

P. Hehenberger and D. Bradley (eds.), *Mechatronic Futures*,
https://doi.org/10.1007/978-3-031-83571-1_15

these developments, is an associated expansion in networking technology and infrastructure, with faster throughput, and extensive use of wireless communication on Internet-enabled mobile and static devices.

A useful effect of the ready availability of communication and data is the expansion of interaction beyond person-to-person contact and beyond engagement with local devices to the monitoring and control of network-accessible systems wherever they may have application, especially in our homes, offices and factories.

In this chapter, we consider how the current context has given rise to the ideas behind the Internet of Things (IoT), look at how such systems function, and consider the benefits and disadvantages that are likely to accrue from such developments. Beyond the present '*state of play*', we will review key ingredients, applications, ecosystem requirements, potential issues and prospects for a future that is likely to be greatly affected by the IoT.

15.2 Impetus

As noted, our current technological context lends itself naturally towards the extended integration and enhanced data exchange that is central to the Internet of Things. Broad familiarity with communication facilities that are increasingly functional and immediate, raises expectations that other information systems will be equally immediate and responsive. At the same time, increasing use of networked devices can afford savings in service costs, in terms of direct functionality and enhanced capability for remote monitoring. These economic factors will be accompanied by the natural desire to adopt new technologies, as a basis for IoT take-up. Several plausible benefits are reflected in IoT adoption and initiatives across public and private sectors.

15.2.1 IoT Initiatives

The UK Government published a report in 2014 entitled '*The Internet of Things: making the most of the Second Digital Revolution*' [1], while the US Department of Defense issued '*Policy Recommendations for the Internet of Things*' [2], with a view to providing '*… background and policy recommendations to address vulnerabilities (and take advantage of opportunities) related to the increasingly pervasive and semi-autonomous internet-capable devices that make up what is known as the Internet of Things*' [3]. Other countries and administrations have recognised the scope and importance of IoT advances. The government of Singapore '*… has introduced a slew of initiatives as part of its goal to become the world's first smart nation, including a smart nation operating system, Internet of Things scheme targeted at homes, and pilot trials at a designated residential-business estate*' [4].

Nations with developing economies are also rising to the IoT opportunity, '*One of the top most initiatives in the form of Digital India Program of the Government which aims at* 'transforming India into a digital empowered society and knowledge economy', *is expected to provide the required impetus for development of the IoT industry ecosystem in the country*' [5]. In 2014, South Korea published its *Master Plan for Building the Internet of Things* (IoT), subtitled '*that leads the hyper-connected, digital revolution*' [6]. By 2021, this ambition had developed, such that '*The Seoul Metropolitan Government (SMG) announced its plan to complete building the* 'public IoT network' *across the capital by 2023*' [7].

Since these initial steps at government level, industry has taken notice and responded with market activity as well as collaborative initiatives such as the Australian IoT Alliance: '*Our vision is a* 'Data Smart Australia' *and our mission is to advance society through trusted accessible automated data*'. They provide '*a space where organisations large, small and in between work together to raise IoT awareness, capability, impact and trust*' [8]. Such national and organisational perspectives consistently present the prospect of IoT developments as positive steps that will enhance the welfare of the population and the economic benefit of the country. Before reviewing the likely positives and negatives, let us consider the key ingredients for IoT.

15.2.2 Key Elements

Three key IoT elements were identified by the UK Government publication [1] as: Communication; Integration; and Data analysis. First, the present and evolving communication infrastructure, is made up of existing wired network facilities, supplemented with wireless technologies: Wireless LAN (WLAN) [9], Bluetooth [10], GPRS (GSM) mobile telephony standards [11] and emerging standards for 'near-field' and close-proximity device interaction [12].

An essential ingredient for IoT is integration. The scope for IoT and the extent of its penetration will depend upon the consolidation of differing systems and standards, to ensure a technical ecology with '*local*' systems talking to each other and interacting with '*upper level*' systems. The final key ingredient, data analysis has two roles. Firstly, analysis is a means of monitoring and managing the quality of interaction between devices (e.g., for fault detection), and secondly, data analysis is a value-added ingredient that provides insight into usage and performance. (e.g., for targeting bandwidth and premium enhancements). The prospect is that integrated systems support widely distributed data gathering as well as centralised synthesis and analysis of data, to produce useful insights that serve to optimise or monetise services (cf. Fig. 15.1).

Data has proved to be the lifeblood of many developments in recent technologies and IoT devices are a source of data both through their role as monitors of their environment and as insight on user interaction. The facility to assemble extremely large datasets, e.g., through upstream reporting from many small IoT devices, has coincided with major developments in machine learning and artificial intelligence. A direct benefit from collection and analysis of natural language data (very large

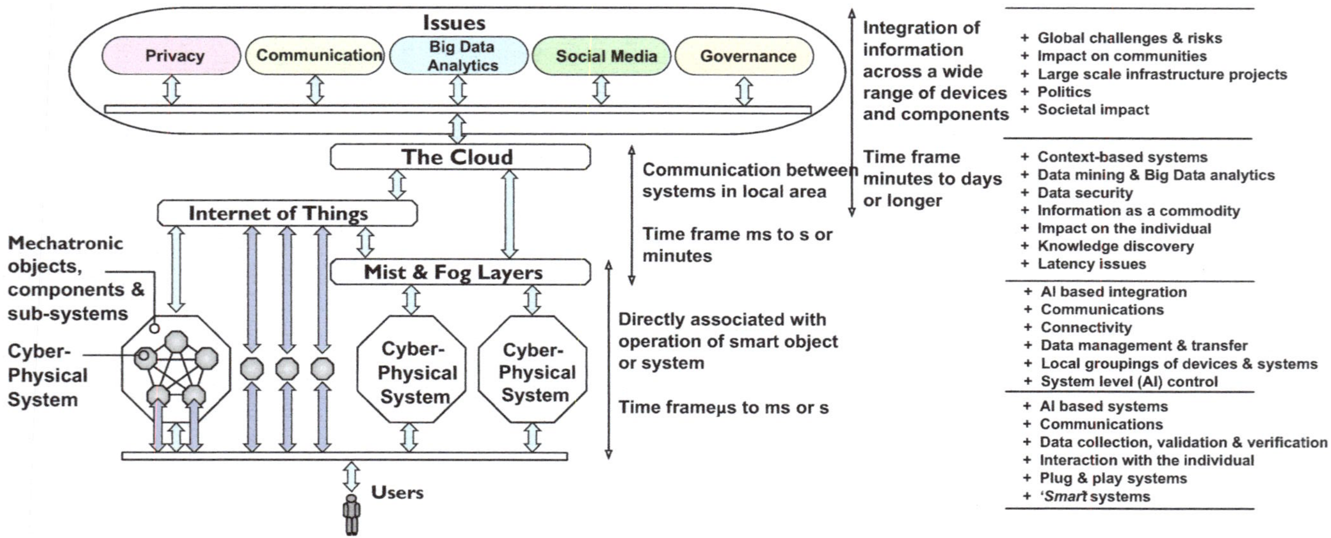

Fig. 15.1 IoT key elements and functional layers (after Bradley et al. (various))

corpora) has resulted in enhanced user computer interaction whereby users can speak to their devices, to elicit actions and vocal responses in return. This development has greatly influenced the adoption of small-scale systems, so-called smart speakers that interface to other domestic facilities, such as central heating and home surveillance. These small devices act as '*thin clients*', with relatively small-scale local processing, and interact with Cloud-based applications that can provide the heavy-duty processing, analysis, and response.

15.2.3 Applications

Here, five core sectors were identified by the UK government publication [1]: Home automation; Agriculture; Energy; Healthcare; and Transport. These sectors were singled out as most likely to afford applications with significant economic potential and appeal. Aspects of home automation should have wide appeal and potential penetration, and this sector has visibly expanded in the intervening years since the government report. So called '*smart home*' technologies '*refer to devices that provide some degree of digitally connected, automated, or enhanced services to building occupants*' [13].

There are two levels of application to homes. Firstly, at the level of individual dwellings and secondly, in the wider context of building management systems. In the first case, smart speakers (such as Amazon's Echo and Google's Nest) and other devices (such as televisions and various '*white goods*') imbued with virtual assistants have grown in popularity and in the scope of their application. A key feature of these domestic systems is the convenience and ease of use afforded by voice interaction with the virtual assistant. Indeed, this must be a prime motivator in the adoption of domestic IoT.

A second factor encouraging the adoption of smart home systems is integration with a range of in-home facilities, such as home surveillance, heat and power monitoring and management. Given the vast number of potential users of IoT technologies at home this is a large growth market. Indeed, estimates suggest that by 2018 aspects of these technologies were adopted by 7.5% of households globally with expected revenues of $44.2 billion in that period [14]. Further, '*Revenue in the Smart Home market is projected to reach US$154.4bn in 2024; Revenue is expected to show an annual growth rate of 10.67%, resulting in a projected market volume of US$231.6bn by 2028;* [and] *Household penetration will be 18.9% in 2024 and is expected to hit 33.2% by 2028*' [15].

At the larger scale, smart home technologies can be applied in the coordination and management of multiple interior systems, e.g., by monitoring and controlling internal systems such as air conditioning, temperature, and lighting. This may apply in buildings with multiple occupancy or, indeed across multiple such buildings in a single complex.

In this context, a single point of feedback, monitoring and control may afford greater finesse and economy in operation [16]. Agriculture and other manufacturing contexts, stand to benefit from the introduction of sensor-based feedback and automation as a basis for production and yield management. Indeed, Internet of Things '… *not only increases agricultural output but can also effectively improve the quality of agricultural products, reduce labor costs, increase farmers' income, and truly realize agricultural modernization and intelligence*' [17].

The energy and water sectors were quick to show movement in the direction of IoT through introduction of smart metres, both to display consumption and to report usage to the provider. Easy domestic views of consumption and associated cost is an attractive option, and, in some quarters, such applications are seen as '… *the transition towards sustainable use of resources such as water, electricity and gas*' [18].

Healthcare is a significant application sector for IoT, both for reducing costs and for enhanced patient care. Thus, '*Over the past* 10–15 *years, health care providers have increasingly become connected through the use of mobile computers, tablets, PCs, smartphones, Wi-Fi phones, and communications badges and help them to become far more proactive about Health Services Delivery*' [19].

Significant IoT growth in the transport sector is anticipated through expanded use of in-car sensors, telemetry, and inter-vehicle communication. Ultimately, these features are the basis for automated self-driving vehicles. As insight, '*The global automotive IoT market size was valued at USD 125.05 billion in 2023 and is anticipated to reach around USD 946.29 billion by 2033*' [20] (see Fig. 15.2). Indications are that adoption and growth will increase exponentially in this sector.

Five principal applications of IoT in transportation are identified by Conure Technology Services as: efficient traffic management, automated toll and ticketing, self-driving cars, advanced vehicle tracking or transportation monitoring and enhanced security of the public transport [21]. In each case, we can detect the impact of remote monitoring, data collection and feedback-based control as means towards greater efficiency, utility, safety, or economy.

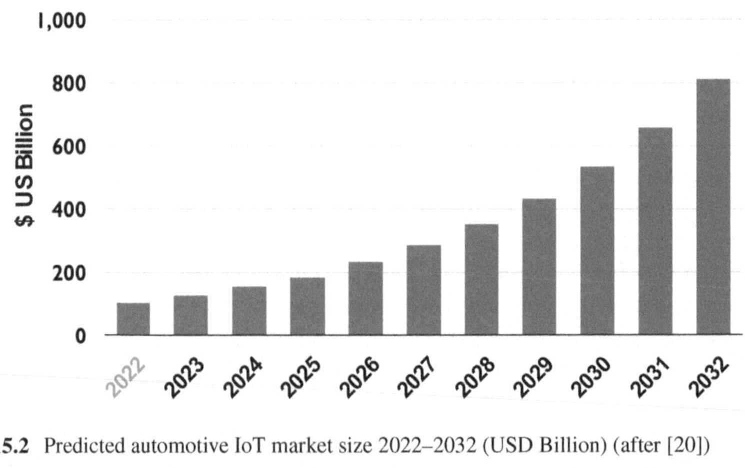

Fig. 15.2 Predicted automotive IoT market size 2022–2032 (USD Billion) (after [20])

According to Allied Market Research, '*The global IoT in transportation market size was valued at $83.25 billion in 2020 and is projected to reach $495.57 billion by 2030*' [22].

The anticipated economic benefits are central to IoT promotion by governments, and we already have relevant devices and technologies in the marketplace that contribute to the adoption and growth of IoT. Home Wi-Fi networks afford a setting for 'smart' devices with in-built communication capabilities, from toasters and kettles, through light switches and multi-room digital music systems, to toothbrushes that rely on end-user mobile apps to report the effectiveness of their use. Mobile telephone or tablet apps often front control and monitoring facilities in the home and support the integration of disparate systems.

Of course, smartphones already support Wi-Fi, Bluetooth and near-field connectivity. With GPS capability and suitable applications, phones seamlessly interact with the local environment, to register their presence (or the presence of the telephone user), capturing localised data for the user, while reacting to personalised settings, perhaps based upon time of day or location.

As supplements to the ubiquitous smartphone, we have smart watches, fitness trackers and other wearable devices, including clothes with in-built sensors. As features of IoT development, wearable devices often engage the data processing and communication functionality of telephones or tablets. The direction of smart clothing development is evident from the following overview.

> *Smart clothing is the fusion of textiles and technology, designed to make your life just a little more quantifiable. A wearable can be found for just about everything, from your head to your toes, and can function in various ways. Sports clothes intended for athletes are equipped to monitor heart rate, muscle activity, speed, and even breathing patterns. On the other end of the spectrum, smart gloves for motorcyclists allow users to control music, media, and smartphones through gesture control, while solar-powered jackets provide wearers with enough power to charge a phone* [23].

As more devices becoming '*smart*' and able to register their status with upper-level systems, we may expect increases in device self-monitoring for performance reporting, and fault detection. Cost benefits and service enhancements may arise through insights on system demand and automated capture of system performance.

15.3 Technology Adoption and Direction

As indicated in our overview of applications (Sect. 2.3), many of the sectors that were identified early in the promotion of IoT have realised significant growth. This is particularly evident in the domestic context, as well as in transport, agriculture, and health care. One area not singled out in the earlier UK government paper is manufacturing. This sector is represented principally by private enterprise. While government may encourage and expect growth and development in manufacturing, governmental engagement is largely hands-off. Despite this '*light touch*', evidence suggests that manufacturing has been greatly influenced by developments in IoT.

In illustration, a 2023 survey commissioned by Ubisense '… *reveals that more than three-in-five manufacturers (62%) have embraced IoT technologies in their manufacturing or assembly processes*' [24]. Furthermore, the survey indicates that the manufacturing sector is increasing adoption of IoT technologies: '… *over two-in-five manufacturers (43%) are currently leveraging real-time location systems (RTLS). This statistic highlights the increasing importance of precise asset tracking and management within modern manufacturing environments*' [24].

This direction should not surprise, since manufacturing is well-placed to benefit directly from the enhanced monitoring, feedback, and resultant efficiency in production management.

Similarly, health was recognised early as a fertile prospect for IoT, seen largely as a means of reducing the increasingly high provision cost of this largely public service. Thereby, the costly realm of healthcare may expect to benefit substantially from remote diagnosis and treatment, as well as through operational enhancements. While in the medium-term, we may look forward to more accessible, efficient, and cheaper health care, the adoption of innovative technologies within the public health service is largely driven by product developments that arise from private enterprise, with costs set at a profitable level for the manufacturers and product re-sellers. In consequence, the major take-up of novel IoT health care technologies is likely to occur in the private sector, where adoption costs can be addressed through income from patients' health insurance.

The promises of economy, efficiency, and convenience, combined with the potential penetration of Internet of Things across all sectors of industry, transportation, health care, agriculture, defence, and the domestic context, we might expect widespread and speedy adoption of these technologies. Yet, historical indications of technology adoption, at the company and individual levels, suggest that adoption is rarely, if ever, a linear process. Indeed, most theories of technology adoption favour a bell curve model of take-up over time [25]. For example, the Technology Acceptance Model [26] identifies five adoption stages characterised in terms of the individuals responsible for making the adoption decision (Fig. 15.3).

Where are we on this curve in relation to IoT adoption? No definitive answer seems possible. In part, this uncertainty arises because IoT technology is not yet

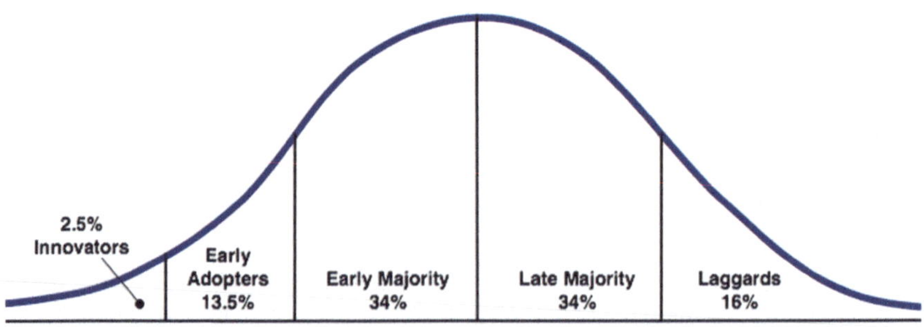

Fig. 15.3 Technology acceptance model (after [25, 26])

fully standardised, and neither is it established in all possible domains of application. Nevertheless, my sense is that we are no further along than adoption by the early majority. This implies that more than half of the adoption lifecycle remains.

15.4 Latest Advances

Can we discern any immediate advances in the field of IoT? Will there be any dramatic shift in scope or performance? Conventional wisdom on the development of technology, particularly computer-based technology, anticipates regular advances in performance due to incremental improvement in processing performance and simultaneous reduction in costs. One clear advance is the application of edge computing as a means of enhancing localised processing and reduction in network traffic. Thus '*Edge computing is a game changer for the IoT. It allows IoT devices to be more independent, storing, processing, and analysing data locally instead of just sending it to a centralised server. This can improve the effectiveness of existing IoT devices and make new devices and deployment topologies possible*' [27].

Other visible influences on the improvement, and thereby, likely penetration of IoT, are machine learning, artificial intelligence (AI) and advances in big data processing. As noted above, the impact of machine learning and big data analysis is already evident in the evolution of voice input and output as a significant step forward in the interface between human and computer.

Since IoT devices are also a rich source of insight on user interaction, there is considerable future scope for data analysis, machine learning and feedback with a view to automating routines and for anticipating user expectations. In related developments, recommender systems seek to extrapolate user preferences based upon past behaviour and choices [28]. These supplemental technologies can be regarded as core ingredients within the larger context of modern Artificial Intelligence [29].

15.5 Issues

In keeping with the previous edition of this chapter, we may characterise emergent or incipient issues likely to arise from the increasing adoption of Internet of Things. Varieties of these issues include: Usability; Reliability and robustness; Availability; Locus of control; Privacy; Integrity and responsibility; and Security.

15.5.1 Usability

Developments in widely used technology, in common with most interactive systems, will create difficulties through poor interface design or inherent system complexity, whereby misunderstandings and incorrect system usage arise. Usability is always a concern and may still be an issue for new instances of interactive IoT systems. On the other hand, IoT adoption may be spurred by the trend towards '*invisible integration*'. By means of connectivity and native '*intelligence*', domestic and commercial devices may become inherent and unseen, with little requirement for user engagement. Thereby, aspects of IoT may become invisible in their usual operation. Progress in both voice input/output and conversational systems suggest that we are heading this way and may ensure that IoT technology will add little overhead to usability with connected devices.

15.5.2 Reliability and Robustness

As we become more dependent upon co-operating systems, so the risks increase. Greater complexity often results in multiple points of failure. Assurance of reliability becomes more urgent because the likelihood and potential impact of failure or error is greater. Moving towards the inevitable situation in which complex integrated systems become mission or life critical, requires the highest levels of robustness in design and reliability in operation. Insights on the predictable performance of individual and combinations of systems, as well as their failure rates and any anomalous characteristics, become key in any SWOT analyses that should precede decisions on technology adoption.

The status of individual devices, inter-linked devices and communication links will affect the robustness of centralised and decentralised services. Reliability is also influenced by capacity and performance level. Thus, quality of service may be as critical as device failure. If there is contention over available resource, some services, and some users, will lose out.

15.5.3 Availability

As suggested above, availability is often determined by reliability and robustness. Limits on system capacity, however they arise, usually mean that some services are curtailed for some users. Ultimately, benefits and limitations may not be equally spread. Without equal service provision (and availability) for all, we have a new setting of '*haves*' and '*have nots*'. A capitalist approach would suggest differing service options at different costs. This is one method of living with the inherent limitations of the technology.

15.5.4 Locus of Control

By its distributed nature, IoT has significant scope for data gathering, data consolidation and data synthesis. The ownership and use of information on end-user interaction, behaviour and preference can be sensitive issues. Where does control of such data reside? Presumably, with the owners of the technology. While some legal considerations already apply to the collection and use of personal data, e.g., data protection laws such as GDPR, the cross-border nature of networked systems and location of data storage can obscure the situation, doubtless at a loss to the user rather than the technology owners. We can find such issues in relation to current data-gathering points, such as search engines. When interacting with third-party systems, individuals generate data that has value, but that data does not belong to the user. Neither do they have influence over how 'their' information may be used. Perhaps the information will be used to benefit end-users, optimise services and minimise costs.

In response, one may argue that any '*added value*' arising from the synthesis of data depends upon the aggregation of many data sources. This is achieved solely through the technical facilities of the technology controller, who, unlike the individual who has contributed data, is able to materialise this value. Since data aggregation requires authorised access to data that derives from individuals or the systems and devices belonging to those individuals, this leads to the issue of privacy.

15.5.5 Privacy

As with other varieties of interactive system, user engagement with IoT generates data on individuals' behaviour. With concern for individual privacy, some people may wish to withhold or limit such information, particularly in contexts where the data can be tied to a specific person. Will integrated systems that support exchange and consolidation of data accommodate such delicacy for individuals? The trend may be to use captured data for the direct benefit of those who own or manage the systems, or for those willing to pay for privileged access to data, for instance, as a means to target adverts or identify sales opportunities. There may be organisations monitoring developments with a view to protecting the rights of individuals as technology users, but recognising new areas of concern and addressing them in law takes time and always follows behind the technical progress that enables any infringement of privacy.

15.5.6 Integrity and Responsibility

The issues discussed above, raise questions about the responsibility of service providers to the end users. Indeed, the likelihood that IoT systems will grow in their use and dependence on Artificial Intelligence leads us to wonder whether we can depend upon the integrity of those in control of system performance. Will they accept full responsibility for the consequences of such progress? The increased reliance on AI-based automated decision-making may lead to social, legal, and moral challenges. These are likely to arise from two emerging side-effects of IoT. The first is the impact on employment, with AI-based systems able to replace large sections of the workforce. The second is the absence, for the most part, of explanation facilities tied to automated decision-making [30]. Open discussion of the legal and social implications is becoming more frequent, but answers and resolutions seem absent so far [31]. Recent attempts to control the scope of AI (through agreement or legislation) seem bound to fail but are worthy of attention in the meantime. This may in time impact on the potential growth and expanded application of IoT.

15.5.7 Security

IoT security is a major area of concern. The strength of any set of complex components is no greater than the weakest point in that set. Ironically, the interconnectivity that underpins much of IoT's utility is also the means whereby unauthorised actors may seek to harm or misuse those facilities. The desirable and potent feature of interconnectivity across devices, often from different manufacturers, relies upon open standards in communication. Thereby, attacks on networked systems use standard communication protocols to reach their target. The objective in such unauthorised actions may be intrusion—e.g., for reconnaissance or looking to exfiltrate data, or harm, seeking means to misdirect and impair system operation.

Examples of contamination of existing networked devices already attest to the risk and likelihood of attempted remote exploitation of accessible systems and the facilities they control. Foreign malware has been found on US SCADA (control) systems with a naturally growing anxiety over risks to infrastructure. Meanwhile, '*unfriendly*' nation-states are often cited as perpetrators of security exploits.

In the United States, a demonstration under Project *Aurora*, simulated an attack on the US power grid and dramatically illustrated the risks of vulnerability to attack, with a $1 Million diesel-electric generator destroyed as a culmination to the experiment [32].

By their nature and modes of operation, interconnected devices may be vulnerable to illicit remote influence, as evidenced by exploits such as misdirecting Internet-enabled fridges to send spam emails [33]. Of course, the risks can be increased because of poor decisions, such as wireless-linked cameras and routers with unchanged default passwords. If devices are installed without change to pre-set access passwords, they are acutely vulnerable to remote access and the associated risk of misdirected devices or theft of personal data.

Denial of service attacks, seeking to disable a networked service and deny access to authorised users, are a real threat to any facility that permits remote access [34]. Often, these attacks overwhelm the networking feature of the system, e.g., through a distributed botnet attack [35], and thereby disable the service. In addition to loss of service utility, there may be data loss and associated negative impact on the organisation's reputation.

Vulnerability to attack is a key measure of system security. Clearly, widespread acceptance of increasingly interconnected and automated systems, such as drones and self-driving cars, will depend upon perceived safety and reliability [36]. While there are moves to address endemic security issues in the realm of IoT [37], the frequency of significant data breaches is further indication that interconnections between systems harbour weaknesses as well as strengths. Efforts to allay such concerns continue apace [38], but are inherently reactive, often in response to previously unrecognised risks.

15.6 The Future Internet of Things

We are still early in the penetration and acceptance of IoT. In the coming years, the functionality and scope are likely to expand considerably. The rise in functionality via Cloud and edge computing, especially employing generative AI systems (such as ChatGPT) and growing ease of user interaction will empower end users with extended smart home accessories. As noted above, the increasing ease of access to 'new' content through systems based on machine learning and big data, will impact upon conventional means of production. This may affect employment both in the Arts and in more scientific occupations, including computer programming. Ethical concerns must rise with increased reliance on AI-based automated decision-making, and inevitably lead to social, legal, and moral challenges.

Should we be happy with IoT prospects or concerned at the threats that may well emerge? As with any march of technology, the optimal response is to embrace the positive and, wherever possible, oppose the negative effects of technical innovation. Often, this requires that we weigh up the benefits, usually in terms of utility and economy, against the drawbacks, often in terms of social inequality and disruption. In concluding, I refer the reader to the chapter title, and underline the difference between the Internet of Things promising '*a better connected world*'—in which the interconnections make the world better—and merely promising '*a better-connected world*'—in which, simply, connectivity is enhanced. We must hope (and strive) for the former but can, at least, expect the latter.

References

1. The internet of things: making the most of the second digital revolution—a report by the UK Government Chief Scientific Adviser, www.gov.uk/government/uploads/system/uploads/attachment_data/file/409774/14-1230-internet-of-things-review.pdf. Accessed 23 August 2024
2. www.securityinformed.com/insights/co-5188-ga-sb.13927.html. Accessed 23 August 2024
3. www.hsdl.org/c/abstract?docid=799676. Accessed 23 August 2024

4. www.zdnet.com/home-and-office/networking/singapore-unveils-plan-in-push-to-become-smart-nation. Accessed 23 August 2024
5. www.csc.gov.in/digitalIndia. Accessed 23 August 2024
6. www.onem2m.org/images/files/News/KOREA-IoT-Master-Plan.pdf. Accessed 23 August 2024
7. www.itu.int/hub/2021/07/seoul-to-implement-city-wide-public-iot-network-by-2023. Accessed 23 August 2024
8. www.iot.org.au/page/AboutUs. Accessed 23 August 2024
9. www.nibusinessinfo.co.uk/content/wireless-networking-standards. Accessed 23 August 2024
10. www.bluetooth.com/specifications. Accessed 23 August 2024
11. Salih AA, Zeebaree SR, Abdulraheem AS, Zebari RR, Sadeeq MA, Ahmed OM (2020) Evolution of mobile wireless communication to 5G revolution. Tech Rep Kansai Univ 62(5):2139–2151
12. www.mouser.co.uk/applications/rfid-nfc-introduction/. Accessed 23 August 2024
13. Sovacool BK, Furszyfer Del Rio DD (2020) Smart home technologies in Europe: a critical review of concepts, benefits, risks and policies. Renew Sustain Energy Rev 120:109663
14. Shin J, Park Y, Lee D (2018) Who will be smart home users? An analysis of adoption and diffusion of smart homes. Technol Forecast Soc Change 134:246–253
15. www.statista.com/outlook/cmo/smart-home/worldwide. Accessed 23 August 2024
16. Daissaoui A, Boulmakoul A, Karim L, Lbath A (2020) IoT and big data analytics for smart buildings: a survey. Proced Comp Sci 170:161–168
17. Xu J, Gu B, Tian G (2022) Review of agricultural IoT technology. Artif Intell Agric 6:10–22
18. Fettermann DC, Borriello A, Pellegrini A, Cavalcante CG, Rose JM, Burke PF (2021) Getting smarter about household energy: the who and what of demand for smart meters. Build Res Inform 49(1):100–112
19. Chiuchisan I, Costin HN, Geman O (2014) Adopting the internet of things technologies in health care systems, 2014 IEEE International conference & exposition on electrical & power engineering, pp 532–535
20. www.precedenceresearch.com/automotive-iot-market. Accessed 23 August 24
21. www.conurets.com/5-major-applications-of-iot-in-transportation. Accessed 23 August 2024
22. www.alliedmarketresearch.com/IoT-in-transportation-market. Accessed 23 August 2024
23. www.wearables.com/collections/smart-clothing. Accessed 23 August 2024
24. www.ubisense.com/a-rapid-increase-in-iot-adoption-manufacturing-iot-in-2023. Accessed 23 August 2024
25. Rogers EM (1995) Diffusion of innovations, 4th edn. Free Press, New York
26. Davis FD (1986) A technology acceptance model for empirically testing new end-user information systems: theory and results, doctoral dissertation, Sloan school of management, Massachusetts Institute of Technology
27. www.run.ai/guides/edge-computing/edge-computing-in-iot. Accessed 23 August 2024
28. Jannach D, Manzoor A, Cai W, Chen L (2021) A survey on conversational recommender systems. ACM Comput Surv (CSUR) 54(5):1–36
29. Zhang J, Tao D (2020) Empowering things with intelligence: a survey of the progress, challenges, and opportunities in artificial intelligence of things. IEEE Internet Things J 8(10):7789–7817
30. Vyas B (2023) Explainable AI: assessing methods to make AI systems more transparent and interpretable. Intl J New Med Stud Int Peer Rev Sch Indexed J 10(1):236–242
31. Hagerty A, Rubinov I (2019) Global AI ethics: a review of the social impacts and ethical implications of artificial intelligence, arXiv preprint arXiv:1907.07892
32. www.wired.com/story/how-30-lines-of-code-blew-up-27-ton-generator/. Accessed 23 August 2024
33. Arthur C (2014) Help! My fridge is full of spam and so is my router, set-top box and console, The Guardian, www.theguardian.com/technology/2014/jan/21/fridge-spam-security-phishing-campaign. Accessed 23 August 2024

34. Tyagi V, Saraswat A, Bansal S (2023) An analysis of securing internet of things (IoT) devices from man-in-the-middle (MIMA) and denial of service (DoS). CRC Press, In Smart Cities, pp 337–357
35. www.bbc.co.uk/news/technology-24608435. Accessed 23 August 2024
36. Karthikeyani R, Karthikeyan E (2023) A review on distributed denial of service attack. Asian J Res Comput Sci 16(4):133–144
37. Alladi T, Chamola V, Sikdar B, Choo KKR (2020) Consumer IoT: security vulnerability case studies and solutions. IEEE Consum Electron Mag 9(2):17–25
38. Sarker IH, Khan AI, Abushark YB, Alsolami F (2023) Internet of things (IoT) security intelligence: a comprehensive overview, machine learning solutions and research directions. Mobile Netw Appl 28(1):296–312

Part IV
Hard Technologies

Chapter 16
Bioinspired Design in Mechatronics: EcoMechatronics as a Pathway to Sustainability

Maki K. Habib

Abstract This chapter explores the synergy of bioinspired design principles within Mechatronics to drive sustainability and innovation. It introduces interdisciplinary Mechatronics, discusses its adaptation toward EcoMechatronics, and presents the significance and methodology of this research. The chapter then discusses the importance and objectives of incorporating bioinspired designs, detailing both non-bioinspired and bioinspired paradigms and their comparative synergies. The research methodology section outlines the structured design process, emphasizing sustainability, ecosystem-based management, and applying circular economy principles. Additionally, the chapter highlights the integration of advanced and smart materials, technologies, and computational tools in facilitating effective bioinspired designs. Comprehensive case studies illustrate the practical applications of bioinspired design in environmental monitoring and medical devices. The chapter concludes by examining the best general practices in Mechatronics design and best practices of bioinspired design in Mechatronics, including pitfalls and research challenges, offering insights into the future potential of bioinspired approaches in sustainable engineering design.

Keywords Bioinspired design · Nature · Mechatronics · EcoMechatronics · Sustainability · Innovation · Circular economy · Environmental monitoring · Smart materials · Ecosystem

16.1 Introduction

Mechatronics is an interdisciplinary field that synergizes knowledge from various fields and disciplines, such as mechanical engineering, electronics, computer science, and control engineering, with open-ended knowledge to develop intelligent, reliable, compact, and automated processes, systems and products. This field emphasizes

M. K. Habib (✉)
The American University in Cairo, Cairo, Egypt
e-mail: maki@aucegypt.edu

P. Hehenberger and D. Bradley (eds.), *Mechatronic Futures*,
https://doi.org/10.1007/978-3-031-83571-1_16

creating synergized and adaptable solutions with added values that enhance functionality and performance through the various contributing domains. Mechatronics promotes innovation and addresses complex engineering challenges by synergizing multiple disciplines, perspectives, depth, and powerful insight [4]. Recent advancements in Mechatronics include the development of smart systems and automation technologies that improve efficiency and effectiveness in various applications, such as manufacturing, healthcare, aerospace, transportation, smart factories, and environmental monitoring [2, 3, 11, 21].

Mechatronics has evolved from focusing solely on mechanical and electronic integration to incorporating advanced fields such as robotics, artificial intelligence, and sustainable engineering. This evolution reflects a growing awareness of the need to develop energy-efficient, environmentally friendly, and resource-efficient technologies. Integrating smart materials is essential in this evolution, providing adaptive and responsive characteristics to engineered systems. EcoMechatronics extends Mechatronics principles to create sustainable solutions that minimize energy consumption, reduce waste, and optimize resource use, aligning with the circular economy principles and sustainable development [4, 6, 13, 22]. Emerging trends in EcoMechatronics include integrating renewable energy sources, sustainable materials, and lifecycle analysis in the design and development of Mechatronic systems [12, 19]. The application of circular economy principles and ecosystems is central to EcoMechatronics, as highlighted by [5], who emphasizes the importance of designing systems that are not only efficient but also sustainable within their environmental context [5].

This chapter explores integrating bioinspired design principles in Mechatronics to enhance sustainability and innovation. The increasing environmental challenges and the need for sustainable engineering solutions necessitate a shift toward EcoMechatronics [6]. This study uses bioinspired design to provide practical insights and methodologies for developing sustainable Mechatronics systems. This exploration addresses the critical need for sustainable development in engineering, highlighting how bioinspired approaches can lead to innovative solutions that are both functional and environmentally friendly [8, 21].

The chapter adopts a comparative analysis of bioinspired and non-bioinspired design paradigms, followed by detailed case studies to illustrate the application and benefits of bioinspired design in Mechatronics. This approach involves identifying biological analogies, extracting principles, and applying them to engineering problems to develop innovative solutions. The methodology emphasizes a systematic and iterative process that integrates principles from biology into engineering design to create sustainable and efficient systems [9]. For instance, adapting nonlinear control for landing in bioinspired ornithopters (a type of flying machine) has demonstrated the target benefits of integrating biological principles into the mechatronic design [10]

This chapter aims to contribute to understanding, using, and effectively synergizing bioinspired design into Mechatronics, providing useful insights and methodologies for sustainable engineering practices. It seeks to highlight the potential of bioinspired design to address complex engineering challenges and enhance the sustainability of Mechatronics systems [4]. The contributions include the development of new design frameworks, demonstrating innovative applications, and identifying best practices and pitfalls in bioinspired design [4, 7, 17, 20]. Moreover, the design and kinematic analysis of an 8 DOF serial robot manipulator for electrostatic spray painting demonstrates the applicability of these principles in real-world scenarios [7].

The chapter is structured as follows: an introduction to Mechatronics and EcoMechatronics, a theoretical foundation comparing non-bioinspired and bioinspired design paradigms, and a detailed explanation of the research methodology, applications, and case studies demonstrating bioinspired approaches. Additionally, the chapter discusses best practices, potential pitfalls, and challenges in bioinspired design, providing insights for future sustainable engineering practices. Finally, a conclusion summarizing the findings and contributions.

16.2 Theoretical Foundation

16.2.1 Non-Bioinspired Design Paradigms

Non-bioinspired design paradigms in Mechatronics typically involve traditional engineering approaches focusing on functionality, performance, and efficiency without explicitly drawing inspiration from biological systems. These paradigms include top-down design, bottom-up design, structured design, and object-oriented design, each with its advantages and disadvantages [15, 16, 18]. See also Table 16.1.

1. Top-down design, starting from a high-level solution and refining it through successive layers of detail, is structured and systematic, ensuring clearly defined stages from concept to implementation. While it can be rigid and less flexible,

Table 16.1 Comparison of non-bioinspired design paradigms

	Paradigm	Approach	Strengths	Achieving the target	Limitations/ weaknesses	Integration potential
1	Top-down design	Starts from a solution and refines it	Structured, systematic	Clearly defined stages	It can be rigid, less flexible	Combined with bottom-up
2	Bottom-up design	Starts with fundamental principles	Innovative, exploratory	Incremental development	May lack direction	Integrated with top-down
3	Structured design	Maps inputs and outputs systematically	Clear, organized	Detailed mapping	Complex, time-consuming	Enhanced with object-oriented
4	Object-oriented design	Focuses on interacting objects	Modular, reusable	Encapsulation of functionality	Challenging to implement	Combined with structured

potentially limiting innovation, this approach can be combined with bottom-up approaches to allow for flexibility and innovation within a structured framework [1].

2. Bottom-up design begins with fundamental principles and builds to a complete system. It encourages innovation and exploration through incremental development, allowing for continuous improvement. Nevertheless, it may lack direction and coherence without a strong guiding framework. Integrating with top-down approaches can provide structure and direction while maintaining flexibility [15].
3. Structured design involves systematically mapping inputs and outputs, leading to clear and organized designs. Detailed mapping ensures that all aspects of the design are considered and integrated, but this approach can be complex and time-consuming, potentially slowing down the design process. Structured design can be enhanced with object-oriented design to improve modularity and reusability [16].
4. Object-oriented design focuses on the interaction of objects, making them modular and reusable, which facilitates maintenance and upgrades. Encapsulating functionality within objects allows for clear and manageable development stages. However, it can be challenging to implement, particularly in complex systems. Combining it with structured design can provide systematic development and modularization [16].

16.2.2 Bioinspired Design Paradigms (BID Paradigms)

Bioinspired design paradigms utilize principles and strategies from biological systems to address engineering challenges, aiming to create sustainable and innovative solutions. These paradigms include problem-driven BID, solution-driven BID, biomimicry, bioreplication, hybrid bioinspired design, system-of-systems bioinspired design, and the information-processing approach [4, 9]. Table 16.2 summarizes these paradigms.

1. **Problem-Driven BID (Top-Down)**

Problem-driven BID starts with an engineering problem and searches for suitable biological analogies. It is targeted and specific, identifying biological solutions directly addressing the engineering problem. However, it requires deep biological knowledge. This approach can be integrated with solution-driven approaches to enhance the innovation process. For example, a mosquito stinging mechanism inspired the development of polymer needles [2].

2. **Solution-Driven BID (Bottom-Up)**

Solution-driven BID begins with fascinating biological phenomena and explores potential applications. It is innovative with broad applications, promoting exploring and adapting biological solutions to various engineering problems. Nonetheless, it may be less systematic. Combining it with problem-driven approaches can provide

Table 16.2 Summarizes bioinspired design paradigms

	Paradigm	Approach	Strengths	Achieving the target	Limitations/Weaknesses	Integration potential
1	Problem-driven BID	It starts with an engineering problem	Targeted, specific solutions	Identification of biological solutions	Requires deep biological knowledge	Combined with solution-driven
2	Solution-driven BID	It starts with biological phenomena	Innovative, broad applications	Exploration and adaptation	It may be less systematic	Combined with problem-driven
3	Biomimicry	Mimics nature's forms and processes	Sustainable, holistic	Implement of nature-inspired designs	It can be complex to implement	Integrated with Bioreplication
4	Bioreplication	Directly replicates biological structures	Accurate, functional	Advanced manufacturing techniques	Requires advanced manufacturing techniques	Combined with biomimicry
5	Hybrid bioinspired	Combines multiple paradigms	Comprehensive, versatile	Integration of multiple approaches	Complex, interdisciplinary	Enhances versatility and adaptability
6	System-of-systems	Designs multifunctional solutions	Adaptive, flexible	Multifunctional and integrated designs	Requires integration of various principles	Improves adaptability and functionality

a more directed application. Examples include Velcro inspired by cockleburs and gecko tape inspired by gecko feet [21].

3. **Biomimicry**

Biomimicry focuses on sustainability by mimicking nature's forms, processes, and ecosystems. It is sustainable and holistic, aiming to implement nature-inspired designs that promote sustainability. This approach can be complex but can be integrated with bioreplication for accurate and functional designs. An example is the Eastgate Building in Harare, which mimics termite mound ventilation for energy-efficient cooling [9].

4. **Bioreplication**

Bioreplication directly replicates biological structures to achieve specific functionalities. It is accurate and functional, utilizing advanced manufacturing techniques to replicate biological features and function. However, it requires advanced manufacturing techniques. This approach can be combined with biomimicry for comprehensive sustainability solutions. For instance, replicating the iridescent elytra of the Ash borer beetle to develop decoys for pest control.

5. **Hybrid Bioinspired Design**

Hybrid bioinspired design combines multiple bioinspired approaches to create comprehensive and versatile solutions. This approach harnesses the strengths of various paradigms to address complex engineering challenges. However, it can be complex and interdisciplinary, requiring coordination across different fields [9].

16.2.3 Comparison of Non-Bioinspired and Bioinspired Approaches

Non-bioinspired and bioinspired approaches differ fundamentally in their starting points, primary focus, and processes. While non-bioinspired approaches typically begin with engineering principles and focus on functionality and efficiency, bioinspired approaches start with biological principles and aim for sustainability and innovation.

16.3 Research Methodology

16.3.1 Design Process

The design process for bioinspired design paradigms involves a structured approach to problem-solving that includes identifying biological analogies, extracting principles, and applying them to engineering problems. This process emphasizes a deep understanding of biological systems and engineering requirements. The methodology is iterative, often requiring multiple design, prototyping, and testing cycles to refine and optimize the solutions. Smart materials are integral to this process, offering dynamic properties that can be tuned and adapted throughout the design iterations.

1. **Identifying the Starting Point**
- Problem-Driven BID. It begins with an engineering problem that needs to be addressed.
- Solution-Driven BID: Starts with an interesting biological phenomenon or solution looking for an application.
- Biomimicry: Focuses on mimicking nature's forms, processes, and ecosystems to solve problems sustainably.
- Bioreplication. It involves directly replicating biological structures for specific functionalities.
- Hybrid Bioinspired Design. Combines elements from various bioinspired paradigms.
- System-of-Systems Bioinspired Design. Designs multifunctional solutions integrating various bioinspired principles.
- Information-Processing Bioinspired Design. Uses computational methods to model and simulate biological systems.

2. **Extracting Principles**

For each starting point of the available BIDs, a detailed analysis and abstraction of the underlying biological principles or mechanisms are necessary. This step highlights the importance of extensive research and collaboration with biologists. Understanding the biological functions and how they can be translated into engineering solutions is a joint effort that enriches the process [5].

3. **Applying Principles to Engineering Problems**

The extracted principles are then creatively applied to engineering problems. This step requires innovation to adapt the biological strategies to practical engineering solutions. Computational tools and simulations often aid in this process to predict and optimize design performance [21].

4. **Prototyping and Testing**

Based on the engineered solutions, prototypes are developed and subjected to rigorous testing to evaluate their performance. This iterative process is crucial for refining the design and improving functionality [8].

16.3.2 Synergy with Mechatronics

Integrating bioinspired design principles into Mechatronics can enhance the sustainability and innovation of engineering solutions. Smart materials are crucial in this integration, providing adaptability and functionality inspired by biological systems. This synergy involves applying biological systems' adaptive, efficient, and sustainable strategies to address complex engineering challenges [22].

1. **Adaptive Systems**

Biological systems are inherently adaptive and capable of responding to changing environments. Integrating this adaptability into mechatronic systems can improve their robustness and efficiency.

2. **Efficiency and Resource Optimization**

Bioinspired designs often exhibit high efficiency and optimal resource utilization, critical traits in sustainable engineering. These principles can be integrated into Mechatronics to develop systems that minimize energy consumption and reduce waste [19].

3. **Sustainable Materials and Processes**

Biological principles can guide the use of sustainable materials and processes in Mechatronics. For instance, the self-cleaning properties of certain plants can inspire the development of surfaces that reduce maintenance needs [5].

16.3.3 Role of Sustainability and Ecosystems in Bioinspired Design

Bioinspired design emphasizes sustainability by mimicking natural ecosystems' efficient and waste-free strategies. This approach seeks to create engineering solutions that are not only functional but also environmentally friendly and resource-efficient. The principles of the circular economy and ecosystem-based management are central to this methodology [5].

1. **Circular Economy Principles**

Bioinspired design aligns with the principles of the circular economy by promoting using renewable resources, recycling, and waste minimization. This approach encourages the development of products with minimal environmental impact throughout their lifecycle [12].

2. **Ecosystem-Based Management**

Integrating ecosystem-based management strategies into engineering design ensures that the solutions are sustainable and resilient. This involves understanding ecosystem interactions and designing systems harmonizing with these natural processes [19].

3. **Sustainable Development Goals**

Bioinspired design contributes to achieving Sustainable Development Goals (SDGs) by promoting sustainable practices in engineering. This approach supports the development of technologies that address global challenges such as climate change, resource depletion, and environmental degradation [20].

16.4 Bioinspired Design Paradigms: Outlines and Characteristics

This section provides a comprehensive overview of each bioinspired design paradigm, highlighting its key features, requirements, strengths, weaknesses, limitations, and possible integration with other bioinspired designs. See Table 16.2, summarizing the available BIDs with more details and examples.

1. **Problem-Driven Bioinspired Design (Top-Down)**

The problem-driven BID paradigm starts with an engineering problem and searches for suitable biological analogies to address it. This approach involves identifying biological systems that have evolved to solve similar problems and abstracting their principles to apply to the engineering context. This approach requires a deep understanding of the engineering problem and the biological systems that can offer solutions. It provides targeted and specific solutions that directly apply to the problem [2]. However, it requires extensive biological knowledge and may be limited by the availability of relevant biological analogies. To overcome its limitations, this BID approach has the potential to be integrated with solution-driven approaches to enhance the innovation process.

2. **Solution-Driven Bioinspired Design (Bottom-Up)**

The solution-driven BID paradigm begins with an attractive biological phenomenon and explores potential engineering applications. This approach involves observing a biological system, understanding its mechanisms, and finding engineering problems that could benefit from them. It promotes innovation and broad applications, encouraging exploring and adapting biological solutions to various engineering challenges. However, it may be less systematic and might lack direction without a specific engineering problem in mind [21]. Combining it with problem-driven approaches can provide more directed applications and enhance effectiveness.

3. **Biomimicry**

The biomimicry paradigm focuses on sustainability by mimicking nature's forms, processes, and ecosystems. Unlike the solution-driven BID, which starts with an interesting biological phenomenon and then finds applications, biomimicry begins with the intent to solve an engineering problem sustainably. It involves identifying sustainable solutions by examining how nature solves similar problems and mimicking these strategies. This approach seeks to implement nature-inspired designs that promote sustainability, such as energy efficiency, waste reduction, and resource optimization. Biomimicry can be complex as it requires a comprehensive understanding of natural forms and processes and how to replicate them in engineering contexts. Despite its complexity, it offers holistic and sustainable solutions. Integrating biomimicry with bioreplication can enhance its accuracy and functionality, making it more practical for engineering applications. An example is the Eastgate Building in Harare, which mimics termite mound ventilation for energy-efficient cooling.

4. **Bioreplication**

The bioreplication paradigm involves directly replicating biological structures to achieve specific functionalities. This approach uses advanced manufacturing techniques to replicate the complex details of biological features and their functions. Bioreplication is highly accurate and functional, providing precise engineering solutions that mimic biological structures. However, it requires sophisticated technology and can be resource-intensive [4, 5, 9]. Combining bioreplication with biomimicry can lead to comprehensive sustainability solutions that are both functional and environmentally friendly.

5. **Hybrid Bioinspired Design**

The hybrid bioinspired design paradigm combines the strength of multiple bioinspired design approaches to create comprehensive and versatile solutions. By integrating different paradigms, this approach harnesses the strengths of each to address complex engineering challenges effectively. Hybrid bioinspired design is comprehensive and versatile but can be complex, requiring interdisciplinary knowledge and coordination [4, 5, 9]. This approach can address the limitations of individual paradigms, offering a more robust and adaptable solution.

6. **System-of-Systems Bioinspired Design**

The system-of-systems bioinspired design paradigm aims to create multifunctional solutions by integrating various bioinspired principles into a cohesive and adaptive system. Unlike hybrid BID, which combines different paradigms to solve a single problem, the system-of-systems approach designs systems with multiple interacting components inspired by different biological principles to achieve overall system adaptability and multifunctionality. This approach focuses on designing systems that can adapt and respond to changing environments and conditions by exploring multiple bioinspired principles. It is highly adaptive and flexible but requires comprehensive integration of multiple principles and a deep understanding of harmonizing them [14]. This paradigm is particularly useful for complex, dynamic engineering problems requiring multiple functionalities and adaptability. An example is designing an adaptive infrastructure system where different components (e.g., water management, energy distribution) are inspired by different biological principles to work together as a cohesive, responsive system.

7. **Information-Processing Bioinspired Design**

The information-processing bioinspired design paradigm uses computational methods to model and simulate biological systems and their principles. This approach systematically applies these principles to engineering problems, using computational tools to optimize and predict design performance. It is systematic and efficient, providing a structured way to translate biological insights into engineering solutions [8]. However, it depends on the availability of computational resources and the accuracy of models.

Table 16.3 compares the available bioinspired design paradigms using their main attributes.

After introducing and analyzing the seven available design paradigms, it becomes clear that the best approach to handle bioinspired design (BID) depends on the specific engineering challenge and the application context. However, several best practices can be outlined:

1. Integrating problem-driven and solution-driven approaches can provide direction and innovation. For example, start with an engineering problem and explore various biological phenomena that could offer solutions.
2. Prioritize sustainability through biomimicry and bioreplication can lead to efficient and environmentally friendly designs. Implementing nature-inspired designs that mimic natural processes can enhance sustainability.
3. Utilizing advanced tools, manufacturing techniques, and computational tools (information-processing bioinspired design) can enhance bioinspired solutions' accuracy, functionality, and optimization.
4. Hybrid bioinspired design, combining multiple paradigms' strengths, can provide comprehensive and versatile solutions that effectively address complex challenges.
5. Employing system-of-systems bioinspired design to create and ensure adaptive and flexible solutions that can respond to dynamic conditions, enhancing the robustness and functionality of the design.

Following these best practices can construct a comprehensive and effective bioinspired design approach, addressing engineering challenges with innovative, sustainable, and adaptable solutions.

16.5 Applications and Case Studies

Before exploring the application of different bioinspired design paradigms, it is crucial to recognize that each paradigm may approach problem-solving from different starting points. Some paradigms begin with a specific engineering problem and seek biological analogies to address it, while others start with a biological phenomenon and identify potential engineering applications. This section will introduce two impressive applications that best suit bioinspired design principles. We will analyze these applications by demonstrating how the first two bioinspired design paradigms can be uniquely applied.

Table 16.3 Comparison of the available bioinspired design paradigms

	Paradigm	Key Features	Requirements	Strengths	Weaknesses	Integration Potential	Examples
1	Problem-driven BID	Targeted, specific solutions	Deep biological knowledge	Directly addresses engineering problems	Requires extensive biological expertise	Combined with solution-driven	Mosquito-inspired needles, Sharkskin-inspired swimsuits
2	Solution-driven BID	Innovative, broad applications	Understanding of biological phenomena	Promotes exploration and adaptation	Maybe less systematic	Combined with problem-driven	Velcro, gecko-inspired adhesives
3	Biomimicry	Sustainable, holistic	Understanding of natural forms and processes	Promotes sustainability	It can be complex to implement	Integrated with Bioreplication	Eastgate building, lotus-effect surfaces
4	Bioreplication	Accurate, Functional	Advanced manufacturing techniques	High precision and functionality	Requires sophisticated technology	Combined with biomimicry	Ash borer beetle-inspired decoys, Iridescent coatings
5	Hybrid Bioinspired	Comprehensive, versatile	Interdisciplinary knowledge	Addresses complex challenges	Complex to manage	Combines strengths of multiple paradigms	Nacre-inspired composites, Biomimetic robots
6	System-of-systems	Adaptive, flexible	Integration of multiple principles	Multifunctional and adaptive solutions	Requires comprehensive integration	multiple bioinspired principles	Adaptive infrastructure systems, Multifunctional robots
7	Information-processing	Systematic, efficient	Computational tools and resources	Predictive and optimized design	Dependent on computational resources	Enhances efficiency and accuracy	Bioinspired algorithms, neural network models

16.5.1 Case Study 1: Problem-Driven Bioinspired Design (Top-Down)

The Problem: Underwater Environmental Monitoring Device

Environmental monitoring involves collecting, assessing, and analyzing data related to environmental conditions. Traditional monitoring devices often suffer from limited mobility, high power consumption, limited deployment time, insufficient sensitivity to certain pollutants, and inability to access remote or hazardous underwater terrains. Bioinspired design approaches offer innovative solutions to these challenges by mimicking nature's adaptive and efficient strategies.

16.5.1.1 When to Select Problem-Driven Bioinspired Design

- When there is a specific engineering problem with defined requirements.
- When traditional approaches have limitations, and nature's solutions offer potential improvements.
- Ideal for problems requiring a tailored, functional design inspired by biological models.

16.5.1.2 Process Assessment and Design Development

(a) **Understanding the Problem**

- Define specific requirements and constraints for effective underwater environmental monitoring.

List of requirements of the presented case:

- Operate on minimal power for long-term deployment in remote areas.
- Detect low concentrations of pollutants accurately.
- Need to withstand harsh environmental conditions.
- Operate autonomously with minimal maintenance.
- Move to access remote or hazardous areas for comprehensive monitoring.
- Specify the access depth and/or the complexity level of underwater terrains.
- **Initial Engineering Design and Assessment**
- Attempt an initial engineering design based on traditional approaches.
- Assess the cost and performance of this initial design.
- Identify limitations and areas where traditional engineering falls short.
- **Search for Biological Analogies**
- Identify fish species known for efficient swimming and adaptability to harsh environments.
- Study their locomotion mechanisms and other relevant features.
- **Extract and Analyze Principles**

- Analyze the principles of locomotion, energy efficiency, and sensory capabilities of the selected fish species.
- Extract specific data such as movement patterns, energy usage, and sensory adaptation.
- Compare the principles and features of different biological species.
- Rank the biological species based on how well they meet the requirements.
- **Shortlisting and Design Development**
- Shortlist the best-matching biological analogies that align with the requirements.
- Develop initial design concepts based on the shortlisted biological principles.
- **Simulation and Performance Comparison**
- Conduct simulations of the different designs to evaluate their performance.
- Compare the performance of the simulated designs against the requirements and key performance indices (KPIs).
- **Prototyping**
- Select the best-performing design from the simulations.
- Develop a prototype of the selected design.
- **Testing and Performance Analysis**
- Test the prototype in various underwater environments.
- Analyze the performance of the prototype against the defined requirements.
- Refine the design based on testing results.

16.5.1.3 Strengths

- Provides a targeted solution addressing specific underwater monitoring needs.
- Utilizes nature's proven strategies for energy efficiency and adaptability.

16.5.1.4 Weaknesses/Limitations

- Requires extensive knowledge of marine biology and robotics.
- Initial development and testing phases may be time-consuming and costly.

16.5.1.5 Recommendations

- Foster interdisciplinary collaboration between marine biologists and robotic engineers.
- Develop educational programs to enhance understanding of both biological and engineering principles.
- Create a repository of biological analogies and their engineering applications for future reference.

16.5.2 *Case Study 2: Solution-Driven Bioinspired Design (Bottom-Up)*

Biological Phenomenon. The Adhesion Mechanism of Gecko Feet

The adhesion mechanism of gecko feet is a remarkable biological phenomenon that allows geckos to adhere to various surfaces using van der Waals forces. This bottom-up approach begins with identifying a biological principle and exploring potential engineering applications.

16.5.2.1 When to Select Solution-Driven Bioinspired Design

- When there is a fascinating biological phenomenon suggesting potential engineering applications.
- Suitable for exploring innovative solutions by applying biological principles to various problems.
- Ideal for broad applicability across different fields, where the biological model provides a unique solution.

16.5.2.2 Process Assessment and Design Development

(a) **Identifying the Biological Function**
  - Study the microstructure and adhesion properties of gecko feet.
  - Understand the underlying principles, such as the role of setae (tiny hair-like structures) and van der Waals forces.

(b) **Initial Engineering Design and Assessment**
  - Consider initial engineering problems where strong, reversible adhesion could be beneficial.
  - Assess the potential applications for gecko-inspired adhesion in various engineering contexts.

(c) **Extract and Analyze Principles**
  - Analyze how the gecko's adhesion mechanism works at a microscopic level.
  - Extract specific data such as adhesion strength, reversibility, and surface adaptability.
  - Compare the principles of gecko adhesion with other biological adhesion mechanisms.
  - Rank the biological principles based on their potential engineering applications.

(d) **Shortlisting and Design Development**
  - Shortlist the best-matching biological principles that align with the identified engineering problems.
  - Develop initial design concepts based on the shortlisted biological principles.

(e) **Simulation and Performance Comparison**
  - Conduct simulations of the different designs to evaluate their performance.

- Compare the performance of the simulated designs against the requirements and key performance indices (KPIs).

(f) **Prototyping**
- Select the best-performing design from the simulations.
- Develop a prototype of the selected design.

(g) **Testing and Performance Analysis**
- Test the prototype in various scenarios to assess its adhesion performance.
- Analyze the performance of the prototype against the defined requirements.
- Refine the design based on testing results.

16.5.2.3 Strengths

- Encourages innovative exploration and adaptation of biological solutions.
- Broad applicability across different fields.

16.5.2.4 Weaknesses/Limitations

- May be less systematic and require extensive iteration.
- Requires a detailed understanding of the biological phenomena.

16.5.2.5 Recommendations

- Develop a comprehensive database of biological phenomena and their associated physical principles.
- Link these phenomena to various potential engineering problems they can help solve.
- Encourage interdisciplinary collaboration to translate biological principles into practical engineering solutions effectively.

16.6 Best Practices, Pitfalls, and Challenges in Bioinspired Design

This section covers the best general design practices in Mechatronics, bioinspired design practices, and pitfalls and challenges in bioinspired design to provide comprehensive guidance for practitioners in the field.

16.6.1 Best General Design Practices in Mechatronics

1. Engage experts from various fields and disciplines to foster innovative solutions, such as engineering, biology, physics, computer science, etc., to bring diverse perspectives that drive innovation and enhance problem-solving capabilities.
2. Conduct extensive brainstorming sessions to understand the problem and define requirements thoroughly to ensure the design addresses all relevant aspects and requirements.
3. Based on the gathered requirements, develop detailed operational scenarios and conceptual designs to help visualize the solution and identify potential challenges early.
4. Employ an iterative design, prototyping, and testing process to refine solutions for continuous improvement and optimization of the design based on feedback and testing results.
5. Prioritize sustainability in design to align with environmental goals. This demands integrating sustainability principles, such as eco-friendly, energy-efficient design, etc., into the design process to ensure that the solutions are environmentally friendly and resource-efficient.
6. Consider the end-user in the design process to ensure practicality and usability based on the end-user's preferences. This ensures the design is practical, user-friendly, and meets the intended purpose.
7. Ensure the design complies with the relevant regulatory standards and guidelines regarding safety, reliability, and market acceptance.
8. Documenting and sharing knowledge supported by knowledge management tools helps preserve insights, facilitate collaboration, and enhance future design efforts.
9. Stay updated with the latest technological and methodological advancements ensures that the design process incorporates the best available practices and innovations.
10. Conduct a lifecycle analysis to evaluate the long-term sustainability of the design, ensuring that sustainability considerations are integrated from inception to disposal.
11. Develop tools to evaluate the environmental impacts of design concepts in the early phases to select the most sustainable options. This will help select and refine designs that minimize negative environmental effects.
12. Perform comprehensive studies on the theoretical bases of ideation methods and their application across different knowledge domains by understanding the theoretical foundations of design methods enhances their effective application and fosters innovative solutions.
13. Design for scalability to ensure solutions can be effectively applied on a larger scale to allow for broader application and impact effectively.
14. Employ computational tools, models, and simulations to predict and optimize design performance to help accurately predict performance and optimize design parameters.

15. Efficient use of resources to minimize waste and reduce environmental impact, such as selecting materials or manufacturing processes, minimizes waste and contributes to sustainability.
16. Conduct thorough and rigorous testing to ensure designs perform as expected in real-world conditions. This helps validate the design and ensure reliable performance under actual operating conditions.
17. Think holistically and act concurrently throughout the development process, ensuring an integrated approach to the design is considered in parallel, leading to cohesive and effective solutions.
18. Establish interdisciplinary development teams covering all aspects, from market/end-user needs to finance/purchasing, researchers from different disciplines, manufacturing, and selling, to engage simultaneously in the development process. Involving all relevant stakeholders from the beginning ensures diverse insights and expertise, leading to more innovative and market-relevant products.

16.6.2 Best Practices for Bioinspired Design Synergized in Mechatronics

1. Recognize the difference between learning about nature and learning from nature. Effective bioinspired design requires understanding and innovatively applying natural principles
2. Understand different biological mechanisms. Recognize that nature evolves various mechanisms to solve real-world problems, offering diverse insights
3. Foster interdisciplinary collaboration. Engaging engineering, biology, and other relevant field experts ensures a holistic approach to problem-solving and innovation.
4. Integrate different bioinspired design paradigms. Combining different paradigms, such as problem-driven and solution-driven approaches, enhances versatility and adaptability in addressing complex engineering challenges. Utilize advanced materials to replicate biological structures accurately.
5. Develop necessary and comprehensive datasets, tools, and databases. Creating and maintaining comprehensive and robust datasets, tools, and databases of biological principles and analogies, along with computational tools, aids in the systematic application of bioinspired design principles
6. Ensure accurate understanding and application of biological principles. Collaborating with biologists and conducting in-depth research into biological systems is crucial for correctly abstracting and applying biological principles to engineering problems. This will help to avoid ineffective designs and applications.
7. Consider bioinspired designs' ethical and financial aspects to ensure their feasibility and public acceptance, especially in sensitive areas like medical applications. Ensure an accurate understanding and application of biological principles to avoid ineffective designs.

8. Implement rigorous prototyping and testing phases to ensure that bioinspired designs perform as expected and can be refined to improve functionality.
9. Develop the targeted bioinspired behaviors, structures, and materials using advanced manufacturing technologies, smart materials, and computational capabilities.
10. Ensure that bioinspired designs selectively integrate the strengths of biological systems without incorporating their weaknesses
11. Understand cognitive mechanisms in bioinspired design
12. Develop tools and techniques that support the cognitive processes involved in bioinspired design, enhancing innovative solutions
13. Explore sustainability solutions in biology. Investigate biological sustainability solutions before developing tools and methodologies to ensure environmentally
14. Facilitate interaction between biology, informatics, and m/nano spaces. Enable innovative solutions by integrating biological principles with advanced information and micro/nanotechnologies.
15. Ensure that bioinspired designs are sustainable throughout their lifecycle, not just in isolated aspects
16. Understand and develop insight into how biological systems interact at component and system levels to inform bioinspired design
17. Ensure that abstraction in bioinspired design considers appropriate levels of abstractions related to all functional aspects to avoid incorrect designs

16.6.3 Bioinspired Design Pitfalls

1. Overly complex designs can hinder implementation and usability.
2. Misapplication of biological principles can result in ineffective designs.
3. Lack of interdisciplinary collaboration can constrain the potential of bioinspired designs.
4. Ignoring user needs may result in impractical and unusable solutions.
5. High costs associated with bioinspired designs must be considered.
6. Inadequate biological knowledge can lead to incorrect applications and ineffective solutions.
7. Insufficient testing may result in designs that underperform in real-world conditions.
8. Navigating regulatory requirements can be difficult and time-consuming.
9. Ignoring ethical considerations can lead to public backlash and legal problems.
10. Designs that cannot be scaled appropriately may have limited applicability.
11. Bioinspired designs can be resource-intensive, requiring careful management.
12. Current technology may not always be capable of replicating complex biological systems.
13. New and innovative designs may face resistance in the market.
14. Protecting intellectual property can be complex and contentious.
15. Balancing performance and sustainability can be challenging.

16.6.4 Bioinspired Design Challenges

1. Develop tools and methodologies that reduce the environmental impact of engineering designs while adding value
2. Combining knowledge from biology, engineering, and other fields requires effective interdisciplinary communication.
3. Systems are complex, and fully understanding them is challenging.
4. Keeping pace with rapid technological advancements is essential for effective bioinspired design.
5. Balancing the costs of bioinspired designs with their benefits requires careful planning.
6. Developing effective prototypes and conducting thorough testing is resource-intensive.
7. Ensuring that designs can be scaled for applications and environments is a significant challenge.
8. Choosing appropriate materials that mimic biological properties can be difficult.
9. Navigating the regulatory landscape for bioinspired designs, particularly in medical applications, is complex.
10. Market acceptance for new and innovative bioinspired designs requires effective communication and marketing.
11. Conducting comprehensive sustainability assessments to ensure long-term benefits.
12. Protecting intellectual property while fostering innovation and collaboration.
13. Addressing ethical implications and ensuring designs do not have adverse effects.
14. Staying updated with the latest research and adapting designs accordingly.
15. Building effective collaborations and networks to support bioinspired design initiatives.
16. Managing public perception and educating stakeholders about the benefits of bioinspired design.

16.7 Conclusion and Future Work

16.7.1 Conclusion

In conclusion, synergizing bioinspired design principles in Mechatronics offers significant potential for enhancing sustainability and innovation. This chapter comprehensively overviews various bioinspired design paradigms, including problem-driven, solution-driven, biomimicry, bioreplication, hybrid, system-of-systems, and information-processing approaches. We have illustrated how these paradigms can address engineering challenges by examining their key features, strengths, weaknesses, and examples.

The practical benefits and challenges of bioinspired design have been illustrated through detailed case studies, demonstrating the effectiveness of these approaches in real-world applications. Best practices have been identified, such as combining problem-driven and solution-driven methods, utilizing advanced tools and techniques, and ensuring adaptability and flexibility through system-of-systems design. Additionally, the chapter has highlighted potential pitfalls and challenges, providing valuable insights for practitioners in the field.

Synergizing bioinspired design principles within Mechatronics not only enhances sustainability but also drives the evolution of Mechatronics toward EcoMechatronics. This evolution marks a new edge in the field, opening up innovative pathways for developing environmentally friendly and resource-efficient technologies, applying ideas abstracted from ecology, and using the principles of the circular economy. By integrating multiple approaches, robust, adaptable, and sustainable solutions can be developed, addressing complex engineering challenges and contributing to sustainable development.

16.7.2 Future Work

Future work in bioinspired design should focus on improving current methodologies and addressing existing knowledge gaps. Key areas for future research and development include:

1. Developing sophisticated computational models and simulation development environments to represent better, develop new models, analyze, predict, and optimize the performance of bioinspired designs. Improved computational tools will enable more accurate translations of biological principles into engineering solutions.
2. Fostering closer collaboration between biologists, engineers, and material scientists to deepen the understanding of biological systems and enhance the application of these insights in engineering design.
3. Addressing challenges related to the scalability and manufacturability of bioinspired designs. Developing new manufacturing techniques that can accurately replicate complex biological structures on a large scale will be essential for widespread adoption.
4. Establishing standardized metrics for assessing and evaluating the performance and sustainability of bioinspired designs. These metrics should encompass environmental impact, resource efficiency, and lifecycle assessment to ensure that bioinspired solutions contribute to sustainable development.
5. Conducting long-term performance studies of bioinspired designs in real-world applications to assess their durability, reliability, and effectiveness. These studies will provide valuable data to refine and improve future designs.

6. Developing educational programs and curricula to train the next generation of engineers and designers in bioinspired design principles. These programs should emphasize interdisciplinary learning and practical applications.
7. Continuously exploring and studying new biological systems that can offer innovative solutions to emerging.

References

1. Braha D, Maimon O (1997) The design process: properties, paradigms, and structure. IEEE Trans Syst Man Cybern Part A Syst Humans 27(2):146–166. https://doi.org/10.1109/3468.554679
2. Christoffersson J, van Noort D, Mandenius C (2017) Developing organ-on-a-chip concepts using bio-mechatronic design methodology. Biofabrication 9:025023. https://doi.org/10.1088/1758-5090/aa71ca
3. Habib MK (2007) Mechatronics: a unifying interdisciplinary and intelligent engineering paradigm. IEEE Ind Electron Mag 1(2):12–24. https://doi.org/10.1109/MIE.2007.901480
4. Habib M, Nagata F (2019) Bioinspired design: creativity and sustainability. Proceeding of 2019 20th IEEE international conference on research and education in mechatronics (IEEE REM 2019), Wels, Austria, pp 1–5. https://doi.org/10.1109/REM.2019.8744140
5. Habib MK (2022). Chapter title: EcoMechatronics and bioinspired design ecology, circular economy, and sustainability. In: EcoMechatronics: foundations, methods, and solutions—challenges for evolution, development & sustainability. (Eds Hehenberger, Habib and Bradley), Springer Nature Switzerland, pp 65–83. https://doi.org/10.1007/978-3-031-07555-1_4
6. Hehenberger P, Habib MK, Bradley D (2022) EcoMechatronics challenges for evolution, development and sustainability. Book, Springer Nature Switzerland
7. Huanca J, Zamora J, Cornejo J, Palomares R (2022) Mechatronic design and kinematic analysis of 8 DOF serial robot manipulator to perform electrostatic spray painting process on electrical panels. 2022 IEEE Engineering international research conference (EIRCON), pp 1-4. Lima, Peru. https://doi.org/10.1109/EIRCON56026.2022.9934104
8. Ilami M, Bagheri H, Ahmed R, Skowronek EO, Marvi H (2020) Materials, actuators, and sensors for soft bioinspired robots. Adv Mater. https://doi.org/10.1002/adma.202003139
9. Lenau TA, Metze AL, Hesselberg T (2018) Paradigms for biologically inspired design. SPIE Proceeding volume 10593, Bioinspiration, Biomimetics, and Bioreplication VIII, Article No. 1059302. Denver, Colorado, USA. https://doi.org/10.1117/12.2296560
10. Maldonado F, Acosta J, Tormo-Barbero J, Grau P, Guzmán MM, Ollero A (2020). Adaptive nonlinear control for perching of a bioinspired ornithopter. 2020 IEEE/RSJ International conference on intelligent robots and systems (IROS), pp 1385–1390, Las Vegas, NV, USA https://doi.org/10.1109/IROS45743.2020.9341793
11. Meza BH, Fernández H, Cornejo J, Palomares R (2021) Mechatronic design and kinematic analysis of delta robot applied on supermarkets for social distancing during COVID-19 pandemic. 2021 IEEE Engineering international research conference (EIRCON), 1–4. Lima, Peru, https://doi.org/10.1109/EIRCON52903.2021.9613165
12. Nnodim CT, Arowolo MO, Agboola BD, Ogundokun RO, Abiodun MK (2021) Future trends in mechatronics. IAES Int J Robot Autom (IJRA) 10(1):24–31. http://doi.org/10.11591/ijra.v10i1.pp24-31, https://ijra.iaescore.com/index.php/IJRA/article/view/20356/12936
13. Ñope-Giraldo RM, Illapuma-Ccallo LA, Cornejo J, Palacios P, Napán JL, Cruz F, Palomares R, Cornejo-Aguilar JA, Vargas M (2021) Mechatronic systems design of ROHNI-1: hybrid cyber-human medical robot for COVID-19 health surveillance at wholesale-supermarket entrances. In: Proceedings of the IEEE 2021 Global medical engineering physics exchanges/pan American

Health care exchanges (GMEPE/PAHCE), Sevilla, Spain, 2021, pp 1–7. https://doi.org/10.1109/GMEPE/PAHCE50215.2021.9434874

14. O'Rourke J, Seepersad C (2011) A methodology for identifying factors that contribute to the sustainability of bioinspired engineered systems. ASME 2011 International Mechanical Engineering Congress and Exposition, IMECE2011-62605, pp 845-857, November 11–17, 2011, Denver, Colorado, USA. https://doi.org/10.1115/IMECE2011-62605
15. Salustri FA, Venter RD (1994) A new programming paradigm for engineering design software. Eng Comput 10(95–111):1994. https://doi.org/10.1007/BF01200180
16. Spitas C (2010) Analysis of systematic engineering design paradigms in industrial practice: a survey. J Eng Des 20(4):343–363. https://doi.org/10.1080/09544820903437734
17. Tantawi K, Nasab A, Potter L, Tantawi O, Wilson N, Henrie A, Sirinterlikci A, Kaplanoglu E (2023) Mechatronics education—current and future trends. In: 2023 11th International symposium on digital forensics and security (ISDFS), pp 1–5. https://doi.org/10.1109/ISDFS58141.2023.10131706
18. Todd R (1993) A new paradigm for pre-engineering education: Design and technology education K-12. Proceedings of IEEE frontiers in education 1993 23rd annual conference, pp 405–407. https://doi.org/10.1109/FIE.1993.405501
19. van der Schoor MJ, Göhlich D (2023) Integrating sustainability in the design process of urban service robots. Front Robot A I:10. https://doi.org/10.3389/frobt.2023.1250697
20. Wendland M (2023) An approach to game-based learning, collaboration and design challenges for teaching the design of mechatronic systems. In: Education and New Developments 2023—Volume 2, pp 52–56. https://doi.org/10.36315/2023v2end011, https://end-educationconference.org/wp-content/uploads/2023/06/01_OP_187.pdf
21. Wu Z, Liu J, Yu J, Fang H (2017) Development of a novel robotic dolphin and its application to water quality monitoring. IEEE/ASME Trans Mechatron 22:2130–2140. https://doi.org/10.1109/TMECH.2017.2722009
22. Zhang Y, Mei J, Yan C, Liao T, Bell J, Sun Z (2020) Bioinspired 2D nanomaterials for sustainable applications. Adv Mater Wiley 2020(32):1902806. https://doi.org/10.1002/adma.202070144

Chapter 17
Thin Films Applications in Mechatronic Systems

Liliana-Laura Badita-Voicu and Adrian-Catalin Voicu

Abstract Mechatronics is an interdisciplinary field that includes precision mechanics, electronics and informatics, offering the opportunity to obtain complex mechatronic systems, such as cars, industrial and medical robots, prosthetic systems and artificial organs. Taking into account the main tribological processes that occur during the use of mechatronic systems causing their wear; in recent years, it has been demonstrated that the thin films can help increase wear resistance and extend their life span. This chapter focuses on the use of thin films, which include a wide variety of maintenance coatings, in mechatronic systems. Following a review of mechatronic applications and methods of increasing their life span, a series of researches are discussed for the use of thin films in order to improve the tribological properties of medical and measurement systems. The main research carried out by the authors of this chapter was related to the increase of resistance to mechanical wear, to understand the relationship between the structure of the thin films and the substrates used, influence of the deposition methods and thin films' behaviour in the conditions encountered during the use. Thin metallic and composite films with thicknesses between 50 nm and several hundred μm have demonstrated their potential to be used with the aim of increasing the wear resistance of mechatronic systems that is very important considering the development of various systems that must withstand mechanical, chemical and biological wear factors. The fact that thin film coatings help to increase wear resistance is one of the main reasons why manufacturers of mechatronic systems will be able to design and realize various configurations of systems with superior tribological properties. Thus, in the coming years, an important role in this field will be played by the structural, tribological and applicative characterization of thin films with new chemical compositions and varying thicknesses.

L.-L. Badita-Voicu (✉) · A.-C. Voicu
National Institute of Research and Development in Mechatronics and Measurement Technique INCDMTM, Bucharest, Romania
e-mail: badita_l@yahoo.com

P. Hehenberger and D. Bradley (eds.), *Mechatronic Futures*,
https://doi.org/10.1007/978-3-031-83571-1_17

17.1 Mechatronics

Mechatronics is an interdisciplinary field that includes precision mechanics, electronics and informatics.

The concept of mechatronics is illustrated in Fig. 1.1. In the process of designing and bringing a mechatronic product to life, a comprehensive approach is essential. This diagram highlights the importance of integrating mechanics, electronics and informatics to address the unique challenges associated with mass movement, electrons' movement and information flow. By combining these three disciplines, a holistic understanding can be achieved, enabling the creation of innovative and efficient mechatronic solutions. Completion through manufacturing and processing of the products obtained is the goal of both conception and design activities, where everything is carried out on the basis of an efficient management, in accordance with the needs of the market.

Based on this concept, a mechatronic system is a technical system that integrates, in a flexible configuration, mechanical, electronic and control components with digital informational systems, to generate an intelligent control of movements, in order to obtain a multitude of functions. The block diagram of a mechatronic system is shown in Fig. 1.2.

A mechatronic system is a complex one characterized by extensive integration and enhanced capabilities in contrast to conventional systems. This advancement has paved the way for new opportunities across various industries, including agriculture, construction, machinery manufacturing, healthcare and other equipment-related sectors.

The modern automobile, robots (industrial and civil robots, micro- and nanorobots used in the military industry), numerically controlled machine tools, computer technology, telecommunications technology, biomedical equipment for medical investigations, prosthetic systems for rehabilitation, intelligent transport systems, research equipment, household equipment, office equipment (computers, printers), audio–video equipment (recording systems), modern agricultural machines, etc., are representative examples of mechatronic products [1, 2].

An advantage of the robot-type mechatronic systems is the fact that there is the possibility that they can be used in special environments, where the presence of an operator is not possible or he is subjected to risk.

17.2 Tribological Phenomenon in the Mechatronics Systems

In general, in mechatronic engineering, depending on the types of systems, the tribological phenomenon can be found between the elements and components that come into direct contact with each other. Some of these components can be in motion, being subjected to various static and dynamic surface forces and friction that represent the main tribological phenomena and can influence their functioning [4].

The friction phenomenon, in general, is encountered in the operation of technical and technological systems/equipment, in the operation of classic and intelligent manufacturing processes with or without processing, in physical, chemical and thermal phenomena on materials and on metallographic structures and networks, at different levels, producing the mechanical process of loss of material–wear. This materials destruction is produced by different mechanisms–adhesive wear [5], abrasive wear, third body wear, superficial fatigue wear, corrosion wear and fretting wear [6]–that in some systems can be combined [7].

To minimize or potentially eradicate the occurrence of friction during the operational movement of **mechatronic components** and systems, fundamental principles are employed in the design and assembly of the mechanisms within technical and technological systems. These principles aim to keep friction to a value as small as possible, thereby guaranteeing the optimal performance of these systems.

Engineering solutions for the maximum reduction, to almost zero, of the friction coefficients are currently known, through the approach of the new guides, such as hydrostatic/hydrodynamic guides that completely eliminate the friction coefficients between materials. Until now, other types of guides, specific to the sliding bearings for rotation, have been made, using the magnetic field, using magnetic fluids, etc., valid for high and very high speeds. And these guides, metal/fluid type, have a certain friction, but its value is incomparable with the metal/metal friction.

Over time, due to the stress which actions on the surface of mechatronic components, the different materials from their composition can be subjected to wear. Therefore, the sliding surfaces of mechatronic components must usually be hard, to keep wear to minimum values. Therefore, materials used to manufacture these types of components must be resistant to mechanical degradation and corrosion and have an increased strength-to-weight ratio and hardness. In the case of biomechatronic components, the used materials must also be biocompatible to avoid adverse reactions in the human body and assure the osseointegration, in the case of the implant. All these properties can decrease the particles generation phenomenon, which generally have negative effects on the mechatronic components.

Until now, materials couples have been created, such as treated steel/untreated steel, steel/plastic material, steel/teflon, steel/cast iron, steel/synthetic material—ruby/sapphire, through which it was tried to obtain some friction coefficients as small as possible, until its complete elimination.

Metal composites based on titanium, together with other materials like carbon elements and ceramics offer the opportunity to increase the wear resistance of this metal [8]. TC4 was combined with a graphene film that has a lubricating role, assuring a Gr/TC4 composite material with improved wear resistance [9]. Si3N4 ceramics combined with tungsten was manufactured by gas pressure sintering, in order to improve the abrasive wear resistance [10].

In order to cater to the demands of **biomechatronic components**, such as orthopaedic implants, materials with enhanced mechanical strength and an anti-corrosive composition were developed. These materials are specifically designed to withstand both dynamic and static stresses, which vary depending on the level of patient's activity.

Using materials that resist to the loads placed on the implant's surface, it is possible to remove the wear that can produce fractures, loosening and rejection of physiological reactions of the body associated with its use.

Pure metals, like titanium, chromium, cobalt or molybdenum, together with stainless steels, and cobalt, zirconium or titanium-based alloys are currently used to produce body-tolerant implants, like hip and knee prostheses, with increased mechanical strength and high corrosion resistance. Ceramics is another category of biomaterials that is used due to its low wear rate, but its main disadvantage is the increased cracking possibility.

All these materials—stainless steel, various metallic alloys, polymers and ceramic composites—undergo degradation after 10–15 years of use. This is the main reason why other metals, like nickel and iron, were added to these alloys in order to obtain superior resistance [11].

Additionally, the utilization of composite materials possessing enhanced characteristics, coupled with the requirement for durable mechatronic and biomechatronic components, incorporating anticorrosive compositions employed in various industries, has resulted in the application of thin films composed of materials possessing superior properties on the surfaces of these components.

17.3 Thin Films Used in Mechatronics

Thin films that can be used to improve the wear resistance of mechatronic components must have a high adhesion to their surfaces, high hardness, corrosion resistance, thermal stability and a low friction coefficient to minimize the generation of wear particles, that contribute to the destruction of these components [12].

To date, the utilization of thin films composed of different types of nanoparticles, varying in thickness from 100 nm to several hundred microns, has been employed to enhance several desirable properties such as scratch resistance, corrosion resistance, UV resistance, IR reflection, anti-microbial activity and self-cleaning capabilities (Table 17.1).

Composite thin films, such as nanoparticle-based composite coating, nano-clay-based composite coating, nanocomposite latex emulsions are the most important types of materials that can increase the wear resistance of the mechatronic components. The application of nano-clays in coatings has proven to be highly successful in enhancing corrosion resistance. This has been accomplished by developing inorganic–organic hybrid polymer latexes through the combination of silica with a variety of monomers including butyl acrylate, methyl methacrylate and styrene [14].

Colloidal silica has been utilized to enhance the abrasion resistance of transparent coatings applied to acrylic eyeglass lenses, floor wear films and automotive top-coats [15].

Table 17.1 Nanomaterials thin films for various properties improvement [13]

Property	Nanomaterials thin films
Anti-microbial	CuO; TiO2; ZnO
Gas barrier	Nanoclays
Corrosion	Nanoclays, boehmite
Electrical conductivity, static charge dissipation	ITO, ATO, SnO2
Fire retardant	Nanoclays
IR-absorption/reflection	ITO, ATO, TiO2, In2O3
Magnetic	Fe2O3
Mechanical, scratch resistance	Al2O3; SiO2; ZrO2
Photocatalysis, self-cleaning	TiO2; ZnO
UV stability	TiO2; ZnO; BaSO4; CeO2

In research studies, it has been shown that increase of polymer coatings scratch resistance can be made by inserting metal-oxide nanoparticles (e.g. nano-alumina and nano-silica) into their surface [16]. Also, aluminium silicate nanoparticles can be applied to increase this resistance of automotive coatings.

Boehmite nanoparticles, chromates or composites that possess carboxylic acids or amines functional groups have been used as anticorrosion protective films.

Physical vapour deposition of TiN and TiC thin films on the surface of mechatronic components destroyed by adhesive wear, such as, for example, bearings, valves, gears, fittings, cutting tools, leads to an increase in wear resistance and a decrease in friction coefficient [17].

A superior wear resistance was also obtained after depositing multicomponent nitride (Al-Ti-V-Cr)N thin films with face-centred cubic structure, on the surface of micro- and nano-devices [18].

The longevity of the wastewater treatment pumps, including wear rings, seals, bearings, rotors and casings, was significantly extended by applying an Al2O3 40TiO2 film onto the substrate of a shaft sleeve. This shaft sleeve operates on a steel bearing disc, typically in conditions of dry lubrication.

In the **biomechatronics** domain, this began with shaping a structure that mimicked the bone using moulds made of CoCrMo alloy material (such as in the JUDET prosthesis designed with a “prometal” surface or the LORD type of prostheses whose surface is produced with a coating of spheres), which allows the growth of osteoblasts and implicitly ensures osseointegration at a value between 20 and 40%. [19] Later, the horizontal monolayer, open pore structure was replaced by a multilayer that interconnects the porous structure.

New developments go towards the application of titanium metallic fibres and spheres on titanium alloys stems by diffusion bonding techniques.

Metal–matrix composite (MMC) thin films combined with ceramic particles are another category of materials used for improving the mechanical properties over conventional monolithic alloys.

Other biocompatible alloys and materials with increased hardness that had been used were nitride-based compounds, such as TiN, ZrN, CrN and TiAlN [20], amorphous carbon, calcium-phosphate [21], graphite-like carbon (GLC), diamond-like carbon (DLC) and tantalum (Ta) [22].

Raw nanomaterials, like nanocrystalline materials and nanoparticles can be also used as coatings for prostheses and implants, in order to increase their wear resistance.

Corrosion resistance improvement and adhesion increase have been obtained in the case of biomechatronic components when metals or other materials, like ultra-nanocrystalline diamond [23] and ceramics had been deposited on their surfaces [24].

During the years, the research studies were directed towards the use of multi-layers deposition. Thus, it was demonstrated that hydroxyapatite/Ti alloy [25], Ti/TiC multi-layer [26], Mo/Ni, TiN/Cu and ZrN/W [27], zirconium oxynitride/zirconium oxide [28] can be useful to increase the wear resistance of biomechatronic components.

Inorganic metal oxides have also been used for biomechatronic components, mainly as anti-microbial coatings. The most important ones that can also have a special resistance are ZnO [29], TiO2 [30], TaO [31].

The capacity analysis of micro- and nanostructured thin films, which are deposited on the surface of certain mechatronic and biomechatronic components using physical methods, has been the primary focus of several research projects conducted by the authors of this chapter at the National Institute of Research and Development for Mechatronics and Measurement Technique in Bucharest, Romania. These studies aim to enhance the wear resistance of these components, based on the valuable insights gained from these observations.

A series of researches have been carried out with the aim of increasing the lifetime of some mechatronic components (mainly, components of guidance/positioning equipment) and biomechatronic components (e.g. hip prostheses), using nanostructured metallic thin films.

In this way, it was desired to choose the optimal variants of substrate material—deposited material following the comparative analysis of nanostructured thin films deposited on the surfaces of some samples by different methods, characterized from a tribological point of view and finally tested in real conditions (on these components).

Cr, Ti, Al thin films and Ti + Al multilayers, with thicknesses between 50 and 100 nm were deposited on RUL1, OLC45, C120 and OSC8 steel substrates (materials used in industrial mechatronic components) by electron beam evaporation technique [32], using a Temescal FC-2000 deposition system [33].

First, these metallic thin films were deposited on samples and tribologically characterized, being selected the ones with an increased wear resistance for functional tests in real conditions. The tribological behaviour of two mechatronic components of a guiding/positioning subassembly (Fig. 3.1a), with 10 cm and 14 cm diameters (Fig. 3.1b), was analysed after the functional testing in dynamic and in real-time conditions.

Following the wear testing and tribological characterization of the coated and uncoated conical-type mechatronic components, it was observed that the Cr (50 nm thick), Al (50 nm thick), Ti (100 nm thick) layers and the Ti + Al multilayer (50 nm.

+ 50 nm thick) penetrate through all the irregularities from the surface of steel substrates, creating a good bond between layer and substrate.

After the analysis of the tribological parameters' values, Ti and Cr thin films were the ones that had superior tribological properties, being selected to be tested in dynamic real-time conditions in order to determine the wear resistance of mechatronic components coated with these thin films.

The results obtained showed that nanostructured Ti and Cr thin films increase the wear resistance of the mechatronic components made of RUL1 steel, the Cr thin film having the highest hardness after the wear tests in real conditions, and being useful to protect and increase the lifetime of the tested pieces.

Similar experimental studies have been made for testing TiN, TiO2 and CrN thin films, in order to determine their usefulness for improving the wear resistance of biomechatronic components (femoral heads of hip prosthesis). TiN, TiO2 and CrN thin films, with three different thicknesses—0.5 μm, 1 μm and 1.5 μm—were deposited on biocompatible CoCr and M30NW steel substrates by various physical vapour deposition (PVD) techniques, respectively, cathodic arc (CA) [34] direct current (DC) sputtering [35] and High-Power Impulse Magnetron Sputtering (HiPIMS) techniques [36].

The deposition of all thin films by the three methods mentioned above was performed using a Cathodic-Arc UVN-MGM equipment and a DC and HiPIMS equipment Leybiold Z-550-S [37].

In order to determine the hardness (H) and elastic modulus (*E*r) of deposited TiN, TiO2 and CrN thin films, nanoindentation tests were realized for recording the force–displacement curves and extracting these parameters. These tests were performed using a Hysitron TI Premier nanoindenter, equipped with a Berkovich peak 142.3° [37], applying a maximum compressive force of 0.7 mN (except for the CrN layer–force of 2 mN).

The results obtained showed that DC sputtering is the best method for deposition of biocompatible thin films on metallic substrates, taking into account that obtained TiN and CrN layers on both CoCr and M30NW substrates presented the highest hardness (Fig. 3.3) and elastic modulus.

This demonstrated an increased wear resistance of the femoral heads (components of hip prosthesis—Fig. 3.4) coated with these layers and was tested in real dynamic conditions using an INSTRON 8872 equipment (Fig. 3.5a) by applying a force of 1.5 kN, at a frequency of 5 Hz and movement amplitude of 0.75 kN. [37] Other tests in real dynamic conditions have been made using a MTS Bionix Implants and Prosthetic Components Testing Equipment (Fig. 3.5b) that allows to perform complete flexion/extension, abduction/adduction and internal/external rotation movements of this prosthesis.

The surfaces of the tested uncoated and coated femoral heads of TiN (Fig. 3. 6) and CrN (Fig. 3.7) thin films, characterized using the KLA TENCOR-MICRO XAM-100 system [38], demonstrated that the thin films help to increase the wear resistance of the implant.

Within another project, the authors of this chapter studied and analysed the wear evolution of total hip prostheses components (femoral heads) coated with TiN thin films deposited by pulsed laser deposition (PLD) method and using scratch tests.

TiN depositions on SS316L stainless steel substrate were carried out in a deposition chamber (Fig. 3.8a), with a stainless-steel reaction chamber (Fig. 3.8b), in a dynamic flow of N2 constantly controlled with the help of a regulator mass flow the number of consecutive laser pulses being varied from 5000 to 20,000 pulses.

The adhesion strength of the TiN layers was determined by making scratches (with lengths of 1 cm) with a diamond sphero-conical tip (with a radius of 2 mm), which were drawn at a constant speed over the sample (1 cm/min), at loads of 2.5 N, 5 N, 10 N, 20 N, 40 N, 80 N and 125 N.

After performing the scratching tests and examining the scratch marks with the optical microscope and atomic force microscope (AFM), several types of destructions (main critical events) of the surfaces that appeared as a result of the action of the critical loads were detected: deformations and material detachments (Ec1), detached deposition fragments (Ec2), semicircular deposition cracks (Ec3), material lifting on the edges (Ec4), cutting of the base material (Ec5) and oxidation (Ec6), which represent the failure of TiN coating adhesion in real applications.

It was observed that on all three layers deposited at 5000, 10,000 and 20,000 pulses, critical events such as Ec1, Ec2, Ec3, Ec6 appear, depending on the thickness of the layer, from the action of forces with low values. The critical events Ec4 and Ec5 appear at higher values of the force in the case of layers deposited at 10,000 and 20,000 pulses. This demonstrates the larger hardness of these layers compared to the one deposited at 5000 pulses.

From the analysis of the average roughness values, obtained by characterization using the Nano Laboratory AFM Probe NTEGRA NT-MDT system (Fig. 3.9), it is observed that there is a direct correlation between an increase in the applied force and an increase in the roughness values. This discovery highlights the influence of force on the surface roughness, underlining the significance of controlling and understanding the impact of applied force in surface characterization. This may be due to the fact that as the force increases, the scratches show different depths, and for certain values, the layer exfoliates (thereby resulting an increase in the non-uniformity of the layer). At certain force values and only for certain layers, it was possible to observe that the scratching test also affected the substrate, as a result of the layer's exfoliation.

The scratching tests to which the TiN thin films were subjected demonstrated that these coatings increase the hardness, but there are also cases in which they are affected by exfoliation or scratching.

The essential factor for increasing the hardness of the femoral heads with thin films is to ensure a strongly adherent coating on the substrate. This is a problem that belongs to the processing of the substrate's surface and also to the parameters of the coating process. In the present case, PLD coatings with the maximum number of pulses (20,000 pulses) can represent an alternative technology to ensure the adhesion and scratch resistance of TiN coatings on SS 316L steel and even Ti-6Al-4 V femoral heads.

The solution to address these challenges lies in the utilization of hard multilayer coatings. To achieve optimal results, several conditions must be met. Firstly, it is crucial to establish the ideal roughness of the substrate, additionally, finding the optimal balance between the hardness of the base material and the coating layer or multilayer structure is essential, and lastly, both the base material and the coating should possess similar elasticity modulus. By adhering to these principles, a comprehensive and efficient resolution can be attained for these issues. All parameters can be controlled and varied in order to identify the optimal regime for obtaining the structures and thin films. Fine control of the layer thickness can be achieved by controlling the number of pulses.

The main conclusion of the results obtained from the projects realized by the authors of this chapter is that physical vapour deposition methods, like DC sputtering and physical laser deposition (PLD) are the best methods for deposition of nanostructured thin films with superior mechanical properties that can increase the wear resistance of mechatronic and biomechatronic components.

Thin films integrated into the mechatronic components can protect and increase the lifetime of real mechatronic systems.

Thus, the information obtained from these researches can be used in the future to modernize mechatronic components and increase the quality of various products useful for society.

Even if the use of these thin films and methods has demonstrated the increase in the wear resistance of mechatronic and biomechatronic components, in the future more experimental studies will be carried out to determine the optimal tribological parameters of the surfaces and the technological parameters of the deposition methods to ensure the highest possible adhesion between layer and substrate (facilitating the establishment of double diffusion between layer and substrate) depending on the material and on the deposition method used.

17.4 The Future of Thin Films in the Mechatronics Field

Advanced research on the development and behaviour of new materials, in order to obtain optimized materials with improved physical, chemical, mechanical, electronic characteristics, with specific nano-structural properties and with application possibilities in the most diverse conditions and natural phenomena is the main research direction of materials engineering.

Currently, tribology is the basis for optimizing the operation of all technical and technological systems/equipment, of micro- and nanotechnologies and of optimizing the obtaining of new intelligent materials with superior/improved properties.

The recent discoveries of new materials with "added" molecular and atomic structures lead to credible hypotheses that in the near future we shall have "intelligent materials" that will revolutionize the engineering world by their properties and the way they interact.

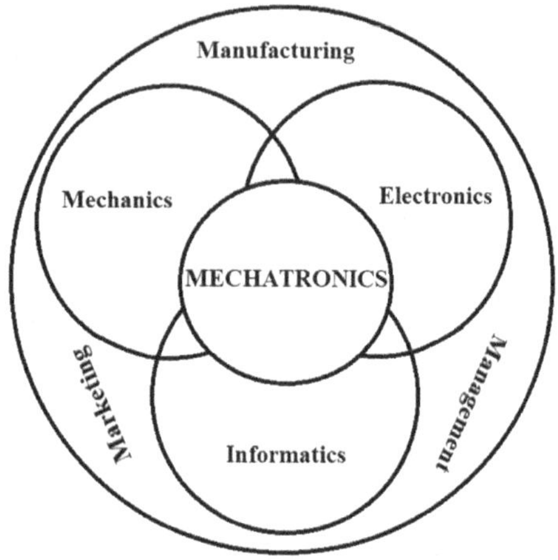

Fig. 17.1 Concept of mechatronics

The structures of these new materials, in advanced concepts, help the "atomic agglomeration" of the discovered chemical element, resettling its nano-structures and creating new, superior complex properties.

Thus, as a virtual structure, in the future, we can mention:

1. Tribology of internal phenomena, from the neural networks of molecules, on a nanometric and subatomic scale;
2. Intelligent tribology of internal microphenomena, from structural/microstructural and nano-structural networks;
3. Atomic and subatomic tribology, regarding the new advanced intelligent materials.

Through these structural approaches, materials engineering will discover in the future, new materials based on the carbon atom, the tungsten atom, the lithium atom, the chromium atom, etc., in the form of a neural network, with "atomic agglomeration" with "concentrations of properties/characteristics" distributed on their different chemical levels and "treated with subatomic phenomena".

The future of the field seems assured by the development of new digital and intelligent high-tech systems and technologies that will benefit, in addition to the contribution of AI software, from micro-nano-technologies and conventional and unconventional micro-nano-processing, on the surface and in the mass with electrons beam, laser, X-rays, UV/IR rays, in 3D and multi-D.

The future use of thin films will contribute to ensure the high-tech and long-lasting level of systems, products and mechatronic components that will develop science and engineering for the new industry, the new knowledge-based economy and the new information society.

In this way, there will be a development of the industry and, implicitly, of the economy, at the national, European and world levels.

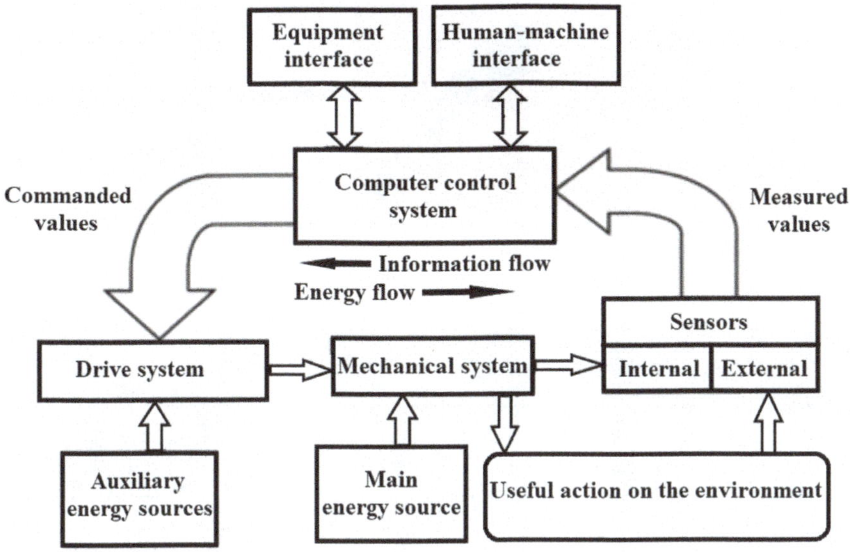

Fig. 17.2 Block diagram of a mechatronic system [3]

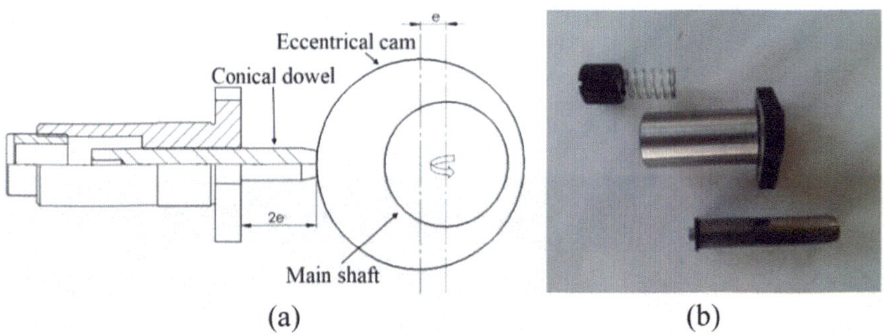

Fig. 17.3 Scheme (**a**) and components (**b**) of the tested guiding/positioning subassembly

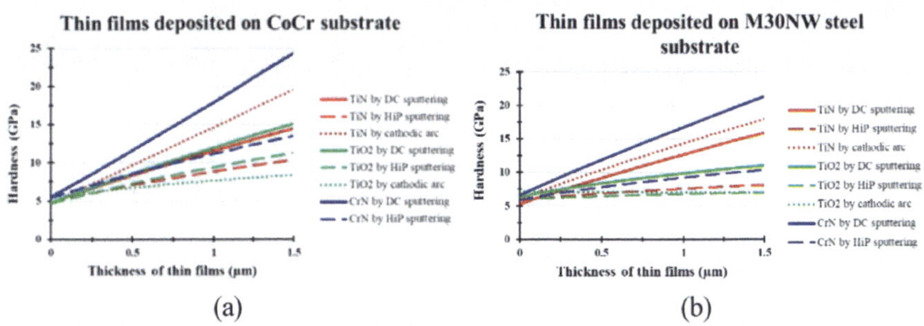

Fig. 17.4 Variation of the hardness depending on the thickness of each type of thin film deposited by DC sputtering, HiPMS sputtering and cathodic arc on (**a**) CoCr and (**b**) M30NW steel substrates

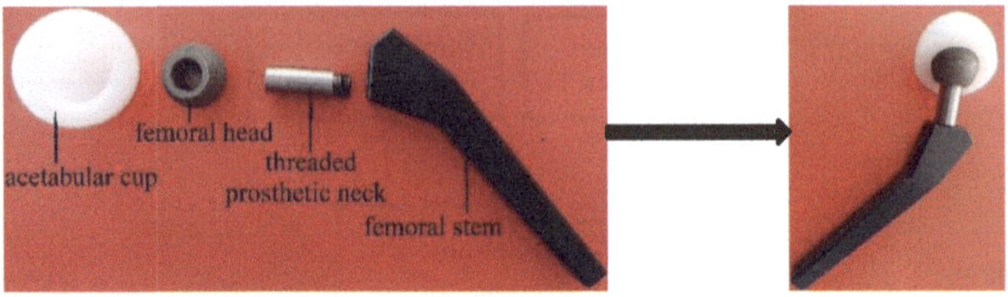

Fig. 17.5 Components of a hip prosthesis

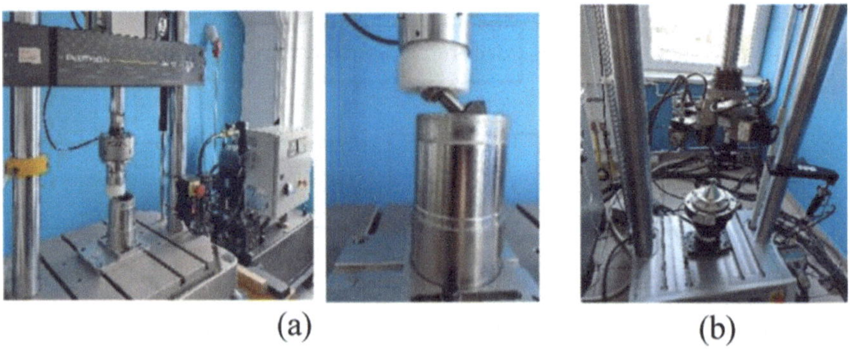

Fig. 17.6 (**a**) INSTRON 8872 system for fatigue wear testing of prosthetic components. (**b**) MTS bionix implants and prosthetic components testing equipment

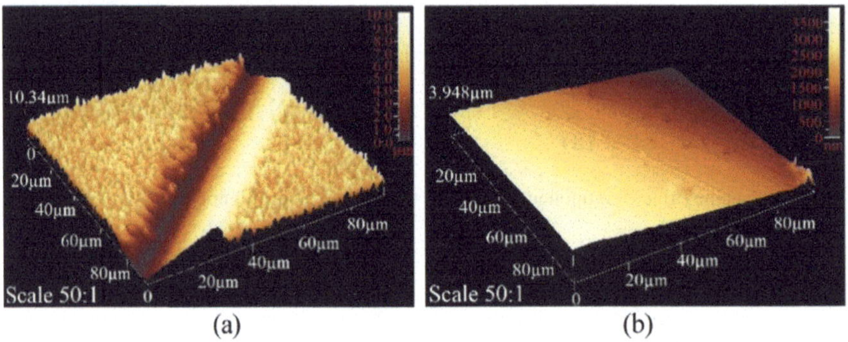

Fig. 17.7 Surfaces of the uncoated M30NW steel femoral heads (**a**) and of the TiN thin films (**b**) deposited on the M30NW steel femoral heads, after the fatigue wear tests (scale 50:1)

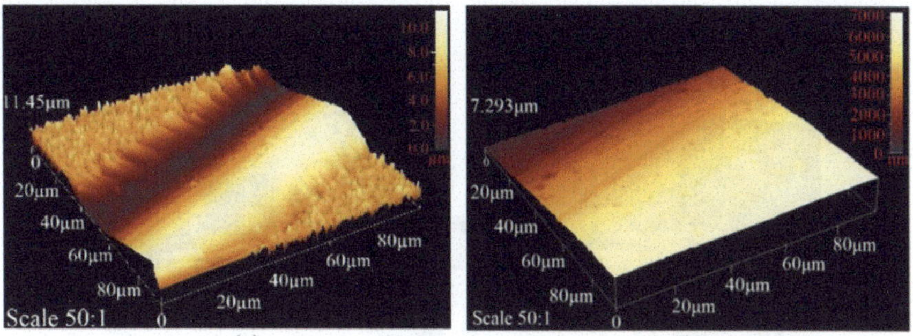

(a) (b)

Fig. 17.8 Surfaces of the uncoated CrCr femoral heads (**a**) and of the CrN thin films (**b**) deposited on the CoCr femoral heads, after testing with MTS system (scale 50:1)

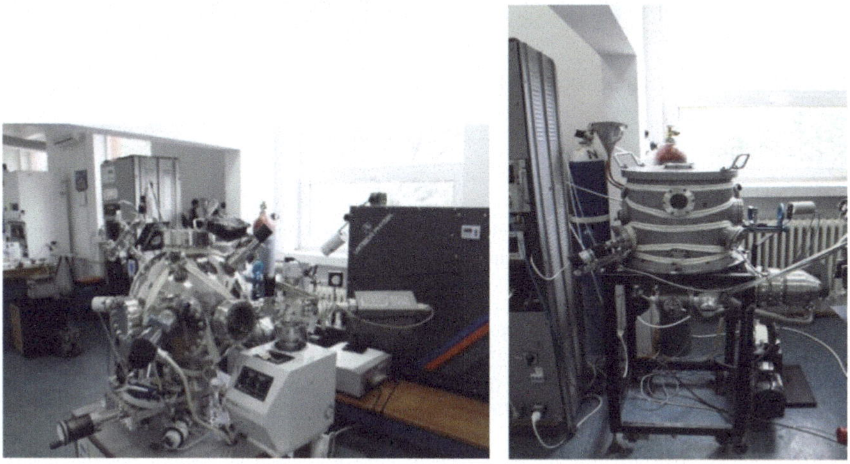

(a) (b)

Fig. 17.9 Laser deposition chamber (**a**) and stainless-steel reaction chamber (**b**)

Fig. 17.10 Atomic force microscope NTEGRA probe nano laboratory

References

1. Bolton W (2018) Mechatronics—electronic control systems in mechanical and electrical engineering, Pearson Educational Limited
2. Bishop RH (2002) The mechatronics handbook. CRC Press
3. Ardeleanu M Gheorghe G, Matei G (2007) Mechantronics. Principles and applications, Agir
4. Elzayady N, Elsoeudy R (2021) Microstructure and wear mechanisms investigation on the brake pad. J Mater Res Technol 11:2314–2335
5. Podgornik B (2022) Adhesive wear failures. J Fail Anal Prev 22(11):113–138
6. Stachowiak GW, Batchelor AW (2006) Engineering tribology, Publisher Butterworth—Heinemann
7. Straffelini G (2015) Friction and wear methodologies for design and control, Springer Cham
8. Ammisetti DK, Kruthiventi SSH (2021) Recent trends on titanium metal matrix composites: a review. Mater Today-Proc 46(19):9730–9735
9. Zhao ZY, Liu MR, Du WB, Bai PK, Hu YB (2023) Tribological behavior of Gr/TC4 matrix composite through selective laser melting. Mater Chem Phys 305:127973
10. Wang LJ, Qiao ZH, Qi Q, Yu Y, Li TY, Liu XJ, Huang ZR, Tang HG, Liu WM (2022) Improving abrasive wear resistance of Si3N4 ceramics with self-matching through tungsten induced tribochemical wear. Wear 494–495:204254
11. Wang L, Wang C, Wu S, Fan Y, Li X (2020) Influence of the mechanical properties of biomaterials on degradability, cell behaviors and signaling pathways: current progress and challenges. Biomater Sci 8:2714–2733
12. Cordill MJ, Kreiml P, Mitterer C (2022) Materials engineering for flexible metallic thin film applications. Materials 15(3):926
13. Fernando RH, Sung LP (2009) Nanotechnology applications in coatings, ACS Symposium Series, 1008
14. Zaarei D, Sarabi AA, Sharif F, Kassiriha SM (2008) Structure, properties and corrosion resistivity of polymeric nanocomposite coatings based on layered silicates. J Coat Technol Res 5(2):241–249
15. Tiarks F, Landfester K, Antonietti M (2001) Silica nanoparticles as surfactants and fillers for latexes made by miniemulsion polymerization. Langmuir 17(19):5775–5780

16. He Q, Wu L, Gu G, You B (2002) Preparation and characterization of acrylic/nano-TiO2 composite latex. High Perform Polym 14(4):383–396
17. Motru S, Hussain N, Khan ZA (2020) Tribological studies of high surface finish ceramic coatings for low friction and adhesive wear resistant. Mater Today Proc 27(3):2208–2212
18. Feng CS, Feng XB, Guan Z, Song HQ, Wang TL, Liao WB, Lu Y, Zhang FX (2022) Nanocrystalline (AlTiVCr)N multi-component nitride thin films with superior mechanical performance. Nanomaterials 12(15):2722
19. Ogiso MJ (1998) Bone formation on HA implants: a commentary. J Long-Term Eff Med 8(3/4):193–200
20. Subramanian B, Muraleedharan CV, Ananthakumar R, Jayachandran M (2011) A comparative study of titanium nitride (TiN), titanium oxy nitride (TiON) and titanium aluminum nitride (TiAlN), as surface coatings for bio implants. Surf Coat Technol 205(21–22):5014–5020
21. Mahmoodian R, Rafieerad AR, Ashra MR, Bushroa AR (2015) Surface characterization and corrosion behavior of calcium phosphate-base composite layer on titanium and its alloys via plasma electrolytic oxidation: a review paper. Mater Sci Eng C 57:397–413
22. Ching HA, Choudhury D, Nine MJ, Abu Osman NA (2014) Effects of surface coating on reducing friction and wear of orthopaedic implants. Sci Technol Adv Mater 15(1):014402
23. Auciello O, Renou S, Kang K, Tasat D, Olmedo D (2022) A biocompatible Ultrananocrystalline diamond (UNCD) coating for a new generation of dental implants. Nanomaterials 12(5):782–796
24. Catledge SA, Fries MD, Vohra YK, Lacefield WR, Lemons JE, Woodard S, Venugopalanc R (2002) Nanostructured ceramics for biomedical implants. J Nanosci Nanotechnol 2(3–4):293–312
25. Sygnatowicz M, Tiwari A (2009) Controlled synthesis of hydroxyapatite-based coatings for biomedical application. Mater Sci Eng, C 29(3):1071–1076
26. Shanaghi A, Chu PK, Rouh Aghdam AS, Xu R (2013) Structure and corrosion resistance of Ti/TiC coatings fabricated by plasma immersion ion implantation and deposition on nickel—titanium. Surf Coat Technol 229:151–155
27. Abadias G, Michel A, Tromas C, Jaouen C, Dub SN (2007) Stress, interfacial effects and mechanical properties of nanoscale multilayered coatings. Surf Coat Technol 202(4–7):844–853
28. Cubillos GI, Bethencourt M, Olaya JJ, Alfonso JE, Marco JF (2014) The influence of deposition temperature on microstructure and corrosion resistance of ZrOxNy/ZrO2 coatings deposited using RF sputtering. Appl Surf Sci 309:181–187
29. Sollazzo V, Pezzetti F, Scarano A, Piattelli A, Bignozzi CA, Massari L, Brunelli G, Carinci F (2008) Zirconium oxide coating improves implant osseointegration in vivo. Dental Mater 24(3):357–361
30. Giavaresi G, Giardino R, Ambrosio L, Battiston G, Gerbasi R, Fini M, Rimondini L, Torricelli P (2003) In vitro biocompatibility of titanium oxide for prosthetic devices nanostructured by low pressure metal-organic chemical vapor deposition. Int J Artif Organs 26(8):774–780
31. Rahmati B, Sarhan AAD, Zalnezhad E, Kamiab Z (2016) Development of tantalum oxide (Ta-O) thin film coating on biomedical Ti-6Al-4V alloy to enhance mechanical properties and biocompatibility. Ceram Int 42(1):466–480
32. Yushkov Y, Oks EM, Tyunkov A, Zolotukhin D (2022) Electron-beam synthesis of dielectric coatings using fore vacuum plasma electron sources (Review). Coatings 12(1):82
33. Badita LL, Gheorghe G, Bratu V, Gornoava V, Vocurek M, Zapciu A, Munteanu IS (2017) Mechanical characterization of nanostructured thin films used to improve mechatronic components. Scient Bul Valahia Univ Mater Mech 15(12):48–54
34. Brown IG (1998) Cathodic arc deposition of films. Annu Rev Mater Sci 28:243–269
35. Gobbi AL, Nascente PAP (2013) D.C. Sputtering, in Encyclopedia of Tribology (eds) Wang, QJ & Chung, YW, Springer Science and Business Media, pp 699–706
36. Ehiasarian A (2010) High-power impulse magnetron sputtering and its applications. Pure Appl Chem 82(6):1247–1258

37. Badita LL, Zapciu A, Vitelaru C, Parau AC, Constantin LR, Sobetkii A (2022) Influence of the deposition method on the hardness and elastic modulus of biocompatible thin layers deposited on metallic substrates, In: International conference on reliable systems engineering. Lecture Notes in Networks and Systems (ed) Cioboata, DD, Springer, pp 338–357
38. Badita-Voicu LL, Zapciu A, Vitelaru C, Parau AC, Constantin LR, Sobetkii A (2023) Evaluation of the applicative-tribological capacity of thin films used to increase the lifetime of hip prostheses components. Int J Mechatron Appl Mech 3:174–183

Chapter 18
The Integration of Ecomechatronics and Industry 4.0 Towards Sustainable Manufacturing

Mutaz Ryalat, Marwa Al-Faouri, Hisham El-Moaqet, Natheer Almtireen, and Ghaith Al-Refai

Abstract This chapter explores the integration of ecomechatronics and Industry 4.0, aiming to provide a comprehensive overview of the synergies between these two fields towards sustainability. Ecomechatronics represents a multidisciplinary approach that combines principles of mechanical engineering, electronics, computer science, and environmental science to design and develop sustainable and environmentally friendly mechatronic systems. Industry 4.0, on the other hand, embodies the fourth industrial revolution characterised by the integration of digital technologies into manufacturing processes. The fusion of these two domains has the potential to revolutionise the way we design, produce, and utilise mechatronic systems, fostering sustainability and efficiency. As a case study, a smart plastic injection moulding machine will be designed that demonstrates a successful integration of Industry 4.0 technologies and ecomechatronics towards the environmental and economic benefits achieved through the integration of smart technologies.

Keywords Ecomechatronics · Industry 4.0 · Sustainability · Robotics · Cyber-physical systems · IoT

18.1 Introduction

In recent years, there has been a growing awareness of the need for sustainable and eco-friendly practices in various industries. This includes the field of mechatronics, which focuses on the integration of mechanical, electrical, and software components in systems [1]. With the emergence of Industry 4.0, there is an opportunity to leverage advanced technologies to develop and implement ecomechatronics solutions. These strategies have the potential to enhance energy efficiency, decrease carbon emissions,

M. Ryalat (✉) · M. Al-Faouri · H. El-Moaqet · N. Almtireen · G. Al-Refai
Mechatronics Engineering, German Jordanian University, Amman, Jordan
e-mail: mutaz.ryalat@gju.edu.jo
URL: https://www.gju.edu.jo

P. Hehenberger and D. Bradley (eds.), *Mechatronic Futures*,
https://doi.org/10.1007/978-3-031-83571-1_18

and reduce waste in manufacturing processes [2]. Furthermore, the incorporation of Industry 4.0 technologies into ecomechatronics can also enable the development of intelligent and interconnected systems, leading to improved automation, efficiency, and flexibility. The purpose of this chapter is to investigate the idea of ecomechatronics through the utilisation of Industry 4.0 technology and to examine how it could potentially influence sustainable manufacturing.

We will highlight the key components of ecomechatronics that can be enhanced through Industry 4.0 technology. These include intelligent sensors and actuators, real-time data monitoring and analysis, predictive maintenance, and digital twins. Additionally, we will discuss the challenges that need to be addressed in order to successfully implement ecomechatronics using Industry 4.0 technology. Finally, we will present a case study that demonstrates the successful implementation of ecomechatronics using Industry 4.0 technology in the smart plastic injection industry.

18.2 Ecomechatronics

Mechatronics is a multidisciplinary field that combines mechanical engineering, electronics, computer science, and control engineering to design and create intelligent systems or processes. Intelligent mechatronics refers to the integration of intelligent or smart technologies into mechatronic systems. The addition of intelligence in mechatronics involves incorporating advanced sensors, actuators, controllers, and computational capabilities to improve overall performance, adaptability, and autonomy of these systems [3]. The crucial application of mechatronics has been identified because of its contribution and benefits that have resulted from the integration of its disciplines. Rapid advancements in technology and the increasing complexity of systems are transforming modern industry. These changes lead to higher-quality products, lower production costs, and continuous breakthroughs in manufacturing technology.

Leveraging the power of mechatronics and fostering innovation to develop new solutions while reducing existing costs can pave the way for a significant shift towards more environmentally friendly practices. This approach, focusing on the creation of sustainable and eco-friendly solutions through mechatronics, lays the foundation for a new field known as ecomechatronics [1]. Ecomechatronics aims to create intelligent and sustainable mechatronic systems by considering environmental impact throughout the product life cycle. Ecomechatronics, a burgeoning field, extends beyond traditional mechatronics by incorporating environmental consciousness throughout the life cycle of a system [4]. It encompasses the design, development, operation, and disposal of intelligent mechatronic systems with a focus on minimising environmental impact and promoting sustainability. The focus is on incorporating sustainable principles into mechatronic systems throughout all stages of product lifecycle management (PLM), a topic that has been extensively investigated in [5] and illustrated in Fig. 18.1.

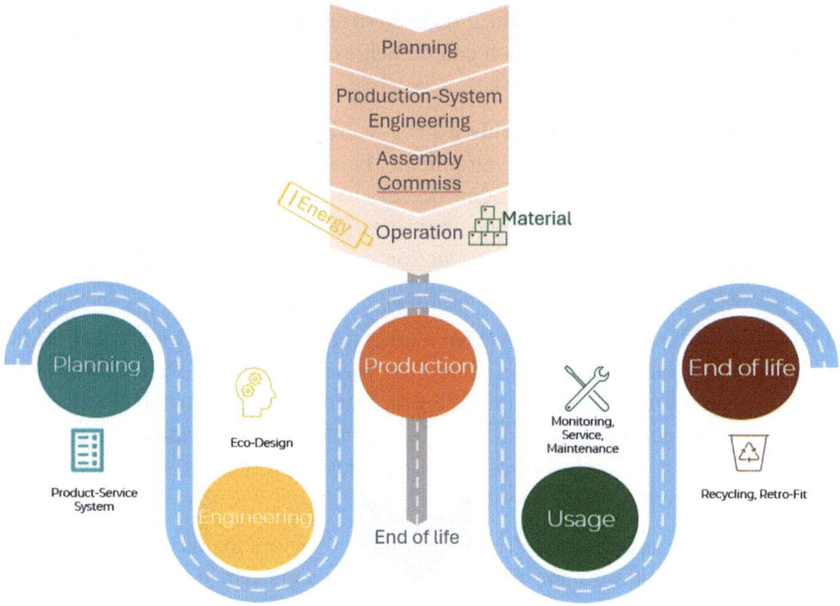

Fig. 18.1 Product lifecycle management (PLM) stages

Ecomechatronics solutions play a role in advancing several of the 17 Sustainable Development Goals (SDGs) established by the UN Department of Economic and Social Affairs [6]. They achieve this by decreasing carbon emissions, reducing waste, enhancing energy and resource efficiency, and optimising product performance across their entire lifespan [7]. In a global scenario where 1,940 million tonnes of carbon dioxide are released every year and 9 billion tonnes of waste are disposed of in landfills annually, the potential influence of ecomechatronics on worldwide sustainable progress is significant [8, 9]. This comprehensive perspective takes into account:

18.2.1 Material/Part Selection

In ecomechatronics, the selection of materials should not only consider their performance and durability but also the environmental impact of the materials utilised. Sustainable materials, such as recycled plastics or biodegradable composites, can help reduce the overall carbon footprint of a product. Additionally, choosing materials that are easily recyclable or have a longer lifespan can contribute to a more sustainable product lifecycle. Industries can advance their environmentally friendly efforts and support a more sustainable future by thoughtfully choosing materials that are in line with their sustainability objectives. Prioritising sustainability in material selection allows companies to reduce their environmental footprint and promote a greener future for the planet, which can be achieved by focusing on the following practices [10–12]:

Recyclability and reusability: This aspect is crucial to take into account when selecting materials for product development. By selecting materials that are easily recyclable or reusable, businesses can decrease waste generation and lower their environmental impact. Moreover, integrating recycled materials into their products can facilitate the circularity of the production process and diminish the reliance on new (virgin) resources.

Durability and longevity: There is also a crucial factor to consider when choosing materials for product development. Choosing materials that are durable and long-lasting can help to reduce the frequency of replacement and ultimately decrease the amount of waste generated. By investing in durable materials capable of enduring damages such as wear and tear, industries can develop products with an extended lifespan, thus fostering a more sustainable future.

Biodegradability and toxicity: There are two key factors to consider when evaluating the sustainability of materials. Biodegradable materials break down naturally over time, reducing the environmental impact of waste disposal. On the other hand, toxic materials can harm both the environment and human health. Choosing materials that are non-toxic and biodegradable can help to minimise the negative effects of product manufacturing and disposal. Furthermore, incorporating renewable materials into product design can also help to reduce reliance on finite resources and promote a more circular economy.

18.2.2 Energy Efficiency

Enhancing energy efficiency is a crucial strategy to cut costs and mitigate environmental damage. Energy efficiency involves accomplishing more with less and attaining similar outcomes with less energy usage. By fine-tuning the systems to reduce energy consumption during their operation, we can substantially decrease waste and decrease our ecological footprint by [1, 13]:

- Utilising low-power components,
- Employing renewable energy sources (e.g. solar, wind),
- Implementing intelligent control systems to optimise energy usage,
- Reducing parasitic losses through efficient design and material selection.

18.2.3 Life Cycle Assessment (LCA)

Life cycle analysis is an important tool for assessing the environmental impacts and sustainability of products and systems. It involves evaluating the entire life cycle of a product, from raw material extraction to end-of-life treatment, to identify and mitigate environmental impacts [14]. In the context of mechatronic systems, integrating sustainability criteria into the design process is crucial. This can be achieved by considering the three dimensions of sustainability (economic, environmental, and social) and using decision support methods to select and calculate indicators for each alternative [15]. Additionally, life cycle costing can be used to assess the economic sustainability of mechatronic systems, complementing the traditional focus on costs [16].

By examining the environmental impacts of power electronic systems, such as energy usage and depletion of resources, it is possible to suggest ways to enhance their sustainability. For instance, in the automotive sector, the role of a mechatronic engineer has evolved. While previously their focus was on designing cars efficiently, the rapid technological advancements in recent years necessitate greater attention to environmental concerns throughout the entire lifespan of a vehicle. Thanks to progress in computational and information technologies, notably the rise of the internet and the expansion of cyber-infrastructure (like shared software tools and services, e-commerce, and online communication), various stages in a product's life cycle—from material extraction and manufacturing to distribution, usage, maintenance, and disposal—can now be monitored, assessed, and managed more efficiently than ever before.

18.3 Industry 4.0

The term Industry 4.0, also known as the fourth industrial revolution, is currently a rapidly developing concept in academia, industry, and society. The driver of the new wave of industrial revolution is considered a wide range of digital technologies that impact manufacturing companies [17, 18]. Industry 4.0 proposes a method to shift from traditional machine-focused manufacturing to digital manufacturing using advanced technologies such as cyber-physical systems, the Internet of Things, autonomous robotics, artificial intelligence, big data analytics, blockchain, cloud computing, and additive manufacturing [19]. At the heart of Industry 4.0 lies a set of foundational technological pillars that drive this transformation [19–22]:

o Cyber-physical systems (CPSs): Cyber-physical Systems (CPSs) denote integrated systems in which physical objects and software are closely intertwined, facilitating advanced interaction through information exchange among various components in various ways.

- Internet of things (IoT): It represents a network that provides communication between 'things' (i.e. objects or devices: physical devices, vehicles, and appliances) embedded with sensors, software, and connectivity capabilities through information and communication technology (ICT) infrastructure, resulting in real-time collection and sharing of data among machines, processes, and complete factory facilities. This data serves as the essential component of Industry 4.0 systems.
- Big data analytics: The substantial volume of data produced by IoT devices necessitates robust tools for processing and analysis. Big data analytics are used to derive insights from this data, recognise patterns, enhance processes, and facilitate predictive decision-making.
- Artificial intelligence (AI): Artificial intelligence is a crucial component of Industry 4.0, transforming the manufacturing sector. AI's ability to process vast amounts of data, identify complex patterns, and learn from experience enables unprecedented levels of optimisation and intelligent decision-making. In manufacturing, AI drives predictive maintenance, minimising downtime by anticipating equipment failures. It optimises production processes, leading to enhanced resource utilisation and reduced waste. AI-powered quality control systems detect defects with pinpoint accuracy, ensuring product consistency. Furthermore, AI facilitates mass customisation, allowing manufacturers to cater to individual customer needs without sacrificing efficiency.
- Cloud computing: Cloud computing provides scalable, on-demand computing resources (large datasets and clusters) and storage needed for Industry 4.0 applications. It facilitates data sharing, remote monitoring, and collaboration across distributed manufacturing networks without extensive on-premise infrastructure.
- Robotics: Highly advanced robots with enhanced sensing capabilities, dexterity, and AI-powered decision-making play a crucial role in Industry 4.0 factories. Industrial robots, cobots, and mobile robots are autonomous, flexible, and cooperative, for industrial automation with the goal of performing production tasks more precisely with minimal human involvement.
- Cybersecurity: Connectivity increases vulnerability. Robust cybersecurity measures are critical to protect Industry 4.0 systems from cyberattacks, safeguarding data integrity, and ensuring the availability of critical manufacturing infrastructure.
- Augmented reality (AR): AR overlays digital information and visualisations onto the real world through mobile devices, tablets, or smart glasses. In manufacturing, AR is used for maintenance, providing technicians with real-time instructions and guidance, remote expert assistance, and equipment visualisation.
- Additive manufacturing (3D Printing): Additive manufacturing revolutionises design and production. It allows for the creation of complex, lightweight components, on-demand manufacturing for customisation, rapid prototyping, and spare parts production, reducing waste and minimising supply chain constraints.

18.3.1 Sustainability in Industry 4.0

Industry 4.0 presents a crucial opportunity to promote social and technological innovation by integrating cutting-edge technological architectures. There are important impacts and benefits that can be realised through the use of Industry 4.0 technologies in manufacturing and different industries around the world. The three main spectrums of the impact of Industry 4.0 are the economic impact, the social impact, and the environmental impact shown in Fig. 18.2. The impact of Industry 4.0 on economic development has been thoroughly researched and recorded. Industry 4.0 technologies and design principles greatly enhance the efficiency of manufacturing and industrial operations. Enhanced profit margin, increased innovation capacity, improved production control, higher productivity and enhanced customisation, reduced labour and material costs, and decreased time-to-market for products are advantages that digitalisation provides industrial organisations [23]. Industry 4.0 also transforms the workplace to support social sustainability. Smart and autonomous systems relieve workers of hazardous or tedious tasks and promote physical and mental health. This leads to higher job satisfaction and a more motivated workforce [20]. The environmental consequences of Industry 4.0 are evident in the implementation of sustainable practices in sectors such as enhanced global resource efficiency and significant waste reduction achieved by embracing novel business strategies and digitally interconnected processes [24]. The well-planned technological development and adoption of Industry 4.0 can result in global resource efficiency gains and big waste discharges as we tend to step towards a virtuous eco-friendly world. From an ecological perspective, Industry 4.0 technologies hold significant promise for environmental sustainability. These technologies enable manufacturers to reduce energy and resource consumption throughout the production and supply chains. Real-time data analysis across processes allows for optimisation, minimising waste, and energy usage [25]. Additionally, Industry 4.0 facilitates data-driven carbon footprint assessments, enabling targeted strategies to minimise CO2 emissions [26]. Furthermore, these technologies promote closed-loop production systems by facilitating product disassembly and material recovery for reuse, recycling, or remanufacturing, contributing to a more circular economy [25, 26].

18.4 Eco-Manufacturing via the Integration of Ecomechatronics and Industry 4.0

In recent years, the global industrial landscape has undergone significant transformations due to continuous technological advancements and innovations. The fourth industrial revolution, often known as Industry 4.0, is a term used to characterise the current trend of intelligent automation, digitalisation, information and communication technologies. This revolution is transforming the manufacturing industry by integrating diverse technologies and disciplines, such as mechatronics, to create

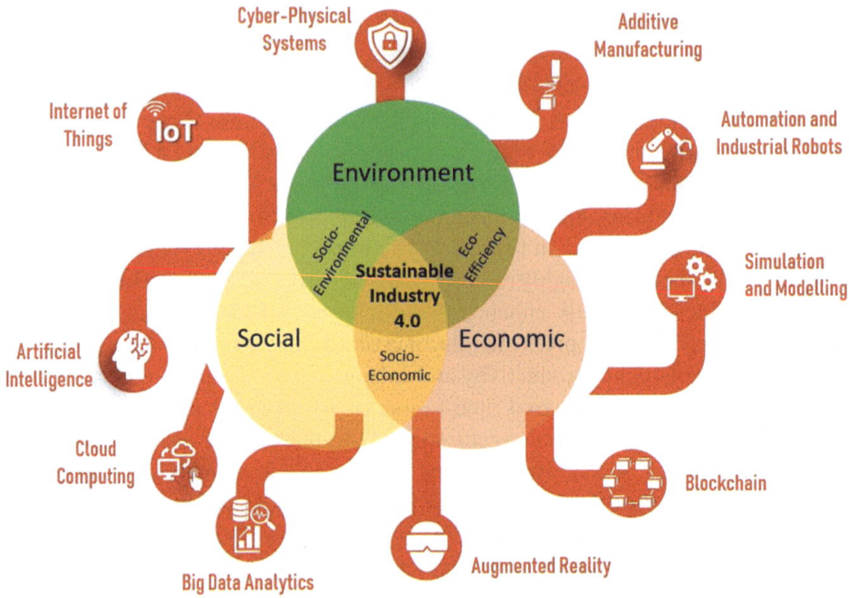

Fig. 18.2 Framework of sustainable industry 4.0

smart factories that achieve superior levels of automation, precision, productivity, and flexibility. The integration of Industry 4.0 and ecomechatronics serves as a catalyst for sustainable manufacturing covering a variety of sustainability objectives shown in Fig. 18.3. Manufacturing organisations have a huge impact on the three pillars of sustainability: economic, environmental, and social. The implementation of Industry 4.0 principles and ecomechatronics in manufacturing processes can contribute significantly to the creation of sustainable value. Using Industry 4.0 technologies, such as cyber-physical systems, Internet of Things, and data-driven decision-making, ecomechatronics enables the creation of intelligent and interconnected systems that improve automation, efficiency, and flexibility in manufacturing and industrial processes as well as serve the sustainable objectives of industries. The compatibility of Industry 4.0 and mechatronics is evident, as mechatronics provides the necessary hardware and software components for the implementation of Industry 4.0 technologies.

One of the key pillars of Industry 4.0 is the development of smart factories, which are efficient, secure, and sustainable. The integration of mechatronics principles within the smart factory concept plays a crucial role in achieving this goal. The integration of Industry 4.0 and mechatronics provides several advantages in manufacturing. Firstly, improved automation and efficiency in production processes can be achieved by implementing intelligent systems that can monitor and optimise manufacturing processes. For instance, predictive maintenance systems can detect potential equipment failures and schedule maintenance before the equipment

Fig. 18.3 Contributions of mechatronics and Industry 4.0 towards sustainability

breaks down, reducing downtime and increasing productivity. Secondly, improved product quality and customisation can be achieved by implementing smart systems that can monitor and control product quality in real time. For example, sensors can detect defects in products and trigger corrective actions to ensure that customers receive only high-quality products. Lastly, reduced production costs and waste can be achieved by implementing intelligent systems that can optimise energy consumption, reduce material waste, and improve supply chain management. It is evident from the earlier sections that both ecomechatronics and Industry 4.0 play a role in advancing sustainability in manufacturing.

18.4.1 Examples of How They Work Together for Eco-Manufacturing

Energy Efficiency

- Ecomechatronics designs inherently energy-efficient motors, drives, and systems. There is a growing demand to create industrial machines that are more energy-efficient, reliable, flexible, and accurate to meet market needs for increased

productivity while reducing costs in a sustainable manner. The mechatronic design process offers a structured methodology for precise motor selection and the development of optimised motion profiles. By carefully designing these trajectories, undesirable effects such as vibration, noise, and excessive stress are minimised in mechanical and electronic components [27, 28].
- Industry 4.0 sensors track energy use patterns in real time. AI-powered analytics identify areas for improvement and optimise processes to minimise waste. Industry 4.0 enables data-driven decision-making, energy system monitoring, and autonomous energy consumption optimisation, thus enhancing energy efficiency. Energy data needs to be gathered and combined from smart metres, sensors, and other energy devices and then incorporated into production management [29].
- In smart grid application, control of energy flow, from one or multiple energy sources such as capacitors, batteries, or the grid, is an integrated ecomechatronics and I4 approach that focuses on designing distributed control strategies to dictate energy flow in energy storage elements [22, 29]

Material Optimisation

- Ecomechatronics promotes the use of recycled, recyclable and less toxic materials. The field of soft robotics [30] recognises the need for sustainable practices and proposes effective recycling of elastomeric materials to reduce environmental impact.
- Industry 4.0 helps track material flows, ensure efficient use, and minimise material waste throughout the supply chain.
- In the context of Industry 4.0 and 3D printing technology, the material optimisation process is used in the initial stages of the design process. The current FEA software, which serves as an ecomechatronic tool, offers the ability to adjust the parameters of various materials for the purpose of optimisation [31].

Predictive Maintenance [19, 32]

- Using intelligent IoT sensors and platforms, ecomechatronics can perform real-time data collection and comprehensive condition monitoring to enable predictive maintenance.
- Industry 4.0 using advanced big data analytical tools (such as data mining, statistical analysis, and predictive analytics) can process data collected from sensors and devices to monitor and predict machine failures, overloads, or any other issues. When combined with AI which is another Industry 4.0 core has the capability to automatically detect abnormalities in the initial phases of production. This feature allows for predictive maintenance, reducing unexpected failures and preventing breakdowns that lead to unnecessary resource use and waste.

Another field of applications involves process optimisation, where Industry 4.0 provides real-time process visualisation. By integrating ecomechatronics with AI, it can identify inefficiencies that result in saving resources and enhancing modifications.

18.5 Case Study: Smart Plastic Injection Moulding Machine

The chosen smart plastic injection moulding machine in the case study was utilised to illustrate the benefits of incorporating ecomechatronics and Industry 4.0 for sustainability purposes. The initial phase of the design was introduced in [33]. The novel model introduced in this study is an IoT-enhanced Smart Desktop Injection Machine, a small and energy-efficient plastic injection moulding device that integrates Internet of Things (IoT) technology and cloud computing. This integration offers users the ability to remotely monitor and control processes, conduct real-time data analysis for optimising operations, utilise advanced automation features, and smoothly integrate into Industry 4.0 settings.

Figure 18.4 shows the smart plastic injection moulding machine. The main structure measures 1246 mm in length, 290 mm in height, and 250 mm in width, with two fixed plates connected by two tie rods. This structure was assembled on a base table that also houses the electric motors, the control unit, and other mechanical components including (as shown in Fig. 18.4 with numbers: (1) the barrel, (2) the injection screw inside the barrel, (3) the nozzle plate, (4) the sliding plate, (5) the ball screw, (6) linear joints, (7) springs, and (8) the mould. Furthermore, the structure was made of Aluminium 7075, which has excellent mechanical properties such as ductility, strength, toughness, and fatigue resistance. The tie rods were made of 30 mm rolled steel, which can withstand the high stresses caused by clamping and injection forces.

The machine operates according to a clearly defined sequence. The process starts with the barrel being heated to the designated temperature. Once the temperature is reached, the raw material is introduced into the barrel to facilitate melting. Subsequently, the sliding plate holds the mould in place. The sliding plate then moves the nozzle plate and barrel back to the correct position to inject the molten material into the mould through the nozzle. Each injection cycle distributes the required dose of material. The sliding plate retracts by reversing the motor's direction, whereas the nozzle plate and barrel are restored to their initial positions by the springs' forces.

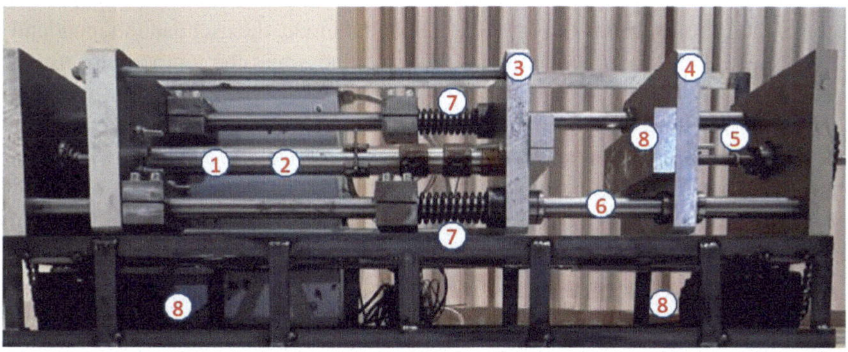

Fig. 18.4 Smart injection moulding machine

18.5.1 Predictive Maintenance

Predictive maintenance has become increasingly advanced with the integration of machine learning and Internet of Things (IoT) technologies. In the context of an injection moulding machine, data collected from sensors, such as motor RPM and heater temperature, can be processed using machine learning algorithms. These algorithms can analyse the patterns and trends in the data to detect anomalies and predict potential failures before they occur. By incorporating Arduino IoT devices, which can collect and transmit sensor data in real time, the injection moulding machine can be connected to a centralised monitoring system. This system can apply machine learning models to the collected data, enabling proactive maintenance strategies. Through the use of Arduino IoT, predictive maintenance becomes more efficient and effective, as it allows continuous monitoring of data, remote access, and timely decision-making, ensuring optimal performance, reducing downtime, and prolonging the life of the machine. Furthermore, the Arduino IoT devices that are used offer additional features for predictive maintenance of the injection moulding machine.

Using this IoT platform, you can interface multiple devices with each other and permit them to exchange real-time data. Besides this, you will be able to monitor and control the data from anywhere using a simple interface. For instance, we incorporated additional sensors such as encoders, vibration sensors, and current sensors into the Arduino platform to gain valuable insight into the condition of different machine parts. By monitoring the vibration levels or current draw of various motor components, it is possible to detect early indications of mechanical deterioration or electrical problems.

The connectivity and data processing capabilities of Arduino IoT devices allow predictive maintenance algorithms to be implemented directly on the machine. This edge computing strategy enables instant analysis of sensor data, reducing delay, and enabling prompt response to crucial maintenance requirements. This cloud enables the creation of a separate dashboard to display historical data, and then using big data analysis and AI, predictive maintenance is implemented. It also improves the machine's ability to independently recognise and deal with possible malfunctions. To monitor the whole process, the Arduino IoT Cloud interface (GUI) for manual and automatic control is shown in Fig. 18.5. In manual mode, the user can independently set the speed and direction of motors, thus the blades and the temperature in each zone.

Figure 18.6 shows the real-time temperature monitoring in each zone of the machine. The machine was successfully operated and integration with the IoT cloud platform allowed remote monitoring and control. The Arduino Cloud dashboard provided an intuitive interface for managing the machine's operation and monitoring temperature variations.

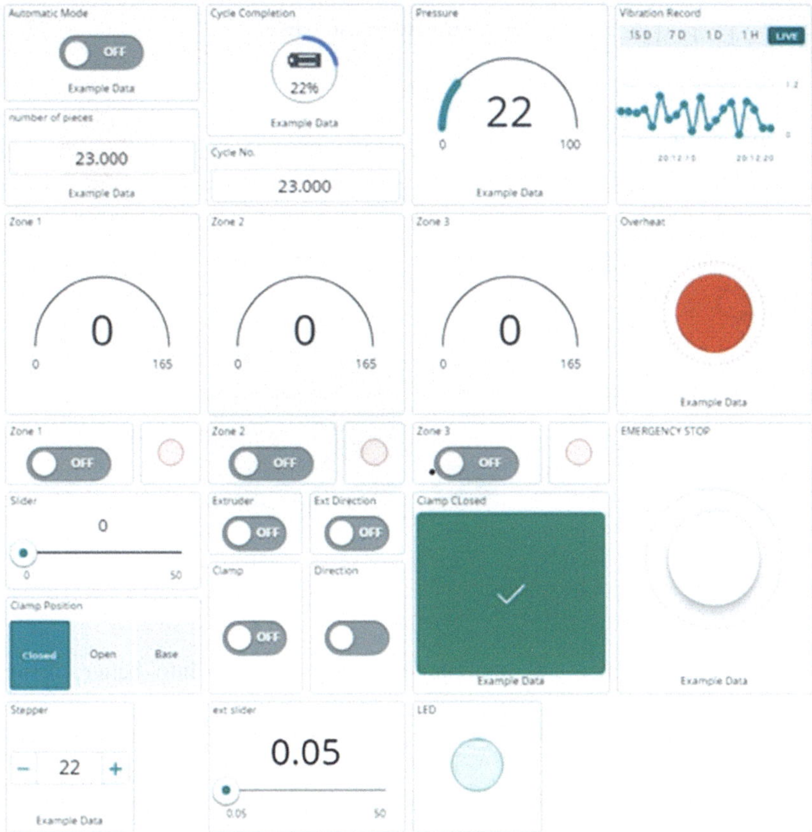

Fig. 18.5 Plastic injection moulding process dashboard

18.6 Conclusion

With increasing concerns about global warming and climate change, as well as high energy and power prices, it is essential that the environmental community actively participates in research and development activities. This is to explore technology that can help reduce emissions and pollution and reduce energy and material waste by focusing on green technology and environmentally friendly design. Also, it is important to focus on generating knowledge of best practices for sustainable development. The various research and developments in the area of ecomechatronics under the umbrella of Industry 4.0 discussed in the previous section have shown good promises and indications for the future. In general, the integration of ecomechatronics and Industry 4.0 has the following sustainable benefits:

- Reduced environmental footprint: lower emissions, waste, and resource depletion.

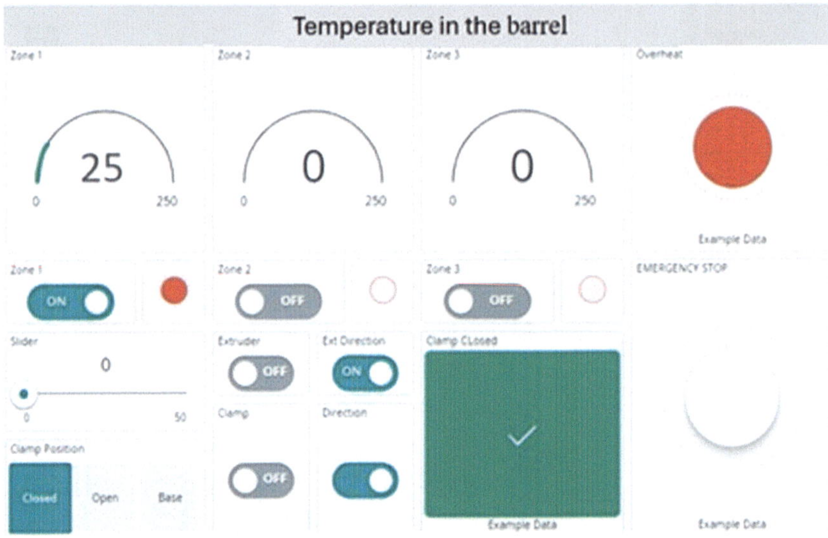

Fig. 18.6 Temperature on all zones dashboard

- Cost savings: energy and material efficiency translate into better financial performance.
- Improved product lifecycles: easier repair and remanufacture extend the use of the product.
- Enhanced innovation: data-driven insights foster new eco-friendly designs.

References

1. Hehenberger P, Habib MK, Bradley D (2022) EcoMechatronics challenges for evolution, development, and sustainability. Springer, EcoMechatronics
2. Afolalu SA, Ikumapayi OM, Abdulkareem A, Soetan SB, Emetere ME, Ongbali SO (2021) Enviable roles of manufacturing processes in sustainable fourth industrial revolution—a case study of mechatronics. Mater Today Proceed 44(1):2895–2901
3. Merzouki R, Samantaray AK, Pathak PM, Bouamama BO (2013) Intelligent mechatronic systems: modeling, control and diagnosis. Springer, London
4. Rassolkin A, Tammi K, Demidova G, HosseinNia H (2022) Mechatronics Technology and Transportation Sustainability. Sustainability 14(3):1671
5. Hehenberger P (2022) A holistic and sustainable view on the product creation process for mechatronic systems. In: Hehenberger P, Habib MK, Bradley D: EcoMechatronics challenges for evolution, Development, and Sustainability, pp 13–28. Springer
6. UN Department of Economic and Social Affairs, https://sdgs.un.org/goals
7. Simpson E, Whiting K, Carmona LG (2022) EcoMechatronics: re-envisioning innovation for sustainability. In: Hehenberger P, Habib MK, Bradley D: EcoMechatronics challenges for evolution, development, and sustainability, pp 13–28. Springer

8. McGlynn B (2022) Mechatronic applications in respect of sustainability and climate change. In: Hehenberger P, Habib MK, Bradley D: EcoMechatronics: challenges for evolution, development and sustainability, pp 29–48. Cham: Springer International Publishing
9. Annual global emissions of carbon dioxide 1940–2023, https://www.statista.com/statistics/276629/global-co2-emissions/
10. Kishawy HA, Hegab H, Saad E (2018) Design for sustainable manufacturing: approach, implementation, and assessment. Sustainability 10:3604
11. Machado CG, Winroth MP, Ribeiro da Silva EHD (2020) Sustainable manufacturing in industry 4.0: an emerging research agenda. Int J Prod Res 58(5):1462–1484
12. Rockizki J, Peggy Z (2013) Material selection for eco-design. In Smart product engineering: proceedings of the 23rd CIRP design conference, Bochum, Germany, March 11th-13th, (pp 875–883). Springer Berlin Heidelberg
13. Pellicciari M, Berselli G, Leali F, Vergnano A (2013) A method for reducing the energy consumption of pick-and-place industrial robots. Mechatronics 23(3):326–334
14. Merschak S, Diallo TM, Hehenberger P (2022) Life cycle sustainability assessment as a decision-making tool for the design of mechatronic systems. Int J Prod Lifecycle Manag 14(2–3):142–173
15. Iskandirova M, Blecha P, Holub M, Brad´aˇc F (2014) Eco-design of mechatronic systems. In Mechatronics 2013: recent technological and scientific advances (pp 65–72). Springer International Publishing
16. Iskandirova M, Blecha P, Holub M, Dudarev I (2014) Assessing the impact of mechatronic systems on the environment. In Proceedings of the 16th IEEE international conference on mechatronics-mechatronika, pp 706–710
17. Zheng T, Ardolino M, Bacchetti A, Perona M (2021) The applications of Industry 4.0 technologies in manufacturing context: a systematic literature review. Int J Prod Res 59(6):1922–1954
18. Lasi H, Fettke P, Kemper HG, Feld T (2014) Hoffmann, M: Industry 40. Bus Inform Syst Eng 6:239–242
19. Oztemel E, Gursev S (2020) Literature review of industry 4.0 and related technologies. J Intell Manuf 31:127–182
20. Bai C, Dallasega P, Orzes G, Sarkis J (2020) Industry 4.0 technologies assessment: a sustainability perspective. Int J Prod Econ 229:107776
21. Ryalat M, ElMoaqet H, AlFaouri M (2023) Design of a smart factory based on cyberphysical systems and internet of things towards industry 4.0. Appl Sci 13(4):2156
22. Frank AG, Dalenogare LS, Ayala NF (2019) Industry 4.0 technologies: implementation patterns in manufacturing companies. Int J Prod Econ 210:15–26
23. Grybauskas A, Stefanini A, Ghobakhloo M (2022) Social sustainability in the age of digitalization: a systematic literature review on the social implications of industry 4.0. Technol Soc 70:101997
24. Javaid M, Haleem A, Singh RP, Suman R, Gonzalez ES (2022) Understanding the adoption of industry 4.0 technologies in improving environmental sustainability. Sustain Oper Comput 3:203–217
25. Stock T, Seliger G (2016) Opportunities of sustainable manufacturing in industry 4.0. Proced CIRP 40:536–541
26. Awan U, Sroufe R, Shahbaz M (2021) Industry 4.0 and the circular economy: a literature review and recommendations for future research. Bus Strategy Environ 30(4):2038–2060
27. da Silva AM (2015) Mechatronics design process with energy optimization for industrial machines, Doctoral dissertation, Marquette University
28. Cusimano G, Casolo F (2021) An almost comprehensive approach for the choice of motor and transmission in mechatronic applications: Torque peak of the motor. Machines 9(8):159
29. Nwulu N, Damisa U (2003) Energy 4.0. Concepts and applications; AIP Publishing: College Park, MD, USA. 208p
30. Partridge AJ, Chen HY, Le NH, Xu C, Eichorn H, Pulvirenti E, Rossiter J (2022) ReRobot: recycled materials for trustworthy soft robots. In 2022 IEEE 5th International conference on soft robotics (RoboSoft). pp 148–153

31. Ali MH, Yerbolat G, Amangeldi S (2018) Material optimization method in 3D printing. In 2018 IEEE International conference on advanced manufacturing (ICAM). pp 365–368
32. Classens K, Heemels WM, Oomen T (2021) Digital twins in mechatronics: from model-based control to predictive maintenance. In 2021 IEEE 1st International conference on digital twins and parallel intelligence (DTPI). pp 336–339
33. Ryalat M, Alawamleh H, Elmoaqet H, Almtireen N (2023) Mechatronics design and implementation of a smart plastic injection moulding machine. In 2023 11th International conference on control, mechatronics and automation (ICCMA). pp 352–357

Chapter 19
Integrated Manufacturing: The Future of Fabricating Mechatronic Devices

Matthew P. Shuttleworth, Nicholas Fry, Rob Richardson, Jordan H. Boyle, and Robert W. Kay

19.1 Introduction

This chapter explores current and future technologies for manufacturing mechatronic devices, in particular, robotic systems. A robot consists of electronics, actuators, and sensors within a self-contained mechanical structure giving them the ability to perform tasks and interact with the physical world around them. There is no doubt that robots have the potential to revolutionise a wide range of sectors [1] and in recent years have seen greater adoption worldwide. But there remain many barriers to widespread use, including public perception of robots and challenges around physically and computationally integrating them into organic and changeable real-world environments. Additionally, due to the complex and multi-disciplinary nature of the engineering requirements, robots are often costly to design and manufacture. Improvements in manufacturing techniques for robots enable a deeper level of integration between the various electrical and mechanical sub-systems. This in turn reduces the economic barriers while also leading to smaller, more robust, and more dexterous robots capable of negotiating challenging environments where their use could have a greater impact.

M. P. Shuttleworth (✉) · N. Fry · R. Richardson · R. W. Kay
University of Leeds, Leeds, UK
e-mail: M.P.Shuttleworth@leeds.ac.uk

J. H. Boyle
Delft University of Technology, Delft, NL, Netherlands

P. Hehenberger and D. Bradley (eds.), *Mechatronic Futures*,
https://doi.org/10.1007/978-3-031-83571-1_19

19.2 Traditional Manufacturing Methods

Conventionally, bolts and other fasteners are used to integrate the various components into the chassis of mechatronic devices. The use of fasteners to secure all parts limits robustness, as anything that has been assembled can also be disassembled [2]. The joints between components are common weaknesses, and as devices get smaller the space taken up by fasteners becomes increasingly significant. Assembly of these complex devices is also a significant challenge; once a product has been designed, it must undergo a rigorous 'design for assembly' process to ensure it is suitable for manufacturing. This introduces additional costs, delays, and greatly limits the designer's freedom. For example, the shape of a mechanical case must be simple enough that it can be released from a mould; circuit boards (both rigid and flexible) can only be populated when held flat and must incorporate off-the-shelf sensors in specific packages (physical cases) and footprint sizes (the size and shape of the components connectors). Manufacturing processes dictate the way we design and produce items, imposing limitations on what is possible. We predict that advances in Additive Manufacturing (AM) will enable a new integrated manufacturing method that allows fabrication around mechatronic components. This could reduce the need for post-fabrication assembly and the use of fasteners, yielding rapid production of robust devices.

19.3 Current Methods: The Rise of Direct Digital Manufacturing

Traditionally, manufacturing was either manual, making it slow, and labour-intensive, or automated by simple robot operations, making it inflexible and costly to set up requiring large product numbers to justify the expenditure. The integration of computerisation and manufacturing is helping to change the economic patterns that have defined manufacturing since the Industrial Revolution. The convergence of different technologies such as 3D Printing with automation and digital design is allowing us to produce designs of unparalleled complexity, in less time and with reduced labour than ever before.

19.3.1 Automation

The design of a production line revolves around the need to make it as efficient as possible. Adjacent manufacturing operations should be located as physically close as possible, and the movement of human workers and parts carefully choreographed to minimise wasted actions and time. While the philosophy behind assembly lines has not changed since they were introduced, the technology that powers them has, and it brings along its own set of advantages and challenges.

Through the realisation of paradigms such as Industry 4.0 and 5.0, factories are becoming increasingly automated, smarter, and more efficient [3, 4]. Where once a human worker had to load parts into a machine, a robot arm can now be coupled with a computer vision system that identifies a specific part and places it in the jig itself. This can increase the speed and efficiency of the process and is often safer, removing the need for a human to work closely with dangerous machinery.

For low volume manufacturing, Computer Numerically Controlled (CNC) machines have had a significant effect. The first CNC machines were traditional machine tools, such as milling machines and lathes, with the addition of computerised control. Over time this has expanded to include laser cutters, 5-axis milling machines, and even 3D printers. CNC gives huge advantages as machines can quickly take a digital model of a part and turn it into a high-precision physical part. This pivots from the highly manual processes that were traditionally used for small production runs, reducing the time and cost to produce parts, without impacting and potential improving the quality. Additionally, there is an increased flexibility from the automation of these machines, as modifying the digital design file, can directly change the final part being produced.

19.3.2 Additive Manufacturing

Additive Manufacturing (AM) is a disruptive technology that was first developed in the 1980's but has burst into the public consciousness in recent years, as consumer desktop 3D printers become cheaper, more readily available, and more approachable by non-technical users. Additive Manufacturing and 3D printing are interchangeable terms to describe a range of technologies. The ISO/ASTM standard 52,900:2021 defines AM as *'The process of joining materials to make parts from 3D model data, usually layer upon layer, as opposed to subtractive manufacturing and formative manufacturing methodologies'* [5].

There are different methods of depositing material in AM processes, but in all cases the first step is to digitally slice the computer model into layers. These layers are then converted to machine-specific commands and physically laid down, one after the other, gradually building up a three-dimensional part. Thinner layers give finer detail, but also increase print time as more layers are required. Deposition technologies can be split into four broad categories [6], although the ISO/ASTM standard breaks these down further [5].

First there are extrusion-based machines. These feed semi-liquid material out of a moving nozzle, which then solidifies to form a hard layer. Many different materials have been used, including clay, chocolate, and concrete [7–9], although thermoplastics are by far the most common, particularly in consumer-grade 3D printers. While commonly referred to as Fused Deposition Modelling (FDM), this is a trademark of Stratasys®. The generic process, and what we refer to in this chapter; is Fused Filament Fabrication (FFF).

Second, there are techniques that bind particles of powdered material together, either using glue or a heat source (such as lasers). In this method, the part is constructed inside a vat of powder. To form a layer, a roller sweeps a thin film of powder across the top of the print bed. The print head then traces out the contours of that layer selectively solidifying the powder. The bed is then lowered by one layer thickness and the sequence repeats until the full part has been built.

Third, vat-photo polymerisation is a term which refers to solidification of a liquid resin using a light source. Lasers, UV lights, and even digital projectors can be used. Stereolithography (SLA or SL) was the first form of AM to be developed and is widely used in industry due to the high resolution afforded by light-based processing. An SL machine includes a vat of resin with a build plate inside. At the start of a build, the plate is positioned just below the surface of the resin while a laser traces out the layer, curing the resin. The plate then moves away by the height of one layer and the resin either flows or is wiped across, before repeating the process.

Finally, there is sheet lamination, a process in which each layer is formed by cutting the shape from a thin sheet of paper, plastic, or even metal. These sheets are bonded together using an adhesive or ultrasonic welding. The paper-based approach produces parts with poor mechanical properties but can be combined with inkjet printing to give full-colour models.

The layer-by-layer nature of AM enables new design freedoms as very complex shapes; even parts that cannot be manufactured using traditional techniques can often be realised (e.g. nested spheres). There are, however, some limitations to the geometry of the parts that can be manufactured. The use of layers introduces a 'stair step' effect on sloped and curved surfaces (Fig. 19.1) as the shape must be digitised into discrete layers. Furthermore, each layer must be supported by a layer underneath, which means that any overhanging or free-floating sections must be supported. Support structures can be printed in the same material as the part and designed to be cut, snapped, or machined off, or if the equipment allows; a different material is used that can be washed, peeled, or dissolved away.

Although the impact of AM technology is sometimes exaggerated, it has undeniably had a dramatic effect on the way products are manufactured and will continue to do so. Since the Industrial Revolution, economies of scale have defined the structure of the manufacturing industry [10]. When a greater number of products are

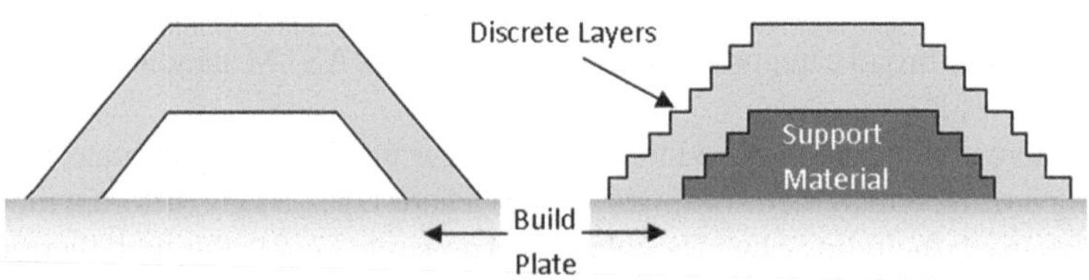

Fig. 19.1 Effect of layering, **a** subtractive machining yields a smooth surface, **b** AM part has a stepped surface and requires support material under the overhang

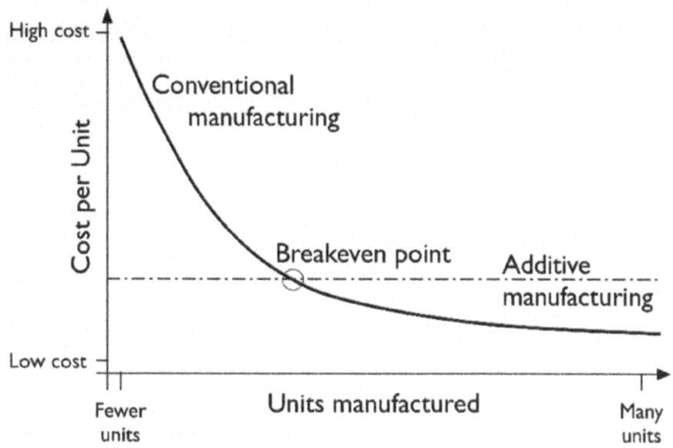

Fig. 19.2 A comparison of traditional and Additive Manufacturing cost models

produced, cost per unit decreases. Unfortunately, this traditionally requires a large initial investment for manufacturing to become cost-effective and therefore leads to large scale, centralised, mass manufacturing. Since AM requires minimal tooling and set-up, the price per unit levels off very quickly and is often modelled as a flat line when compared with traditional manufacturing cost structures [11], as illustrated in Fig. 19.2. This makes AM a very interesting prospect option for low volume manufacturing. Traditional methods are more cost-effective for large quantities, but the breakeven point will improve as 3D printers become more capable, cheaper, and faster.

AM will also significantly impact the economies of scope. This term refers to the fact that if one set of equipment, processes, or materials can be used to make different products then the unit cost falls as the investment is split between more items [10]. 3D printers are extremely versatile in the geometries they can produce and there is little time or cost penalty incurred when swapping between parts and they can even produce different objects side by side on the same build plate.

The abilities of 3D printers are so different from traditional manufacturing that they may allow new approaches to production that are not practical or possible with current manufacturing methods. For example, the layer-based approach allows very complex geometries that could not be machined or cast due to object geometry limiting tool access or mould removal. Internal features can also be created, without the need to split the part into multiple pieces. This in turn leads to parts that can be lighter and require less assembly, all while reducing material waste and lead times.

Additive Manufacturing is predominantly used as a rapid prototyping and rapid tooling technology, allowing engineers to greatly reduce the design cycle of a new product and its manufacture by iterating through design variations within hours rather than days or weeks. Increasingly, however, AM is expanding into Direct Digital Manufacturing (DDM), whereby AM technology is used to manufacture, final production parts, rather than prototypes. While most 3D printers available are still aimed at prototyping or tooling, in some; particularly high value sectors such as aerospace and medical, AM is being used to shorten supply chains, produce optimised parts, and manufacture custom components [12].

The rest of this chapter will focus on currently emerging technologies that will unlock the full potential of AM to revolutionise mechatronics manufacture. The far-reaching consequences of this revolution are explored in the final section.

19.4 Future Technology

19.4.1 Embedding Parts: Towards Non-Assembly Structures

Joining components with fasteners tends to impose a set way of building mechatronic devices which relies on specific materials and fastening techniques and often results in a design which is inherently limited in terms of size, geometry, and robustness [2]. The layer-by-layer construction of AM gives the unique opportunity to access the enclosed internal geometry of a part. This gives rise to the possibility of embedding components (motors, sensors, electronics) within object structures—to become an inherent part of the structure. This approach can introduce major benefits when creating complex mechatronic systems. Consider a humanoid robotic hand, for example. Researchers built two versions, the first of which used traditional assembly and included 60 parts (40 of which were fasteners) in each finger [13]. For the second version, each finger was a single part manufactured using a hybrid AM method called Shape Deposition Manufacturing (SDM). SDM is a layer-based manufacturing method which involves a cycle of depositing material followed by precision shaping using subtractive methods. This approach allows different materials to be incorporated in a part, and even for components to be embedded inside the structure. Sensors were integrated into each finger, creating a reliable interface while protecting the sensitive electronics [13]. The joints and finger pads are made of a compliant material, while the link sections are a stiff polymer. This compliance greatly increases the robustness of the gripper; researchers have even released a video of it being repeatedly hit with a baseball bat, with no resulting damage [14] demonstrating its durability. The SDM process was also used to create devices mimicking the impressive agility of a cockroach [15] and the climbing of a gecko [16].

The ability to make integrated joints is key to simplifying assembly [17] and allows unique mechanisms to be created. For example, the process has enabled ball and socket joints to be redesigned to limit their motion in certain directions by changing their internal geometry—something that is not possible using traditional manufacturing methods [18].

Producing non-assembly joints (working mechanisms that require no assembly) is difficult, and support material typically must fill small gaps between parts [17] required to give good tolerances for the joints. This, however, makes the support removal problematic. Traditional joints have been redesigned to result in close tolerances and simple support removal [17], but this work must be extended to enable a wider range of non-assembly joints.

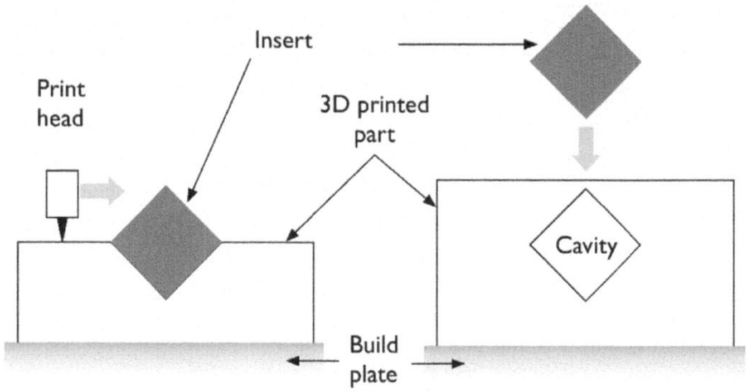

Fig. 19.3 Use of inserts in 3D printing **a** print head collides with protruding insert, **b** insert cannot be embedded until the build is as tall as the insert, but by that point the cavity could be enclosed

Embedding inserts during the 3D printing process has given some interesting results, with components such as motors, circuit boards, and gears being inserted to create, for example, a moving model of the stereo lithography machine that printed the parts [19], a remote-controlled buggy [18] and a robotic hand [20]. These inserts could also change the mechanical properties of the object. For example, metal could be inserted to increase the stiffness of a part.

There are challenges to overcome, however. Existing 3D printers are designed to work with planar horizontal layers, so the top of the part being built must always be flat to avoid collisions with the print head or recoating wiper arm (see Fig. 19.3a). This requirement prevents the insert from being embedded until the build level is as high as it is. However, if the upper geometry of the part is convex, then the cavity will be enclosed by the time the build level is high enough to shield the component (Fig. 19.3b).

To overcome these issues, shape converters can be used (Fig. 19.4). These are additional parts that fit around the component, converting its geometry into one with vertical sides and a horizontal top that can be inserted without protruding [21]. These allow embedding but remove some of the possible advantages by adding extra process planning and assembly steps. The shape converter must be designed (it could have a complex shape) and manufactured, before being manually attached to the component for insertion into the main build. In more recent work, the ability to print using transparent materials has enabled printing around embedded optical elements to create displays and illumination, or even customised optoelectrical sensors [22]. This required infrared emitters and receivers to be embedded during the printing process. These were embedded by pausing the printer and manually inserting the component and a shape converter; the same labour-intensive technique that was developed a decade before [19]. The creation of fully integrated robotic mechanisms is being held back by a lack of suitable embedding processes that allow inserts to be seamlessly built around.

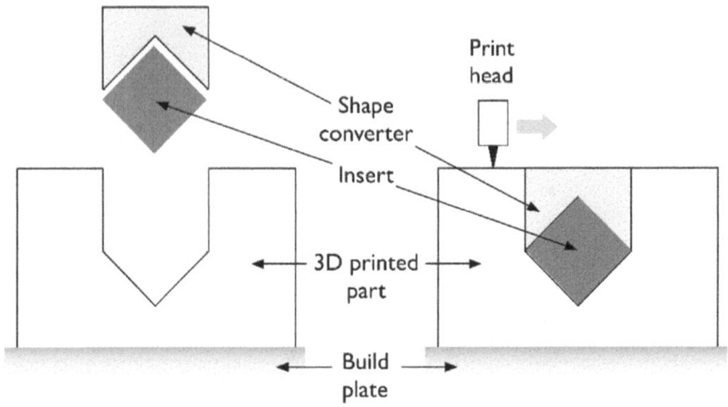

Fig. 19.4 A shape converter is assembled with the insert, converting the geometry to vertical sides and horizontal top, removing collisions

19.4.2 3D-Printed Functionality

Including sensors, actuators, and electronics in 3D-printed structures will turn them into functional mechatronic devices. There is little reason to 3D print versions of parts that are easy to acquire and could simply be inserted, yet the integration and customisation potential of printing certain components does bring unique advantages that will be valuable in many cases. For example, a method to print speakers in any shape has been developed [23]. This allows the component to be totally integrated with the device it is in, allowing much greater design freedom, as the casing no longer needs to fit the shape of a standard speaker cone. By changing the shape of the speaker, the focus of the sound can also be adjusted [23]. A cylindrical speaker could be used to make an ultrasonic sensor capable of detecting obstacles all the way around a robot. Or a single object could have many individual speakers giving very directional sounds which could be used for artistic displays or communication [23].

The ability to print soft and transparent structures allowed researchers to develop novel optoelectronic sensors [22]. The electrical components can be situated near the surface of the part and externally wired to a control circuit, while sensing can be achieved using light signals channelled through printed optical conduits, avoiding the need for conductive material within the part.

By embedding infrared (IR) emitter and receiver pairs into a part many different sensors can be made [22]. Using an Objet multi-material 3D printer, flexible, clear material can be used to guide the light [22]. Deforming this light guide changes the reflections, allowing movement to be detected. Switches, sliders, encoders, and accelerometers have all been made using different configurations; for example, a button is made by guiding IR light from the emitter to the receiver with a cantilever beam. When the user presses the button down it deforms the light guide, bending it away for the receiver. The receiver now registers a much lower signal and can be used to trigger a different electrical circuit. Using this technique customised, accurate sensors, and user interface devices can be easily, quickly, and repeatedly incorporated into any product [22].

The resistance of electrical traces can also vary in response to external stimuli such as force or temperature. By integrating meandering traces into flexible printed parts, the deformation or bend radius of a single piece flexure can be measured [24] giving a form of feedback to mechatronic devices. The same traces could also be used to produce 3D antennae, saving valuable space by wrapping them around the structure, or providing unique radiation patterns for directed communications [25].

19.4.3 Integrating Multiple Materials

Extending 3D printing processes to work with multiple materials opens a huge range of possibilities for the technology, and while it is available for some methods, there are still challenges to overcome before this becomes a mainstream manufacturing technique.

The ability to combine different materials into one part would allow the mechanical and aesthetic properties to be adjusted. This can increase functionality as compliance could be included in different sections, or different colours or surface finishes could make AM more appropriate for final production runs as opposed to prototypes [26]. If materials with special properties, such as electrical conductivity, could be incorporated, then we would be much closer to an all-in-one process to fabricate functional mechatronic devices.

The Stratasys® Objet range of 3D printers are possibly the most capable multi-material printers available at the time of writing [27]. They can be used to print parts with combinations of flexible, hard, transparent, and full-colour materials. The printer uses the Polyjet process, somewhat like that of an inkjet printer, in which tiny droplets of liquid build material are deposited from holes in the print head. The print head also includes an ultraviolet light that passes over each layer to cure it and solidify the polymer before laying down the next layer. This jetting process means that different materials can be combined in a controlled way to create functional gradients of materials. In a process such as SDM each material is deposited individually with a clear interface between the sections. In contrast, Polyjet allows gradual material transitions that remove the weakness caused by a sudden change in materials. The material properties can even be precisely adjusted at different points throughout a model. This is an ability that is only possible with multi-material 3D printing. Adjusting the compliance, elasticity, and damping of a part through its structure could have applications in many fields. For example, surgical devices may need to be strong and stiff in one section to aid handling or insertion while having a soft end effector to reduce tissue damage. Achieving this with one integrated part would also facilitate waterproofing and sterilisation. In robotics, flexible joints actuated by shape memory alloy wire have been created, showing the potential of integrated non-assembly joints [28].

The Polyjet process, while giving many advantages, also limits the materials that can be used. As the polymers are cured using UV light, each material must be created to work in the printer, giving rise to Stratasys'® 'rubber-like' and 'ABS-like' descriptions of its materials. They share similar properties but are not the same as the traditional materials engineers are familiar with using. One of the biggest challenges, and opportunities, for 3D printing is increasing the range of materials that can be reliably used.

Malone and Lipson created the Fab@Home project, in which they designed a multi-material 3D printer that is open source and low cost [29]. They used this to demonstrate that by using different materials and processes during a build they can produce working electromechanical devices. The system uses a syringe to deposit a wide range of liquid, gel, or paste materials and a molten-extrusion head to deposit thermoplastics and solder. This enabled the creation of flexible circuit boards, strain sensors, electromagnets, electromechanical relays, electroactive polymer relays, and even batteries in different shapes. While creating functional electromechanical parts is a huge step for Additive Manufacturing, the Fab@Home system is low-cost, low-resolution, and labour-intensive, as processes must be carefully planned, and the various materials are manually supplied to the machine. These components were functional but cannot match the performance of their traditionally manufactured counterparts. Process planning is a difficult step as multi-material Additive Manufacturing gives new ways of making parts, so no set way of working has been established. This will continue to be challenging as different technical processes will affect the design and planning stage.

In recent years, the consumer FFF market has seen a significant shift from basic machines that require extensive technical expertise to produce acceptable parts to technologically sophisticated devices that require little user interaction. This has brought with it an array of consumer printers with the ability to switch materials mid-layer by retracting the feedstock, inserting a different one, purging the nozzle until clean, and then finally continuing the print.[1] This advancement has not only enabled the extensive hobbyist community to explore a wide range of potential applications such as soft coating or flexible elements in rigid parts, but also lowered the barrier to entry for research and development. The downside to this approach is the added time and wasted material when switching feedstocks; however, it appears that further advancements in the form of tool-changing printers with discrete tools for each material that do not require purging between materials are on the horizon.[2]

[1] Bambu Lab. *Bambu Lab X1 Series* @ bambulab.com/en/x1 (accessed September 2024)

[2] Prusa Research. *Original Prusa XL* @ www.prusa3d.com/en/product/original-prusa-xl-semi-assembled-single-toolhead-3d-printer/ (accessed September 2024)

19.4.4 3D-Printed Electronics

One of the most keenly awaited developments in AM is the ability to print electronics and conductive traces. This would enable functional mechatronic devices to be directly manufactured in one process. Novel designs that are not constrained by the need to include a flat printed circuit board would be possible. Wires could also be replaced with conductive tracks, removing the difficulty of routing wires through complex and tight spaces, and reducing the size and weight of the resulting device.

Batteries and transistors have been made via 3D printing [29–32], but there is a long way to go before these parts can compete with the efficiency and price of established processes. In the future many electrical components may be printed from their basic materials, allowing a huge catalogue of parts to be used without the need for each to be kept in stock. Despite this, merely printing conductive connections and inserting pre-made electronic components would still open up a huge range of possibilities. For the foreseeable future, at least, this will be the preferred method of incorporating electronics into an AM part as the mass manufacturing processes for electronic components are so well optimised.

The main options for printing conductive connections are depositing conductive inks, extruding solder, extruding conductive thermoplastic filament, embedding wires, or selectively plating metals. Commercial aerosol jet printers are available and have been demonstrated to produce a circuit on the wing of a UAV model constructed via 3D printing [22]. The process uses a mist generator to atomise conductive ink, then aerodynamically focusses it, using a sheath gas, to create a fine flow. Feature sizes less than 15 μm are possible and the approach shows great promise for high-resolution circuits [33].

Conductive silver inks have been used to Direct Write (DW) onto parts produced by both FFF and vat-photopolymerisation. In these approaches, an air-backed syringe was used to trace a circuit design onto the surface of the component [34, 35]. By hybridising the processes and switching between AM and DW mid-production embedded electronics were also demonstrated. The main drawbacks of these approaches; however, were the inability to use solder due to the low thermal resistance of the printable materials, and the reduced conductivity of silver inks when compared with conventional electronics.

Embedding wires is another method of incorporating conductive paths into the AM build process and is appealing due to the potential of low resistance, multicore, or shielded wires, along with low cost. Wire is difficult to embed, however, as it does not directly adhere to the object and needs to be in tension to manipulate. Wire has been successfully embedded into thermoplastic parts either by using ultrasonic agitation to locally heat the wire and surrounding plastic and pushing the wire just below the surface, or by co-feeding a wire into the plastic extradite as the part is being built up. This has been demonstrated with wires down to 50 μm in diameter and has the advantage of using copper wire, granting it excellent electrical characteristics [36–38].

An alternative approach demonstrated by Esfahani et al. was to selectively plate copper onto the surface of a 3D-printed polyetherimide (PEI) component [25]. Through multiple stages of chemical treatment followed by circuit patterning with an ultraviolet laser the group were able to produce conductive circuitry over a range of conformal and flexible parts. The PEI also has the advantage of good thermal stability allowing electronics to be directly soldered to the circuit using traditional solder reflow technologies, bringing the process closer to compatibility with existing technologies. Furthermore, the conductivity of plated copper was shown to 64% that of bulk copper, much higher than many conductive inks or polymers. This method was also shown to be useful for a range of mechatronic applications, offering capabilities including 3D antennas, flexible sensors, and solid-state actuation [24].

At a consumer level, FFF printers are the most common so there is great interest in conductive materials for this process. Researchers have used low melting point alloys, in the form of solder, as filament for FFF printers, which allows circuits to be printed [39]. This is cheap, readily available and can be used in unmodified FFF printers. However, it is difficult to control the feeding and cooling rates to give good results. There is much excitement around new conductive filaments that can be used in FFF printers. The filaments are made by mixing a thermoplastic, such as PLA with a conductive material. Graphene and nano-composite materials have been shown to provide the best results, but the resistance is still relatively very high. If short traces and low currents are used, then functional devices and circuits can be successfully created using this method. The fact that this conductive material can be deposited with the same process as the main build material simplifies the task of embedding the tracks within the part and means many printers will require no custom hardware. This will greatly speed up the process of adoption. While these conductive materials are far from ideal and the resolution of FFF does not approach that of aerosol jet or standard PCBs, the availability of this process to a large and inventive community will drive a rapid push-in innovation as individuals attempt to solve a wide variety of problems and use it to manufacture their own designs. This will be unlikely to pose a threat to companies interested in using 3D circuits for a competitive edge; however, they would be wise to pay attention to this unofficial research, as it is likely to be where novel uses and applications are explored.

19.5 Future Impact

While many individual experiments have confirmed the great potential for using AM to manufacture end-use mechatronic devices, combining these technologies is yet to happen. This section will look at the possibilities opened up by technologies such as embedding components and printing electronics along with challenges that engineers will need to face to realise them.

19.5.1 Fully Integrated Mechatronic Devices

If discrete components could be combined with multiple materials and conductive tracks, entirely new types of mechatronic devices could be created. Traditional motors and bearings could be inserted alongside integrated, compliant AM joints, adding damping and robustness. Conductive tracks could connect actuators to a 3D-printed battery whose shape has been customised to fit in the gaps between other components, allowing the device to be smaller. The conductive tracks could follow complex paths around the structure of the material, removing the need for a flat PCB and allowing wires to be protected. These tracks may then lead to a 3D-printed socket allowing simple connections to different modules. All of this would be integral to the mechanical structure of the device. Assembly would happen during manufacture, eliminating joints in the casing, clips that can snap, or screws that could come loose. Removing the need for assembly could speed up manufacturing, reduce labour costs, and help provide consistent quality. The monolithic structures that could result from fully integrated manufacturing would be ideal for harsh environments where toughness or sealing is important.

One must consider the whole life cycle of a product, however. It may be necessary to include access panels for maintenance and repair. Traditional inspection methods may be impossible so new processes will be required. Tight integration of multiple components and materials will present a particular challenge at the end of the product's life when the materials should be recovered and recycled. Some AM processes use materials such as thermoplastics that can be re-melted, but others use chemical reactions that are not as simple to reverse.

While inserting components that are made in other ways is likely to be the most efficient method in the near future, it is possible that some components will benefit from 3D printing. Custom sensors, for example, may be a good candidate for Additive Manufacturing since modifying the shape to fit exactly where they are needed could be very useful. Or it may be that integrating basic but cheap sensors is best done through AM, while more complex parts are manufactured in an external process and embedded. This trade-off between the benefits of different manufacturing processes will have to be closely examined for each component to find the most efficient process.

19.5.2 Designing Differently: The Integrated Design Process

Fully integrated manufacturing methods will require innovative design processes. Additive Manufacturing enables part geometry that could not be produced any other way; be it moulded or machined and frees the designer from considering the complexities of multi-part moulds or tooling requirements. However, the procedure is not yet as simple as just pressing print. The process planning step is still vital for checking if the design is suitable for AM. Currently, designers often try to minimise the support structures needed to build a part, but the introduction of multiple materials and components will add further considerations. Engineers and designers must ensure there is a good dialogue as these new AM technologies are developed, ensuring the effects of any design for manufacturing rules are reduced. Separate Computer-Aided Design (CAD) packages are currently required for mechanical and electrical systems, but this will need to change to enable tighter component integration. It is important that user-focused, intuitive software is developed alongside new integration technologies to enable the designer to consider a holistic mechatronic design.

Communication between the CAD package and the printer is also important. Since the mid-80 s when the STL was developed by 3D systems, it has been the standard file format for rapid prototyping and AM systems. This specifies the surface geometry of a 3D object, but does not include any information on the materials, colours, or textures used. This information is no longer sufficient for modern multi-material printers. To address this issue new file formats such as the Additive Manufacturing File Format (AMF) or 3D Manufacturing Format (3MF) have been developed. These are backwards compatible with STL, but also add features such as being able to specify multiple materials, colour, and geometry with higher fidelity. 3D printer manufacturers and CAD software developers must now take the lead and add compatibility to their products. The ability to create a streamlined workflow from conception to fabrication will be important for integrated design.

19.5.3 Automated Additive Manufacture and Assembly

One possible route for the future is towards fully automated manufacture and assembly of mechatronic devices. This could be as simple as using robotics to facilitate the manual steps described in earlier sections involved with embedding components and shape converters. Better still would encompass multiple processes such as DW or wire embedding to provide the interconnects between the discrete devices, allowing a full device to be produced without operator interaction [40]. This could have a significant impact on repeat manufacture, whether it be multiples of the same device such as in swarm robotics or mass customisation whereby each subsequent device is tailored to a unique application.

The use of a robot arm to both 3D print and place sub-assemblies for a quadcopter was demonstrated by Kosmal et al. [41]. In their hybrid AM robot workcell, a single robot arm was fitted with multifunctional tool that could perform both FFF-based printing and electromagnetic pick and place. The main issue that is still present, however, is the ability to connect components electrically. Kosmal et al. overcame this by building sub-assemblies by hand with push-in connectors that were orientated to suit the robot arm, but this contravenes the idea of fully autonomous assembly and is an area that would benefit further research.

Automated assembly also lends itself to improved sustainability [42] through one of three potential avenues. Firstly, tighter integration would likely result in reduced component size and quantity, minimising the amount of raw material required in the first place. Secondly, if designed with it in mind, devices that are assembled can also be disassembled. This would allow automated assembly systems to effectively recycle devices that were no longer needed and repurpose the components into new devices. Finally, rather than full disassembly for recycling, a damaged device could be partially broken down and repaired before being retuned in a functioning state.

19.6 Conclusions

Additive Manufacturing is having a huge impact on the manufacturing industry. However, only when it is combined with embedded parts, multiple materials and integrated electronics will its full potential to improve mechatronic and robotic devices be realised. AM removes the need to rely on the economy of scale and immediately gives a good economy of scope when manufacturing devices. Soon mass customisation will be available in a huge range of products and industries, which will be particularly important for robotics. There are many jobs that would benefit from automation and the use of robots, but creating a single mass-market robot platform to meet these needs is only beginning to be realised, and small numbers of devices are expensive to build, limiting the market for robotics. If an enhanced 3D printing machine could be created that included abilities such as embedding functional components, printing joints, and connecting electronics then customised robots, designed for specific applications could be much more readily available. This method of construction could also make the robots more robust which is extremely important for moving robots from laboratory-based experiments to real-world applications.

The process is also made cheaper and simpler, by minimising the assembly required, reducing material waste compared to subtractive methods, and reducing the skills operators require. As 3D printing does not require a long or an expensive set-up, the lead times for parts could be greatly reduced. Combined with minimal assembly, on-demand manufacturing could be used in new applications such as in surgery (custom implants or tools could be created), or in search and rescue (a robot could be produced to fit a particular void or move on a specific surface). Combining these innovations could also be the next step required for self-replicating 3D printers, such as the RepRap project [43]. Currently, these printers can produce 50% of their own parts (not counting nuts and bolts). Using the ideas of Integrated Manufacturing these fasteners could be removed and the progress in printing electronics may soon help these machines reach the goal of 100% printable parts.

While there is great potential for Additive Manufacturing, many challenges have been identified. Improved methods for embedding parts are a critical area that needs development, and printing electrical connections reliably with low resistance will be vital for mechatronic and robotic devices. Inherent issues such as discreet layers and the requirement for support material also need to be mitigated. Finally, combining all the technology together, with intuitive software and established process planning methods will finally bring about Direct Digital Manufacturing and Integrated Mechatronic Manufacturing.

References

1. Willetts D (2013) Eight great technologies, Policy Exchange. 57
2. Bailey SA, Cham JG, Cutkosky MR, Full RJ (2000) Biomimetic robotic mechanisms via shape deposition manufacturing, Robot Res, 403–410
3. Adel A (2022) Future of industry 50 in society: human-centric solutions, challenges and prospective research areas, J Cloud Comput Adv Syst Appl, 11(1) https://doi.org/10.1186/s13677-022-00314-5.pdf
4. McKinsey, Co (2022) What are industry 4.0, the fourth industrial revolution, and 4IR? www.mckinsey.com/featured-insights/mckinsey-explainers/what-are-industry-4-0-the-fourth-industrial-revolution-and-4ir. Accessed September 2024
5. ISO/ASTM (2021) 52900:2021(en) Additive manufacturing—general principles—fundamentals and vocabulary, in Terms and definition, www.iso.org/standard/74514.html#:~:text=This%20document%20establishes%20and%20defines,into%20specific%20fields%20of%20application. Accessed September 2024
6. Barnatt C (2016) 3D printing, 3rd edition, CreateSpace independent Publishing Platform
7. Chan SS, Pennings RM, Edwards L, Franks GV (2020) 3D printing of clay for decorative architectural applications: effect of solids volume fraction on rheology and printability. Addit Manuf 35:101335
8. Lanaro M, Forrestal DP, Scheurer S, Slinger DJ, Liao S, Powell SK, Woodruff MA (2017) 3D printing complex chocolate objects: platform design, optimization and evaluation. J Food Eng 215:13–22
9. Xiao JZ et al (2021) Large-scale 3D printing concrete technology: current status and future opportunities. Cement Concrete Compos, 122
10. Cotteleer M, Joyce J (2014) 3D Opportunity—additive manufacturing paths to performance, innovation, and growth. Deloitte Review
11. Ruffo M, Tuck C, Hague R (2006) Cost estimation for rapid manufacturing—laser sintering production for low to medium volumes. Proceed Inst Mech Eng Part B-J Eng Manuf 220(9):1417–1427
12. Hassani V (2020) An investigation of additive manufacturing technologies for development of end-use components: a case study. Int J Pressure Vessels and Piping, 187
13. Dollar AM, Howe RD (2006) A robust compliant grasper via shape deposition manufacturing. IEEE-Asme Trans Mechatron 11(2):154–161
14. Guizzo E (2012) iRobot smashes its new robotic hand with baseball bat. [Cited 2024 05/03]; Available from: https://spectrum.ieee.org/irobot-smashes-its-new-robotic-hand-with-baseball-bat
15. Cham JG et al (2002) Fast and robust: hexapedal robots via shape deposition manufacturing. Int J Robot Res 21(10–11):869–882
16. Cutkosky MR, Kim S (1894) Design and fabrication of multi-material structures for bioinspired robots. Philos Trans R Soc Math Phys Eng Sci 2009(367):1799–1813
17. Chen YH, Chen ZZ (2011) Joint analysis in rapid fabrication of non-assembly mechanisms. Rapid Prototyping J 17(6):408–417
18. De Laurentis KJ, Kong FF, Mavroidis C (2002) Procedure for rapid fabrication of non-assembly mechanisms with embedded components. In: ASME 2002 International design engineering technical conferences and computers and information in engineering conference
19. Kataria A, Rosen DW (2001) Building around inserts: methods for fabricating complex devices in stereo lithography. Rapid Prototyping J 7(5):253–261
20. De Laurentis KJ, Mavroidis C (2004) Rapid fabrication of a non-assembly robotic hand with embedded components. Assem Autom 24(4):394–405
21. Liao YS, Li HC, Chen MT (2007) The study of rapid prototyping process with embedded functional inserts. J Mater Process Technol 192:68–74
22. Willis KDD et al (2012) Printed optics: 3D printing of embedded optical elements for interactive devices. Uist'12: Proceedings of the 25th annual Acm symposium on user interface software and technology, p 589–598

23. Ishiguro Y, Poupyrev I (2014) 3D Printed interactive speakers. 32nd Annual Acm conference on human factors in computing systems (Chi 2014), p 1733–1742
24. Shuttleworth MP (2020) A digital manufacturing process for three-dimensional electronics. University of Leeds. p 226
25. Esfahani RN et al (2020) Light based synthesis of metallic nanoparticles on surface-modified 3D printed substrates for high performance electronic systems. Additive Manufacturing, 34
26. Jones RO, IP, BA (2012) Rapid manufacturing of functional engineering components, B. University of Bath, UK, Editor
27. Vaezi M et al (2013) Multiple material additive manufacturing—Part 1: a review. Virtual Phys Prototyping 8(1):19–50
28. Meisel NA, Elliott AM, Williams CB (2015) A procedure for creating actuated joints via embedding shape memory alloys in PolyJet 3D printing. J Intell Mater Syst Struct 26(12):1498–1512
29. Malone E, Lipson H (2009) Multi-material freeform fabrication of active systems. Proceedings of the 9th biennial conference on engineering systems design and analysis—2008, Vol 1, p 345–353
30. Wei TS et al (2018) 3D Printing of customized li-ion batteries with thick electrodes. Adv Mater, 30(16)
31. Zhang MG et al (2020) 3D printing of structured electrodes for rechargeable batteries. J Mater Chem A 8(21):10670–10694
32. Massetti M et al (2023) Fully 3D-printed organic electrochemical transistors. Npj Flexible Electron, 7(1)
33. Paulsen JA et al (2012) Printing conformal electronics on 3D structures with Aerosol Jet technology. In: 2012 Future of instrumentation international workshop (FIIW) proceedings
34. Li J et al (2016) Hybrid additive manufacturing of 3D electronic systems. J Micromech Microeng, 26(10)
35. Medina F et al (2005) Hybrid manufacturing: integrating direct write and stereolithography
36. Espalin D et al (2014) 3D printing multifunctionality: structures with electronics. Int J Adv Manuf Technol 72(5–8):963–978
37. Kim C et al (2021) Intermittent embedding of wire into 3D prints for wireless power transfer. Int J Precis Eng Manuf 22(5):919–931
38. Saari M (2019) Design and control of fiber encapsulation additive manufacturing
39. Malone E et al (2004) Freeform fabrication of zinc-air batteries and electromechanical assemblies. Rapid Prototyping J 10(1):58–69
40. Häfele T et al (2019) Hybrid additive manufacturing—process chain correlations and impacts. 29th Cirp design conference 2019, 84: p 328–334
41. Kosmal T et al (2022) Hybrid additive robotic workcell for autonomous fabrication of mechatronic systems—a case study of drone fabrication. Addit Manuf Lett, 3
42. Javaid M et al (2021) Role of additive manufacturing applications towards environmental sustainability. Adv Ind Eng Polym Res 4(4):312–322
43. Jones R et al (2011) RepRap—the replicating rapid prototyper. Robotica 29:177–191

Part V
Operational Issues and Education

Chapter 20
Project and Engineering Expertise in Mechatronics Futures

Elias Flening

Abstract In the coming decades, what will be the most salient aspect of everyday work for the Mechatronic design engineer? "*Good engineers*" know not to leave the ongoing organizing of project work to project managers, and good project managers know to ensure that all the engineers in their project can be "good engineers" in this way. Based on participatory research in everyday Mechatronic development work, this chapter explores three overarching questions that the future engineer is well advised to consider both today and in ten years. The aim is not to be yet another think-piece about the importance of collaboration, listening or how to hold value-generating workshops. The chapter will focus on you, the Mechatronic practitioner, and what questions to ask *yourself*.

- Firstly, what will be the basis for (your) engineering expertise in the future? *That is, how should one act as a Mechatronics engineer?*
- Secondly, how will perceived project complexity in the real world affect (your) practice? *That is, what will be seen as difficult in projects?*
- Thirdly, and perhaps most importantly, what will the relation between (your) engineering expertise and project complexity (of your projects) be? *That is, how should one act in difficult project situations?*

These three questions are not to be answered directly but should instead serve as containers for more concrete questions of more practical relevance. Otherwise, the chapter *would* simply be another general think-piece. Therefore, the chapter is organized through these three questions. The questions offered are based on research findings from practice-oriented case study fieldwork on micro-features of everyday project work in Mechatronics system development. For each question, a short example or vignette from industry will be used to exemplify what is meant.

E. Flening (✉)
KTH, Stockholm, Sweden
e-mail: flening@kth.se

P. Hehenberger and D. Bradley (eds.), *Mechatronic Futures*,
https://doi.org/10.1007/978-3-031-83571-1_20

20.1 Introduction

Our pursuit of efficient and effective Mechatronic systems development has historically been ambiguous, with the discipline often relying on overly rationalistic assumptions about the actual *process* for producing potential solutions. This opaque backdrop is pointed to by the field's literature which consistently underscores the need to address challenges to Mechatronic industrial practice as identified in a 2012 systematic literature review by Mørkeberg Torry-Smith et al. [1] The challenges raised included:

- "*The absence of a unified understanding of the entire system.*"
- "*The lack of a shared language for conceptual representation.*"
- "*The presence of divergent mental models related to the system, tasks, and design phenomena.*"

  And

- "*The absence of a language conducive to free and open discourse.*"

These issues are not only of current relevance but also historic, echoing the challenges identified by Salminen and Verho [2] who in 1989 articulated similar concerns using the very same phrasing. The enduring nature of these challenges is further corroborated by Jacob Buur in 1990 [3]. The persistent narrative across these discussions indicates a deep-seated belief in the advantages of unity in language and understanding within Mechatronic development, often at the expense of recognizing the benefit of diversity [4, 5].

Question 1: What is the Basis for Engineering Expertise in the Future?

As a Mechatronics engineer, your expertise extends beyond the bounds of your technical knowledge. It encompasses a mastery of the intricate interplay between social and technical elements. This integration is crucial across all engineering disciplines, but it's especially vital in Mechatronics. You experience this every day. Your ability to effectively merge these elements in real-world applications sets the foundation for innovation and efficiency in your projects. This understanding is supported by extensive research indicating that engineering expertise involves a complex blend of social and technical performances, a trait strongly evident in Mechatronics [6, 7].

I like to think about this broad view of Mechatronic practice in terms of a three-part model of engineering expertise (Fig. 20.1). The model pertains equally to you as a Mechatronics engineer, as well as to other project members such as the project manager or team lead. In the case of you being assigned the PM role, or having taken it upon yourself with or (as is often the case) without full formal authority, this model might seem poignant. It argues that expertise is *the ability to enact technical knowledge, through social networks of professional actors, by continually recognizing the present legitimate basis for practice.* There is a causal connection between the three abilities: Understanding *what kind of practice* which is legitimate in any given situation *enables collaboration in networks*, which in turn *enables you to use*

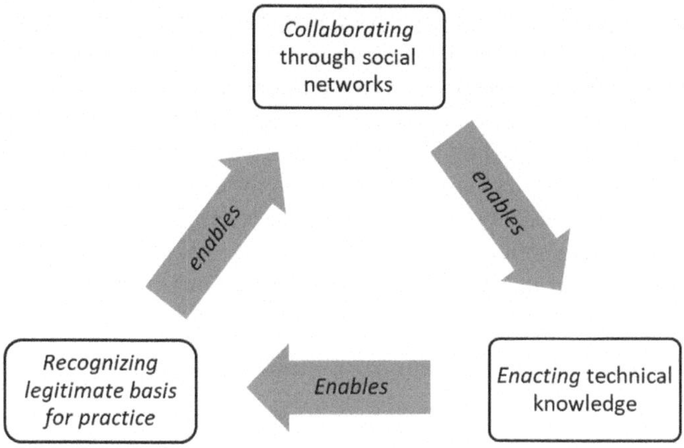

Fig. 20.1 Engineering expertise

your technical knowledge. Of course, having that technical knowledge (both tacit and implicit) also forms the foundation for your being able to act as a legitimate Mechatronics engineer. This closes the loop in the model, making the three abilities interdependent.

Question 1.1: What Specific Technical Skill(S) is of Most Value in This Moment?

This question might seem obvious, but it is specifically when one feels that it is very clear what technical acumen is needed that this question can serve well. If you can clarify which set of skills presently adds value and to whom, then looping through the steps of the model should be easier. However, it is often the case that it *isn't* entirely clear what is needed, or by whom, or for what reason. Keep a light but consistent touch on this question in your mind, because only you will be able to monitor how the answer continuously evolves over various time scales.

Question 1.2: What Systems and People Must I Interact with in This Moment?

> *Sandar, an embedded systems team lead:*
> "I have to think carefully about how to argue about how we setup testing when holding design-meetings for each sprint… if I don't then Brad will shoot of on his own again, because you know, he is impatient…"
> *[sighs]*

Whatever technical skills are presently required, you will rarely have the opportunity/disadvantage (which of these you see depends on your personality) to use them isolated from others, or physically separated from others. Again this probably seems obvious, which is the emergent theme for this chapter, but pertinent to consider in practice.

Usually, this question relates to people rather than systems, but these are increasingly intermingled, and interacting with a system or person is often one thing, such as updating a backlog together on Monday morning or through monitoring pull-requests and the slack-channel (team communication software), ensuring that Brad hasn't done another informal nightly branch-merge simply because "*It works now.*"

Sending Brad a vague and slightly passive-aggressive email with the same intent during the day would probably not have the same effect. So keep this question with you when considering both the *when, who,* and the *how* of the interactions you must engage in to complete the current task. Regardless of whether you are Sandar or Brad.

Question 1.3: What Kinds of Actions Are Considered Legitimate in This Moment?

Eric, an experienced electronics engineer, attends a meeting on tooling for circuit board manufacturing with the internal production team.
The factory is far away and has a long history in the company, so Eric must use their services even though they produce subpar chips for his application. Reasonably concerned, during a budget meeting with the steering group Eric shifts the topic to the quality issues with internal chip production. His interjection, while relevant to the budget work, is met with silence. The meeting proceeds with puzzled and uncomfortable glances.

When switching contexts during your workday, you notice how you treat others, and are treated differently yourself. This is not simply a question of relative status, because while assumptions about you as an engineer *do* matter, you have very limited control over them. What matters is your action choices and whether those decisions strengthen or undermine your perceived legitimacy.

Having conducted planned hardware FEM tests with results motivating immediate action on the level of the product owner and/or sponsor, you might not wait to the next steering group meeting, but quite reasonably call for one. Whether this action is considered legitimate, however reasonable it might be, depends on *how* you call the meeting and *how* you act to present, explain, and recommend action based upon, the FEM test data. Yes, this can be as "*simple*" as the tone of the email and the order and formatting of graphs and tables. Such things are too important to simply be entrusted to the project manager or team lead. So when keeping this question in mind, do so not by considering *if* you should act: as an engineer, you alone often sit on the needed knowledge for the *if*. Rather it is *how* you act, especially if the line is stronger than the projects in your organization.

Question 2: How Will Perceived Project Complexity Affect Everyday Project Work?

You already know that projects are "*complex*", potentially meaning uncertain, interconnected, high levels of dynamic interactions over vastly different time domains, socially implicit and opaque. However, for most intents and purposes, "*complex*" when used in everyday professional situations outside of academia (often broad scope) and domain-specific technical discussions (often narrow scope) simply means "*difficult to do or understand*". This view on complexity in practice is both pragmatic and supported by empirical research on the cognitive nature of complexity [8].

This is what the second question is about, what is difficult for you in your project (i.e. perceived project complexity) is to a large extent a *perceptual exercise*. While thinking about what seems to be "*difficult*", keep "*seeing*" as the metaphor, because the first specific question concerns a core factor for perceived project complexity: size as "*seen*".

Question 2.1: What Part of the Project Seems Big

> *During their daily reporting, Mathilda and her other three project team members opened a backlog that had long been neglected.*
> *They were standing close to a 50-inch screen, higher than the team members.*
> *Many heavily interrelated task entries were in the stack, visualized on screen.*
> *There was a silence and a simultaneous visible squaring of shoulders.*
> *After the meeting, Mathilda sighed and looked away, muttering to herself:*
> "It's always like this, I'm the bottleneck, all of it just feels too big*!*"

As engineers, we often determine the complexity of a project by how large it seems to us—the "*bigger*" it looks, the tougher we assume it'll be. This way of thinking has deep roots in how we are trained to see things as both engineers [9] and human beings [10] (p.51). In my empirical work observing Mechatronics engineers, even traditional "small" projects (small team, known technology, sufficient resources) can be highly complex, especially when it comes to how things work and interact, which isn't always obvious at first glance:

> "*Even if a Mechatronic project has limited ostensive organizational size and low apparent technological novelty and complicatedness it can still be perceived as complex.*" [11].[1]

Question 2.2: What Complexity Dimensions Do I (not) See?

> *Eric was team lead for an embedded systems development team consisting of 4 junior developers. He was very focused on their development as individuals, coaching them systematically while supporting them to maintain an internal unit-test plan together in the team.*
> *While the software-testing went well, this approach blinded Eric—who felt a need to care for the young team members—to scheduling-related concerns raised by the hardware engineers as to their FEM testing of the case material,*

[1] Section 5.3.2.

since all testing had to share computer hardware resources due to testing software licensing limitations.
Eric did not fully understand that the testing software license issue limited the project's ability to schedule its various testing tasks.

In our projects, we often get this nagging feeling that something is off. It might be that one's focus is on a collection of important issues (for good reason), while other less critical (but still important) tasks are left to grow on the wayside. Here we constantly have an opportunity to tap into our professional concern as engineers and ask ourselves what might be quietly breaking while we are not looking?

Question 2.3: What or Who Changes the Burden for the Project the Most Right Now?

Martine, mechanical engineer in an aerospace manufacturing tool project:
"The product owner changed again, and he [*the new PO*] came to me last week, the new guy, and asked me for a functional demo of the system in two week's time for an industry fair he was going to attend. We had to drop everything!"

Of course, sometimes things are not quietly breaking unseen in a corner. Sometimes change is very visible and hard to look away from. While changes of this type might seem obvious, their effects must be monitored, particularly how these effects are *perceived* by not only you but also by the rest of the team.

Question 3: What Will the Relation Between Expertise and Complexity Be?

In the coming decades of Mechatronic development work, both the content of engineering expertise and the ways in which development projects might be seen as complex will change. But there are reasons to believe that the connection between the two will not. As long as the development loop meaningfully includes human elements, the same challenges of differing mental models and language representations will remain. Regardless of the state of the technology involved, you as an engineer will have to relate to three issues between project complexity and your own expertise: how you experience, think about, and act on difficulties (i.e. perceived project complexity).

Question 3.1: When Difficulties Arise, How Are They Experienced?

> *"Practically, everyone who has worked on a project knows the feeling when walking into the project war room and the air is filled with tension. You just know that it is going to be a hot meeting where people for various reasons try to restrain themselves from bursting out in harsh words. It is difficult to explain this feeling."*
>
> Burström & Wilson [12]

This first question about the relationship between your difficulties and competence might seem somewhat odd, but there is strong empirical cognitive science evidence [10, 13–15] for identifying what specific subjective sensory experiences occur in challenging situations. This is summarized in any number of proverbs, as for instance "*go with your gut.*" However, such pithy remarks miss the issue somewhat. The value of looking at how your body reacts to a non-bodily challenge is directly proportional to the specificity you can bring to your analysis. As in question 2.2 above, if you stop at a generalistic answer akin to "*something feels off*" it will be of limited value. But if you can instead identify, say, a pattern for how you physically react in similar situations, this should be seen as valuable and (crucially) actionable information rather than a hindrance. Do not immediately try to soothe or make yourself comfortable but ask yourself: "*What kinds of discomfort can I use as early warning systems for different dangers?*".

Question 3.2: When Difficulties Arise, How Are They Described?

In a conflict-heavy FMEA meeting in which the dev team met with an FMEA-engineer from the line-organization, the meaning and complexity of the word "risk" *stood out.*
The FMEA-engineer tried to "sell" *the method as a service to the team, while the team felt their internal risk-management practice criticised. Arnold, the PM:*
"Hearing how differently the team and the FMEA described risk and the challenges to our development plan, I had to break up the meeting before something ugly happened."[6]

Having established a sense of how a difficult situation is physically experienced in the body, it is of valuable to listen closely to how you and others describe present challenges to your project. What metaphors are used by others? Which do you tend to use intuitively? Do they align or not?

Questions about how challenges are described and felt give clues to how ones team emotionally regulates within the context of these challenges [13, 16, 17]. Furthermore, there are some tentative results that explicitly discussing such questions in a structured format might be of benefit to team performance, lessening the prevalence of negative internal conflict [15].

Question 3.3: When Difficulties Arise, How Are They Acted on

Having spent more than half a year on preparing it, an electronics engineer faced resistance to early integration testing of the electronics at an external facility.

Having had his integration test plan denied, he stormed into a project office management meeting unrelated to the testing, or his project at all, and interrupted with:
"I have to spend so much time on politics every day. Where's your 6-month investigation and data for denying my integration plan?!"

After having considered all these questions, we are still left with the question of *What to do, then?* It is here that we must more than ever be attentive to knee-jerk reactions and intuitive pushing forward. By ourselves, as well as in others, in our projects. Because in not having an understanding of the connection between direct experience, expression, and resulting action in the face of difficulty, we are often faced with what has been called *functional stupidity* [18]: Reactionary behaviour which might be nominally functional, but risky or otherwise sub-optimal.

Provided that the experience and the description of the situation do not indicate a threat to the immediate health and safety of actual people, this question serves as a sanity check.

20.2 Concluding Remarks—A Checklist Approach

In what way should you expect that these nine questions relate to the future of working in Mechatronics development projects? While the above pithy treatment through questions of the mundane reality of development work might seem atemporal and devoid of any forward-leaning discussions of project technology (such as generative AI), there was a thought to this. While all Mechatronic engineers have a very personal and intimate relationship with their own evolving technology landscape, this relationship is quite personal and specific, as for instance expressed in Hillmer about the CAPS-system[1] [19].

Lacking the space and time for a more detailed treatment, I explicitly chose to rely on pre-existing expertise that you have in your own technology environment, and instead focus on the personal and practical everyday treatment of project complexity. Instead of setting and explaining a taxonomy for it—the interested reader is directed to the excellent recent review by Bakhshi and colleagues [20]—and connecting such to general features of everyday Mechatronic work, the chapter will conclude differently. With something that I through my research found several times, a weekly "*sanity*" checklist approach to asking questions. During our hectic weeks, we might find time to stop and consider questions such as those offered above, but to regularly capture the actual value and relevance of our answers can often be more difficult. However, not doing so largely misses the point of the exercise. Sitting down on Friday afternoon, tired and still with important ToDo's in the inbox, it is hard to openly reflect on a blank page.

Therefore, for your future practice of Mechatronic development work, the following checklist is offered. Once a week, preferably by its end, sit down and go through the list. Importantly, you must tailor this list to suit your specific circumstances, both for role and firm, as well as over time. The point is to *maintain a list* while keeping the nine questions offered above in mind.

Start by going through the questions and ask yourself the following:

- Look at your ToDo's from this week, which items do you **need** (rather than want) to close before Friday end? Schedule or delegate each!
- What stole time from me this week? What small step (< 2 min) can I do right now to mitigate this? Set a 120 s-timer and do it.
- What event or issue changed my project (or other significant piece of work) this week the most? (positive/negative/other)
- Has my possible space for legitimate action changed this week?
- What kind of recurring talk (or metaphor) was a hindrance this week?
- Has this kind of talk changed in tone? What, if anything, should I do?
- Who in my team is struggling the most? Is it me?
- What important task is still left from last week's checklist, and why?
- What is the most important action for Monday morning?
- **Finally, add/remove (ir)relevant questions in this checklist.**

As you reflect each week, remember that the value of these reflections lies not just in the answers you find but in the quality of the revisions you bring to your questions. It is the wish and intent of this chapter that you can establish a weekly ritual tailored to the reality of your own Mechatronic engineering practice, both now and for the decades to come.

Acknowledgements Thank you to the editorial team for the discussions on this chapter.

References

1. Mørkeberg Torry-Smith J, Qamar A, Achiche S, Wikander J, Henrik Mortensen N, During C (2012) Challenges in designing mechatronic systems. ASME J Mech Des 135(1):011005. https://doi.org/10.1115/1.4007929
2. Salminen V, Verho AJ (1989) Multi-disciplinary design problems in mechatronics and some suggestions to its methodical solution in conceptual design phase. Intl Conf Eng Des ICED89 1:533–554
3. Buur J, Andreasen MM (1990) A theoretical approach to mechatronics design. Technical University of Denmark, Institute for Engineering Design
4. Trevelyan J (2014) The making of an expert engineer. CRC Press
5. Winberg C (2008) Teaching engineering/engineering teaching: interdisciplinary collaboration and the construction of academic identities. Teach High Educ 13(3):353–367
6. Trevelyan J (2019) Transitioning to engineering practice. Eur J Eng Educ 44(6):821–837. https://doi.org/10.1080/03043797.2019.1681631
7. Wolff K (2018) Researching the engineering theory-practice divide in industrial problem solving. Eur J Eng Educ 45(2):181–195. https://doi.org/10.1080/03043797.2018.1516738

8. Mikkelsen MF (2020) Perceived project complexity: a survey among practitioners of project management. Intl J Managing Projects Bus 14(3):680–698
9. Hyldgaard Christensen S (2015) Engineering identities, epistemologies and values: engineering education and practice in context, 2, 7
10. Lakoff G (2012) Explaining embodied cognition results. Top Cogn Sci 4(4):773–785. https://doi.org/10.1111/j.1756-8765.2012.01222.x
11. Flening E (2023) Mechatronic metaphors : an embodied realist framework of complexity and expertise in mechatronic project work', PhD dissertation, KTH Royal Institute of Technology, Stockholm
12. Burström T, Wilson TL (2018) The texture of tension: complexity, uncertainty and equivocality. Intl J Managing Projects Bus 11(2):458–485
13. Samur D, Lai VT, Hagoort P, Willems RM (2015) Emotional context modulates embodied metaphor comprehension. Neuropsychologia 78:108–114
14. Shelley A (2012) Metaphor interactions to develop team relationships and robustness enhance project outcomes. Project Manag J 43(6):88–96
15. Shelley AW, Maqsood T (2014) Metaphor as a means to constructively influence behavioural interactions in project teams. Intl J Managing Projects Bus 7(4):752–766
16. Gibbs RW (2013) Metaphoric cognition as social activity: dissolving the divide between metaphor in thought and communication. Metaphor Soc World 3(1):54–76
17. Heracleous L, Jacobs CD (2008) Understanding organizations through embodied metaphors. Organ Stud 29(1):45–78
18. Alvesson M, Spicer A (2012) A stupidity-based theory of organizations. J Manag Stud 49:1194–1220
19. Hillmer U (2009) Technology acceptance in mechatronics: the influence of identity on technology acceptance. Springer
20. Bakhshi J, Ireland V, Gorod A (2016) Clarifying the project complexity construct: past, present and future. Intl J Project Manag 34:1199–1213

Chapter 21
Revolutionising Education Through Digital Twins: Case Study and Applications

Esfandiar Khaleghi and Robert Rayner

Abstract This chapter explores the transformative potential of Digital Twin (DT) technology in education, with a particular focus on its application in Science, Technology, Engineering, and Mathematics (STEM) fields. Digital Twins, initially developed for industrial and research applications, are dynamic digital replicas of physical systems, enabling real-time monitoring, predictive maintenance, and operational optimisation. These capabilities provide an unprecedented level of interactivity, allowing learners to engage with complex systems in a way that traditional methods and simulations cannot offer. Digital Twins are shown to enhance operational efficiency, improve decision-making, and provide invaluable hands-on learning experiences. In educational contexts, Digital Twins facilitate immersive and interactive learning by incorporating real-time data from physical entities into digital models, providing students with instant feedback and the ability to simulate real-world scenarios. This chapter outlines the benefits of using Digital Twins, such as fostering deeper understanding, encouraging collaboration, and allowing safe, cost-effective experimentation. Moreover, Digital Twins help bridge the gap between theoretical knowledge and practical application, equipping learners with skills relevant to technology-driven industries. Through an in-depth case study of a miniaturised Connected Autonomous Vehicle (CAV), the chapter demonstrates how Digital Twins can be applied effectively in STEM education. This allows learners to develop skills in areas such as machine learning, sensor fusion, and autonomous vehicle technology. These active learning scenarios promote problem-solving and technical proficiency, preparing students for future challenges in their fields. While implementation challenges, such as high costs and technical complexity, are acknowledged, the chapter argues that the educational benefits far outweigh these obstacles. As Digital

E. Khaleghi · R. Rayner (✉)
University of Hertfordshire, Hatfield, UK
e-mail: Robert.Rayner@member.imeche.org; r.rayner2@herts.ac.uk

E. Khaleghi
e-mail: e.khaleghi@herts.ac.uk

E. Khaleghi
Kingston University London, London, UK

P. Hehenberger and D. Bradley (eds.), *Mechatronic Futures*,
https://doi.org/10.1007/978-3-031-83571-1_21

Twin technology becomes more accessible, its role in revolutionising education will continue to grow, enhancing student engagement and learning outcomes.

21.1 Introduction

In an era where digital transformation is reshaping industries and redefining the boundaries of research, the integration of digital technologies into education is not just a trend but a necessity. The evolution of digital tools from their origins in industrial and research applications to their current role in education underscores a significant paradigm shift. Initially developed to enhance efficiency, accuracy, and productivity in industrial processes and scientific research, these tools are now being leveraged to revolutionise the educational landscape. The incorporation of digital technologies in education offers unparalleled opportunities for interactive learning, real-time data analysis, and the practical application of theoretical concepts, thus preparing learners for the demands of a digital-first world.

This chapter delves into the multifaceted applications of Digital Twin technology, with a particular focus on its transformative potential in the educational sector. By offering dynamic, interactive, and immersive learning experiences, Digital Twins can revolutionise educational practices, particularly in Science, Technology, Engineering, and Mathematics (STEM) disciplines. The chapter also explores the distinct challenges and opportunities Digital Twins hold over traditional simulations.

In examining the specific case studies of two autonomous vehicle teaching aids, this chapter illustrates how Digital Twin technology can be effectively integrated into educational settings. These case studies highlight the practical applications, benefits, and drawbacks of using Digital Twins in teaching complex concepts in engineering and related fields.

21.2 Origins and Evolution of the Digital Twin Concept

The concept of the Digital Twin (DT) originated in the early 2000s and has evolved significantly over the past two decades. This idea was first articulated by Dr Michael Grieves in 2002 during a course on Product Lifecycle Management (PLM) at the University of Michigan [1]. Initially termed the '*Mirrored Spaces Model*', the concept aimed to create a virtual counterpart of a physical system to track and predict its performance throughout its lifecycle. This model eventually evolved into what we now know as the Digital Twin, focusing on representing physical products in a virtual space to enhance their lifecycle management [2].

The Digital Twin concept arose from the need for a more effective way to manage the entire lifecycle of a product, from design and manufacturing to operation and disposal. Grieves's early model included three main components: the physical space, the virtual space, and the data flow between the two. This structure allowed for a bidirectional flow of information, enabling real-time updates and monitoring of the physical system based on its virtual counterpart.

Initially, the implementation of this concept was primarily theoretical, constrained by the technological limitations of the early 2000s. Computing power, data connectivity, and storage capabilities were not sufficiently advanced to fully realise the potential of the Digital Twin. Most notably, David Gelernter imagined the concept as early as 1991, referring to the concept as '*Mirror Worlds*' [3] while Kary Främling et al. in 2003 described the concept as an '*agent-based model*' [4]. However, the foundational idea was set, namely a digital informational construct about a physical system that could mirror the information embedded within the physical system itself.

The term '*Digital Twin*' was first used in a formal context by NASA in their 2010 technology roadmap. NASA described the Digital Twin as [5]:

> '*… an integrated multi-physics, multi-scale, probabilistic simulation of a vehicle or system that uses the best available physical models, sensor updates, fleet history, etc., to mirror the life of its flying twin.*'

This concept was not entirely new to NASA, which had employed similar methods during the Apollo program, using identical space vehicles to mirror each other's conditions and performance [5].

NASA's adoption of the Digital Twin model marked a significant milestone in its evolution, integrating it into their technology roadmaps and space exploration proposals. This demonstrated the practical applications of the Digital Twin in monitoring and predicting the performance and maintenance needs of complex aerospace systems. The US Air Force soon followed suit, applying the Digital Twin for aircraft design, maintenance, and lifecycle management [6].

The application of the Digital Twin quickly expanded beyond the aerospace industry and product lifecycle management encompassing broader industrial applications with the term 'Digital Twin' becoming the accepted technical terminology [7]. The use of Digital Twins began to broaden with applications being seen in various industrial sectors, including manufacturing [8, 9] and healthcare [10, 11].

The expansion of the Digital Twin was driven by advancements in information technology, including improved computing power, enhanced data connectivity, and more sophisticated data storage solutions. These advancements enabled more detailed and real-time data collection, making the Digital Twin a practical tool for various industries.

The Digital Twin's ability to link physical and virtual worlds in real-time offered significant advantages, including real-time monitoring, predictive maintenance, and enhanced system optimisation. This capability was particularly valuable in industries requiring precise monitoring and management of complex systems, such as aerospace, automotive, and industrial manufacturing.

Today, the Digital Twin is recognised as a critical component of Industry 4.0 [12], the ongoing automation and data exchange in manufacturing technologies. The concept has evolved to include not only physical products but also processes and systems, providing a powerful tool for simulation, prediction, and optimisation.

The rise of the Internet of Things (IoT) has further propelled the development and application of Digital Twins [13, 14]. IoT devices provide continuous real-time data from physical systems, enhancing the accuracy and utility of their virtual counterparts. This continuous data flow enables more precise simulations and predictions, improving decision-making and operational efficiency.

21.3 Core Principles and Definitions of Digital Twins

The concept of the Digital Twin has evolved significantly since its inception, leading to various definitions tailored to specific applications and industries. Fundamentally, a Digital Twin is a digital replica of a physical entity, system, or process. This virtual model is continuously updated with real-time data from its physical counterpart, allowing for detailed monitoring, analysis, and simulation. The bidirectional flow of data ensures that any changes in the physical entity are reflected in the Digital Twin, and vice versa, facilitating a dynamic and interactive relationship [15, 16].

A Digital Twin is a dynamic, digital replica of a physical object, system, or process that enables real-time monitoring, simulation, and analysis. This virtual model is created by integrating data from sensors, historical records, and other sources to mirror the exact state and behaviour of its physical counterpart. Through continuous data exchange, a Digital Twin provides a comprehensive view of the physical entity's current condition and performance, allowing for predictive maintenance, optimisation, and enhanced decision-making.

A block diagram representation of the Digital Twin concept model as proposed by Grieves [1, 2] is shown in Fig. 21.1.

Figure 21.1 shows that a Digital Twin consists of three main parts:

Physical Products in Real Space—This represents the actual physical entity or system being modelled, depicted here by a robotic arm. The real space is where physical actions, interactions, and operations occur, providing the source of real-time data.

Virtual Products in Virtual Space—This represents the digital replica of the physical product, existing in a virtual environment. The virtual space mimics the behaviour

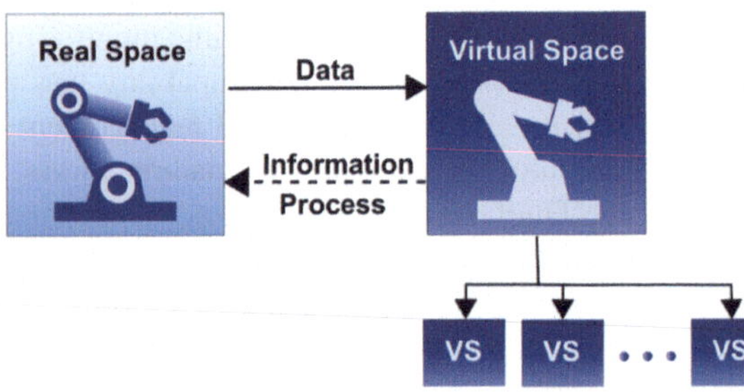

Fig. 21.1 Block of diagram of digital twin (after [15])

and characteristics of the real space counterpart, allowing for simulations, analysis, and optimisations without physical constraints. For a given physical product, it may be possible to represent this device in virtual space using multiple sub-virtual spaces, represented in Fig. 21.1 by blocks labelled '*VS*', each representing the digital replica of a sub-element of the complete physical product.

Connections of Data and Information Processes—This is represented by the arrows between the real space and virtual space, illustrating the bidirectional flow of data and information. Data from the physical product is continuously fed into the virtual model, ensuring it accurately reflects the current state of the real-world system. Conversely, information and insights derived from the virtual model are fed back into the real space, enabling adjustments and improvements. This continuous loop of data exchange allows for real-time monitoring, predictive maintenance, and operational optimisation.

According to Grieves and Vickers [16], Digital Twins can be categorised into different types based on their specific applications and the nature of their physical counterparts:

Digital Twin Prototype (DTP) —This type of Digital Twin represents a prototype of the physical entity. It contains all the necessary information to describe and produce a physical version that duplicates or twins the virtual version. It is used primarily during the design and development phases.

Digital Twin Instance (DTI) —This type of Digital Twin corresponds to a specific instance of a physical product. It remains linked to this specific physical instance throughout its lifecycle, providing real-time updates and enabling precise monitoring and maintenance.

Digital Twin Aggregate (DTA) —This type is an aggregation of multiple Digital Twin Instances. It may not be an independent data structure but a computing construct that queries and analyses data from multiple DTIs to provide insights and predictions.

Digital Twin Environment (DTE) —This is an integrated, multi-domain physics application space for operating on Digital Twins for various purposes, including prediction, simulation, and optimisation.

A Digital Twin has five key characteristics as follows [7, 15]:

Real-Time Data Integration—One of the primary features of a Digital Twin is its ability to receive and process real-time data from its physical counterpart. This data integration allows the Digital Twin to reflect the current state of the physical entity accurately. For example, sensors on a physical machine can continuously send performance data to its Digital Twin, which then updates to show the machine's current status and any potential issues.

Bidirectional Data Flow—The connection between the physical entity and its Digital Twin is bidirectional. Not only does the Digital Twin receive data from the physical entity, but it can also send data back, influencing the physical entity's operations. This capability is crucial for applications such as predictive maintenance, where the Digital Twin can suggest or even implement corrective actions.

Lifecycle Representation—A Digital Twin can cover the entire lifecycle of the physical entity, from design and development through operation and maintenance to eventual disposal. This comprehensive coverage allows for better planning, optimisation, and decision-making at each stage of the lifecycle.

Simulation and Prediction—Digital Twins are equipped with advanced simulation capabilities, allowing them to predict future states of the physical entity based on current and historical data. This predictive capability is valuable for anticipating and mitigating potential issues before they occur, thereby reducing downtime and improving efficiency.

High Fidelity—For a Digital Twin to be effective, it must be a high-fidelity representation of its physical counterpart. This high degree of accuracy ensures that the insights and predictions generated by the Digital Twin are reliable and actionable.

21.4 Digital Twins Versus Computer Simulations: Understanding the Differences

While the terms '*Digital Twin*' and '*computer simulation*' are often used interchangeably in the literature, readers should be wary of this practice. They represent distinct concepts with different functionalities, applications, and levels of complexity. Understanding the differences between a Digital Twin and a computer simulation is essential for leveraging each technology's strengths and capabilities effectively.

As previously discussed, a Digital Twin is a dynamic and comprehensive digital replica of a physical object, system, or process that is continuously updated with real-time data. It is designed to mirror the physical entity's exact state and behaviour, enabling real-time monitoring, predictive maintenance, and optimisation. The bidirectional data flow between the physical and digital entities ensures that changes in one are reflected in the other, providing an interactive and evolving model throughout the physical entity's lifecycle [14, 15].

For instance, a Digital Twin of a Connected Autonomous Vehicle (CAV) includes a corresponding software platform that receives continuous data from the physical prototype vehicle, including GPS position, sensor feedback such as vision camera, LiDAR point cloud, wheel speed, and battery state of charge. This allows car manufacturers to monitor the CAV prototype in real time, predict and optimise vehicle performance characteristics such as the performance of algorithms relating to route planning, object detection, and obstacle avoidance.

In contrast, a computer simulation is a static or dynamic model used to replicate the behaviour of a system based on a set of predefined parameters and initial conditions. Simulations are typically used for analysis, testing, and experimentation under controlled conditions. They are valuable for understanding potential outcomes and behaviours in a virtual environment but do not inherently involve continuous real-time data updates from the physical system.

As an example, an automotive company might use computer simulations to test the aerodynamics of a new car design. Engineers create a virtual model of the car and simulate how air flows around it under various conditions. This helps optimise the car's design for better fuel efficiency and performance before any physical prototypes are built.

Digital Twins and computer simulations differ significantly in their data integration approach. Digital Twins continuously process real-time data from sensors embedded in physical entities, allowing for accurate, up-to-date reflections of the system's state, enabling proactive decision-making and predictive analytics. In contrast, computer simulations typically rely on static data and predefined scenarios, making them suitable for exploring hypothetical scenarios without real-time updates. Digital Twins cover the entire lifecycle of the physical entity, from design to decommissioning, ensuring continuous improvement informed by real-time and historical data. Computer simulations are often limited to specific lifecycle phases, such as design optimisation or stress testing, and may not be updated or maintained long-term. A key feature of Digital Twins is their bidirectional data flow, where data is both received from and sent to the physical system, enabling control and corrective actions. Computer simulations usually operate unidirectionally, lacking real-time feedback, and thus provide limited actionable insights for ongoing operations.

21.5 Digital Twins in Action: Cross-Industry Applications and Benefits

Digital Twin technology is revolutionising the aerospace industry by creating virtual replicas of physical assets, such as jet engines and spacecraft, enabling real-time monitoring and optimisation. GE Aviation [17] and Rolls-Royce [18] use Digital Twins for their aircraft engines, allowing for predictive maintenance, improved performance, and extended engine life by continuously synchronising operational data with their virtual counterparts. Airbus' Skywise platform enhances aircraft maintenance and operational efficiency through real-time data exchange between aircraft systems and their Digital Twins [19]. Boeing applies Digital Twins in its manufacturing processes to improve quality, reduce production time, and minimise costs [20] NASA utilises Digital Twins for spacecraft to simulate performance, enhance mission planning, and extend mission life through real-time monitoring and fault detection. These applications highlight the significant benefits of Digital Twin technology in ensuring safety, efficiency, and cost-effectiveness in aerospace operations [21].

In contrast in the subject area of manufacturing technology, Siemens [22] and Bosch [23] are leveraging Digital Twin technology to revolutionise their manufacturing processes. Siemens uses Digital Twins for factory automation, creating virtual models of production lines that simulate real-time operations using data from sensors and IoT devices on the factory floor. This approach enhances efficiency, enables predictive maintenance, and improves product quality by allowing for the optimisation of production processes and early detection of potential issues. Similarly, Bosch employs Digital Twins in automotive manufacturing to create virtual replicas of production processes, optimising workflows, ensuring precision in manufacturing, and enabling predictive maintenance. Continuous data exchange between physical manufacturing environments and their Digital Twins allows both companies to achieve real-time monitoring and process optimisation, leading to increased efficiency and reduced production costs.

Digital patient twins, sometimes referred to as virtual twins, are highly detailed digital replicas of patients created using data from various medical sources such as imaging, wearables, and electronic health records. Unlike traditional Digital Twins that interact with both physical and virtual spaces, virtual twins in healthcare primarily reside in the virtual space, leveraging real-time data to simulate and predict patient outcomes without direct physical interactions. Examples of Digital Twins in healthcare include Philips' personalised cardiac care models, which enhance heart disease treatment [24] and Siemens Healthineers' [25] aims to create Digital Twins of whole patients to help with the prediction of the onset of illnesses and to create personalised treatment plans accordingly, while Dassault Systèmes' Living Heart Project creates detailed virtual heart models for pre-surgical planning and research [26]. These applications demonstrate the transformative potential of Digital Twins in providing personalised care, improving diagnostic accuracy, and optimising treatment outcomes in the healthcare industry.

21.6 Digital Twins Implementation and Usage: Benefits and Challenges

While both Digital Twins and computer simulations are valuable tools for modelling and analysing physical systems, Digital Twins offer several distinct advantages due to their dynamic, real-time capabilities and comprehensive lifecycle integration. These benefits enhance real-time data integration, continuous lifecycle management, bidirectional data flow, predictive maintenance, high-fidelity modelling, comprehensive simulation capabilities, and improved collaboration and decision-making. Digital Twins exhibit features which make them highly advantageous tools. These advantages include:

Real-Time Data Integration—One significant advantage of Digital Twins is their ability to continuously receive and process real-time data from sensors and other sources embedded in the physical entity. This real-time integration allows for a precise and up-to-date reflection of the physical system's state, enabling proactive decision-making and predictive analytics. This capability enhances the monitoring and optimisation of processes, ensuring that any anomalies or inefficiencies are promptly addressed.

Continuous Lifecycle Management—Digital Twins cover the entire lifecycle of the physical entity, from design and development through operation and maintenance to eventual decommissioning. This comprehensive coverage ensures continuous improvement and optimisation at every stage, informed by real-time performance data and historical trends. The ability to manage and adapt to changes throughout the lifecycle makes Digital Twins invaluable for long-term planning and resource management.

Bidirectional Data Flow—The bidirectional data flow between the physical and digital entities is a hallmark of Digital Twins. Not only does the Digital Twin receive data from the physical system, but it can also send data back, influencing and controlling the physical system based on insights derived from the digital model. This feedback loop enables real-time adjustments and fine-tuning, enhancing operational efficiency and responsiveness.

Enhanced Predictive Maintenance—Digital Twins enable predictive maintenance by using real-time data and advanced analytics to predict equipment failures before they occur. This capability reduces downtime and maintenance costs while improving the reliability and lifespan of the equipment. Predictive maintenance is crucial for maintaining operational continuity and preventing unexpected failures.

High-Fidelity Modelling—Digital Twins provide a high-fidelity, near-realistic digital representation of the physical system. This high degree of accuracy ensures that the insights and predictions generated by the Digital Twin are reliable and actionable. High-fidelity models support precise simulations and analyses, contributing to better decision-making and optimisation strategies.

Comprehensive Simulation Capabilities—Digital Twins offer advanced simulation capabilities that go beyond static models. They can simulate various scenarios and conditions, including the impact of environmental factors, operational changes, and unexpected events. These capabilities allow for thorough testing and analysis, facilitating the development of robust and resilient systems.

Improved Collaboration and Decision-Making—Digital Twins facilitate better collaboration and decision-making by providing a shared, real-time view of the system. Stakeholders can access and interact with the Digital Twin from different locations, improving communication and coordination. This shared access supports more informed and cohesive decision-making processes.

While Digital Twin technology offers numerous advantages, implementing and maintaining these systems is not without challenges. Understanding these drawbacks is essential for organisations to make informed decisions about implementing Digital Twins and to mitigate potential risks effectively. These challenges include:

High Initial Cost and Complexity—Implementing a Digital Twin can be expensive and complex, particularly for large-scale systems. The costs associated with deploying sensors, integrating real-time data streams, and developing sophisticated software models can be substantial. Additionally, the complexity of setting up a Digital Twin infrastructure requires significant technical expertise and resources. This high initial cost can be a barrier for smaller companies or those with limited budgets.

Data Security and Privacy Concerns—Digital Twins rely on continuous data flow between the physical and digital entities, raising concerns about data security and privacy. Sensitive data transmitted from sensors and stored in digital models can be vulnerable to cyberattacks, data breaches, and unauthorised access. Ensuring the security and privacy of this data requires robust cybersecurity measures, which can be both costly and complex to implement.

Data Integration and Management Challenges—Integrating and managing large volumes of data from diverse sources can be challenging. Ensuring data accuracy, consistency, and reliability is critical for the effective functioning of a Digital Twin. Discrepancies or errors in the data can lead to incorrect predictions and analyses. Effective data integration and management require advanced solutions, which can be difficult to achieve consistently.

Maintenance and Upkeep—Digital Twins require continuous updates and maintenance to remain accurate and effective. This ongoing upkeep can be resource-intensive, requiring regular monitoring, calibration, and software updates. Failure to maintain the Digital Twin properly can result in outdated or incorrect models, diminishing their usefulness.

Dependency on High-Quality Data—The effectiveness of a Digital Twin is heavily dependent on the quality of the data it receives. Inaccurate, incomplete, or low-quality data can compromise the reliability of the Digital Twin, leading to incorrect predictions and suboptimal decisions. Ensuring high-quality data is critical but can be challenging to achieve consistently.

Technological and Organisational Barriers—The successful implementation of Digital Twins requires overcoming various technological and organisational barriers. These include the need for advanced analytics, machine learning capabilities, and the integration of different technologies and systems. Additionally, organisational resistance to change and the need for new skill sets can impede adoption. Overcoming these barriers is essential for the effective deployment and utilisation of Digital Twins.

It can therefore been seen that Digital Twins offer dynamic, real-time capabilities and comprehensive lifecycle management that enhance monitoring, predictive maintenance, and decision-making across various industries. Their high-fidelity modelling and bidirectional data flow enable precise simulations and proactive adjustments, fostering improved operational efficiency and collaboration. However, implementing Digital Twins comes with challenges such as high initial costs, data security concerns, and ongoing maintenance demands. Organisations must address these issues to fully leverage the benefits of Digital Twin technology. Building on these advantages and challenges, the next section explores the application of Digital Twins in education, highlighting how this innovative technology can transform teaching and learning processes.

21.7 Digital Twins in Education

As we transition from understanding the general advantages and disadvantages of Digital Twins, it is crucial to explore their specific applications within the educational sector. The integration of Digital Twin technology into education represents a significant advancement, offering dynamic, interactive, and immersive learning experiences that traditional methods and even simulations cannot match. Understanding Digital Twins is essential for educators and institutions aiming to enhance educational practices, particularly in STEM subjects, across all educational levels—from primary schools to higher education and beyond. This technology holds transformative potential by providing real-time, hands-on interaction that enhances learners' understanding of complex concepts. Furthermore, Digital Twins foster improved collaboration, safety, and cost-efficiency in learning environments. Grasping the implications and applications of Digital Twins is vital for leveraging their full potential to revolutionise traditional pedagogical approaches and prepare learners for real-world challenges.

Digital Twins provide learners with access to real-time, interactive learning platforms that can greatly enhance learners' understanding of complex concepts by allowing them to visualise and manipulate digital replicas of real-world systems and processes. This hands-on approach can make abstract STEM concepts more tangible and accessible, fostering deeper engagement and comprehension. In primary education, Digital Twins can help learners grasp fundamental scientific principles by offering interactive experiences that would otherwise be impossible in a traditional classroom setting. By manipulating digital models, learners can explore concepts like planetary movements, gravitational forces, chemical reactions and the functionality of complex machines in an engaging and intuitive manner.

In secondary education, the use of Digital Twins extends to more complex subjects such as physics, chemistry, and biology. Learners can conduct virtual experiments, observe outcomes, and analyse data in ways that traditional labs cannot provide. This not only enhances their understanding of scientific principles but also develops their analytical and critical thinking skills. In higher education, particularly in engineering and medical schools, Digital Twins offer learners the opportunity to work with virtual replicas of sophisticated equipment and systems. This can include everything from automotive engines to human organs, enabling learners to conduct experiments, diagnose issues, and practice procedures in a risk-free environment.

In vocational training and continuing education, Digital Twins simulate real-world scenarios and equipment operation, allowing learners to gain practical experience in a controlled environment.

21.7.1 Benefits of Using Digital Twin in Education

While traditional simulations are important tools for STEM education, Digital Twins offer several distinct advantages that make them superior tools for modern learning environments.

Real-Time Interactivity and Feedback—Digital Twins provide real-time interactivity and feedback, allowing learners to see the immediate effects of their actions on the digital model. This instant feedback loop enhances learning by enabling learners to experiment, make mistakes, and understand the consequences in a controlled environment. This dynamic interaction is a significant improvement over static simulations, which typically do not update in real-time and lack the capability to reflect ongoing changes.

Continuous Learning and Adaptability—Digital Twins are dynamic and continuously updated, reflecting the latest data and changes in the physical system they represent. This adaptability ensures that educational content remains current and relevant, providing learners with up-to-date information and scenarios. Traditional simulations, on the other hand, are often based on static models that do not evolve with new data, potentially limiting their relevance over time.

Enhanced Collaboration and Accessibility—Digital Twins facilitate enhanced collaboration among learners and educators by providing a shared virtual environment. Learners can work together on projects, conduct joint experiments, and share their findings in real-time, regardless of geographical barriers. This collaborative aspect is particularly beneficial in an increasingly globalised educational landscape, where cross-cultural and interdisciplinary teamwork is becoming more common.

Safe and Cost-Effective Learning—Digital Twins offer a safe and cost-effective alternative to physical models and experiments. They reduce the need for expensive equipment, materials, and facilities, while also eliminating the risks associated with hands-on experiments, particularly in hazardous fields like chemistry and engineering. By simulating complex scenarios in a virtual environment, learners can learn and experiment without the potential dangers and costs of physical experimentation.

Transitional Gateway from Simulation to Real World—Digital Twins serve as an effective transitional gateway for learners moving from pure simulations to real-world applications. While simulations provide a controlled environment for understanding basic principles and theories, Digital Twins bridge the gap between these simulations and real-world practice by incorporating real-time data and allowing for interaction with actual physical systems. This transitional role is crucial in preparing learners for real-world challenges by gradually increasing the complexity and realism of the scenarios they encounter.

21.7.2 Challenges of Using Digital Twin in Education

Despite the numerous benefits, there are also challenges associated with the use of Digital Twins in education that need to be addressed, or at least compensated for, to fully leverage their potential.

High Implementation Costs—Implementing Digital Twin technology in educational settings can be expensive. The costs associated with developing high-fidelity models, integrating real-time data systems, and maintaining the necessary infrastructure can be substantial, especially for underfunded schools and institutions. This financial barrier can limit the accessibility of this technology, particularly in regions or institutions with limited resources.

Technical Complexity and Training Requirements—The technical complexity of Digital Twins can be a barrier to their effective use in education. Both educators and learners need adequate training to use these tools effectively, which requires time and resources that some institutions may lack. Ensuring that educators are proficient in the use of Digital Twin technology and can integrate it into their teaching methods is crucial for its successful implementation.

Data Security and Privacy Concerns—As Digital Twins rely on continuous data exchange, ensuring the security and privacy of this data is crucial. Educational institutions must implement robust cybersecurity measures to protect sensitive information from breaches and misuse. This is particularly important in fields like healthcare education, where personal and sensitive data may be involved.

Dependence on Technology Infrastructure—Effective use of Digital Twins requires reliable and high-performance technology infrastructure, including Internet connectivity, powerful computing resources, and advanced software. In areas with limited technological infrastructure, this can be a significant hurdle. Schools and institutions in remote or underserved areas may face challenges in adopting and maintaining the necessary infrastructure to support Digital Twin technology.

21.7.3 Digital Twins in STEM Education: Practical Case Study

While commercial Digital Twin products for academic applications remain limited, their potential is vast and transformative. The challenges of implementation, as detailed in the previous section, and a lack of understanding of the benefits among educationalists contribute to this scarcity. Addressing this lack of awareness is a primary rationale for this book chapter.

One company that has embraced Digital Twins for educational purposes is Quanser, a Canadian-based firm known for creating physical mechatronic products integrated with software to enable Digital Twin functionality. Their products cater to

Fig. 21.2 QCar device showing its key components and features (after [27])

various industrial sectors, including manufacturing and aerospace engineering. One of their most notable Digital Twin products is the QCar device, a 1:10 scale model of a Connected Autonomous Vehicle (CAV), designed to explore technologies in CAV and automotive engineering. An image of this device is shown in Fig. 21.2:

As shown in Fig. 21.2, the QCar is equipped with advanced hardware such as a 2D Light Detection and Ranging (LiDAR) system, multiple high-resolution cameras, a high-performance Graphical Processing Unit (GPU) and Central Processing Unit (CPU), and a 9-axis Inertial Measurement Unit (IMU). Additionally, it includes standard car features such as lights and signals, and connectivity options like USB and HDMI ports.

The QCar integrates with the Quanser Interactive Labs software package, a virtual simulation environment. This software, when used alongside the QCar hardware, allows for the creation of a true Digital Twin. It supports multiple virtual cars, objects, road furniture, pedestrians, buildings, and customisable maps for the QCar. Using the unreal graphics engine, the software provides high-fidelity, mixed-reality experiences with real-time data integration, bidirectional data flow, simulation, and prediction capabilities. Figure 21.3 provides a screenshot of the Quanser Interactive Labs software with multiple cars spawned.

The QCar, when used in conjunction with Virtual Labs software, aligns perfectly with the Digital Twin concept, defined as dynamic digital replicas of physical objects, systems, or processes that enable real-time monitoring, simulation, and analysis. By integrating sensor data and virtual environments, they create accurate digital models that mirror their physical counterparts. This continuous data exchange facilitates predictive maintenance, optimisation, and enhanced decision-making, demonstrating the practical utility of Digital Twin technology. The Quanser Interactive Labs creates an operating environment in which virtual multi-physics actors can exist. It therefore satisfies the definition of a Digital Twin Environment (DTE). In this environment, a singular QCar can exist as a Digital Twin Instance (DTI), or a multiplicity of QCar DTIs can form a Digital Twin Aggregate (DTA) along with DTIs of other environmental components such as smart traffic lights.

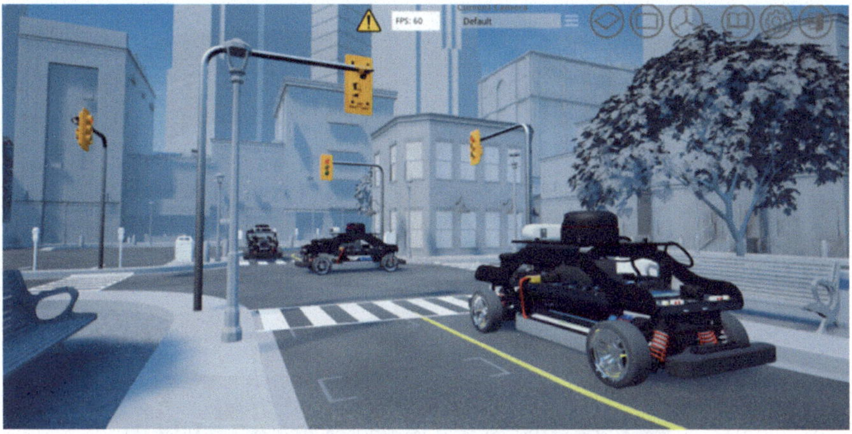

Fig. 21.3 Screenshot Quanser interactive labs (after [28])

The QCar, in conjunction with Quanser Interactive Labs, can be used for teaching activities across various educational levels, although it is primarily designed for undergraduate and postgraduate university and technical college courses. Educators can effectively teach key concepts in CAV technology, such as object recognition, obstacle avoidance, image processing, route planning, deep learning, machine learning, sensor fusion, and the Internet of Things (IoT). Additionally, the QCar serves as a valuable tool for teaching fundamental STEM subjects, including mathematics and physics.

Scenarios can be created using the QCar hardware and Interactive Labs software that embody technical challenges for learners to overcome. For example, a lesson plan could involve using three physical QCars on a physical map, with virtual traffic lights spawned in the virtual space only using the Interactive Labs software. The physical map is also replicated in the virtual space. In this setup, two cars operate autonomously while the third is manually driven by a student using a remote controller. Information on the states of the physical cars is sent back to the Interactive Labs software, allowing them to be replicated in the virtual space, thus creating a Digital Twin. The autonomous QCars, twinned with these virtual cars, will react to the virtual traffic signals as if they were physically present, creating a pseudo-mixed-reality effect.

In this scenario, the human-driven car introduces uncertainty, effectively creating a mobile obstacle that the autonomous cars must avoid and compensate for. This encourages learners to investigate driving concepts such as emergency stops, obstacle avoidance, and blind spot compensation for autonomous vehicles, with the goal of developing algorithms to achieve these effects. This scenario also allows learners to explore how autonomous and human-driven vehicles can coexist on the roads and experience first-hand some of the resultant challenges. During algorithm development, the virtual and physical environments can be decoupled, with the Interactive Labs software initially operating as a pure simulation environment. Once algorithms

Fig. 21.4 A QCar learning activity as viewed in the real world (real space)[1]

have been developed, the hardware and software can be reconnected, reforming the Digital Twin, and allowing for high-fidelity real-world testing with the physical QCars.

By structuring the lesson in this way, numerous learners can develop algorithms for use with the physical QCars, despite their number being far in excess of the number of physical cars present. This provides learners with the opportunity to develop innovative solutions to technical problems through hands-on experience. Figure 21.4, taken at the Connected Autonomous Vehicles Laboratory at the University of Hertfordshire, shows an example of this lesson plan being executed, with the QCars present in the real space.

In Fig. 21.4, the two bottom-most QCars are driving autonomously, while the top-most QCar is human-driven. Notably, the human-driven QCar, due to human driving error, has failed to follow the road correctly and is now positioned across the road, blocking the paths of the autonomous QCars. The autonomous QCars must take evasive action to avoid a multi-car collision. Figure 21.5 shows the corresponding virtual space representation of this scenario as seen in Quanser Interactive Labs, albeit with the cars in slightly different positions.

In Fig. 21.5, the autonomous cars are identified by their yellow and green LiDAR domes, whereas the human-driven car is coloured red. There is continuous bidirectional data flow between the physical QCars and the control computer running the Interactive Labs software. The control computer transmits data to the autonomous QCars, including traffic light colours, the positions of the other QCars, and the desired vehicle trajectory. Conversely, the autonomous and human-driven QCars transmit data back to the control computer, including vehicle state (i.e. speed, steer angle,

[1] Image copyright Esfandiar Khaleghi.

Fig. 21.5 A QCar learning activity as represented in Quanser interactive labs (virtual space) (after [27])

position), RGB & RGBD camera feedback, LiDAR point cloud information, and more, creating a Digital Twin.

The Digital Twin of any of the QCars can also be broken down into sub-virtual spaces, with each sub-virtual space representing the behaviour of a specific aspect of the vehicle. This breakdown can be visualised by examining the control program, written in MATLAB/Simulink, which underpins this scenario and works in conjunction with Interactive Labs. This control program provides the operational logic governing the behaviour of the QCars. An image of this program can be seen in Fig. 21.6.

In Fig. 21.6, coloured, top-level blocks correspond to sub-virtual spaces representing specific aspects of the QCar. Interconnecting arrows show the pattern of data transmission between the sub-virtual spaces. Inside each top-level block are logical commands that receive and process real-time information from the physical cars and coordinate the sending of data back to them.

As part of a teaching activity, learners can be provided with the control program to use as a starting point and then asked to modify it to achieve the operational and behavioural goals required by the activity. Learners can test their modifications using virtual cars within the Interactive Labs software before reconnecting the software with the physical cars for final validation. Such learning activities promote problem-solving, hands-on learning, and computer programming skills. Learners can be placed into groups to work in these task, encouraging the skills of team working and collaboration to be developed. They are inherently active learning activities, as learners actively participate rather than passively observing. Active learning aligns with the constructivist school of thought in education, as proposed by Piaget Vygotsky et al. [30–34], and has been shown to be highly effective in promoting learning. The effectiveness of active learning has been demonstrated in numerous studies, such as the work by Rayner, Kerwin, and Valentine [35].

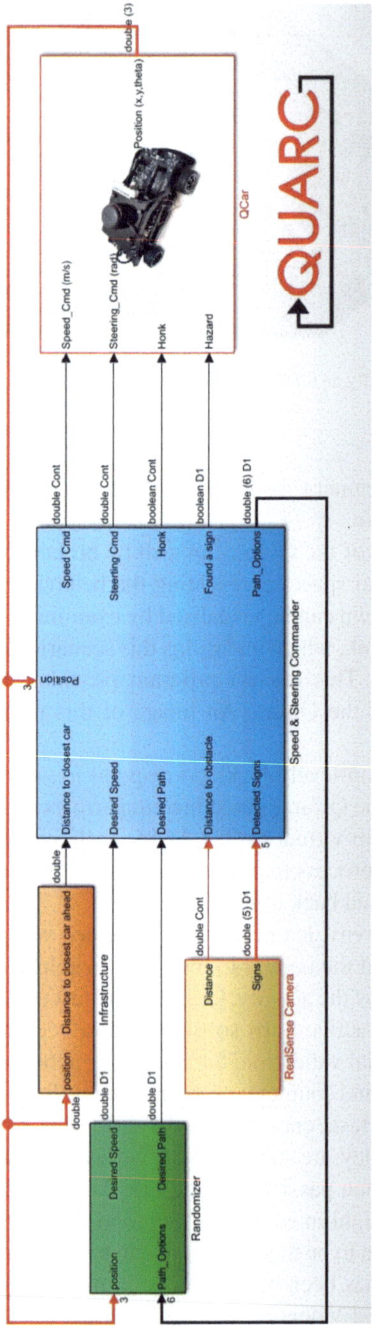

Fig. 21.6 QCar control software showing sub-virtual spaces (after [29])

It can therefore be seen that the Digital Twin comprising the QCars and Quanser Interactive Labs presents substantial benefits for STEM education. The QCar, with its advanced hardware and interactive software, offers real-time interactivity and feedback, enhancing the learning experience by allowing learners to experiment and see immediate impacts on digital models. This dynamic interaction fosters a deeper understanding of complex concepts and ensures that educational content remains current and relevant. Additionally, the collaborative environment facilitated by Digital Twins promotes the development of teamwork skills among learners. The QCar also provides a safe and cost-effective method of experimenting with CAV technology, allowing students to learn and experiment without the risks and costs associated with physical experimentation using a full-scale CAV and real-world environment. By bridging the gap between theoretical understanding and practical skills, the QCar serves as a valuable transitional gateway from simulation to real-world applications.

Despite its great pedagogical usefulness, the implementation of the QCar Digital Twin environment is not without challenges. The greatest challenge in this respect is the amount of practical and financial resources needed to acquire and run the product. The development work that Quanser has carried out to develop the QCar and its accompanying software was undoubtedly complex and time-consuming. This significant effort is reflected in its considerable financial purchase price, which some educational institutions may find prohibitively high. In practical terms, operating the physical QCars requires a large open room. Without careful planning by educational institutions, it is often difficult to find suitable locations to deploy the QCars. Additionally, running Interactive Labs and the accompanying control software requires relatively high-powered computers, even by current 2024 standards. While it is not infeasible for educational institutions to purchase such computers, the cost may prove challenging for more cash-strapped institutions.

For example, the University of Hertfordshire, UK, recently invested in a fleet of five QCars to bolster its teaching abilities in the subject area of Connected Autonomous Vehicles. With a capital cost of nearly £100,000, significant financial planning was required. Practical planning was also necessary to identify a laboratory space large enough to lay out the QCar map, with the map requiring at least 13.5m^2 of floor space. Once a suitable laboratory was identified, considerable work was needed to reorganise the space and move existing equipment to free up sufficient floor space for the map deployment. Additionally, it was a challenge to purchase and install a sufficient quantity of high-powered computers in the laboratory to support teaching activities. However, this extended planning resulted in the creation of a vibrant teaching room where students greatly enjoy the activities, as shown in Fig. 21.7.

Despite these challenges, the benefits of using the QCar Digital Twin environment are clear. Through strategic financial and practical planning, and proper training, educational institutions can overcome these challenges and fully exploit the pedagogical benefits of the QCars.

Fig. 21.7 A teaching activity taking place in the connected autonomous vehicle laboratory at the University of Hertfordshire, UK[2] (Image copyright Esfandiar Khaleghi)

21.8 Conclusion

The integration of Digital Twins in education, exemplified by the use of QCars, provides significant opportunities for rich and vibrant active learning. The QCar, a 1:10 scale model of a Connected Autonomous Vehicle, embodies this approach by offering learners hands-on interaction with advanced automotive technology in a safe, educational environment. This environment exists in both virtual and physical realities, enriched by the bidirectional and continuous data flow between the physical vehicle and the accompanying software. This setup bridges the gap between theoretical knowledge and practical application.

QCars are an excellent example of how Digital Twins can be used for educational purposes and how they can enhance traditional pedagogical approaches. They prepare learners for real-world challenges through immersive and interactive learning experiences, promoting deep engagement and understanding. Most importantly, they provide engaging learning experiences that foster effective learning.

However, the QCars also highlight some of the common challenges faced in implementing Digital Twins, particularly in terms of financial and practical planning factors. Despite these challenges, the potential benefits make the effort worthwhile.

[2] Image copyright Esfandiar Khaleghi.

While the availability of commercially available educational Digital Twin products is currently limited, it is anticipated that a greater range will become available in the future, spanning various subjects and technical disciplines suitable for a wide range of age groups. As educators become more aware of Digital Twins and their significant educational benefits, the demand for such products is expected to increase. Historically, the use of Digital Twins was constrained by the low levels of computing power available to typical users. Today, however, this limitation is far less significant, allowing Digital Twins to transition from the laboratory to the classroom.

Thus, we stand at the beginning of a new era in educational tools marked by the rise of Digital Twins.

References

1. Grieves MW (2005) Product lifecycle management: the new paradigm for enterprises. Intl J Prod Dev 2(1/2):71–84
2. Grieves M (2015) Digital twin: manufacturing excellence through virtual factory replication, Digital Twin Institute White Paper, www.researchgate.net/publication/275211047_Digital_Twin_Manufacturing_Excellence_through_Virtual_Factory_Replication. Accessed September 2024
3. Gelernter D (1993) Mirror worlds: or the day software puts the universe in a shoebox. How it will happen and what it will mean, 1st Ed., Oxford University Press
4. Främling K, Ala-Risku T, Kärkkäinen M, Holmström J (2006) Agent-based model for managing composite product information. Comput Ind 57(1):72–81
5. Shafto M, Conroy M, Doyle R, Glaessgen E, Kemp C, LeMoigne J, Wang L (2010) Modelling, simulation, information technology & processing roadmap: technology area 11 (Draft), NASA, www.researchgate.net/publication/280310295_Modeling_Simulation_Information_Technology_and_Processing_Roadmap. Accessed September 2024
6. Tuegel EJ, Ingraffea AR, Eason TG, Spottswood SM (2011) Reengineering aircraft structural life prediction using a digital twin, Intl J Aerosp Eng, 1–14
7. Jones D, Snider C, Nassehi A, Yon J, Hicks B (2020) Characterising the digital twin: a systematic literature review. CIRP J Manuf Sci Technol 29:36–52
8. Botkina D, Hedlind M, Olsson B, Henser J, Lundholm T (2018) Digital twin of a cutting tool. Proced CIRP 72:215–218
9. Cai Y, Starly B, Cohen P, Lee Y-S (2017) Sensor data and information fusion to construct digital-twins virtual machine tools for cyber-physical manufacturing. Proced Manuf 10:1031–1042
10. Tao F, Zhang M, Liu Y, Nee AYC (2018) Digital twin driven prognostics and health management for complex equipment. CIRP Ann 67(1):169–172
11. Torkamani A, Andersen KG, Steinhubl SR, Topol EJ (2017) High-definition medicine. Cell 170(5):828–843
12. Uhlmann E, Hohwieler E, Geisert C (2017) Intelligent production systems in the era of Industrie 4.0—changing mindsets and business models. J Mach Eng 17(2):5–24
13. Cheng J, Chen W, Tao F, Lin C-L (2018) Industrial IoT in 5G environment towards smart manufacturing. J Ind Inform Integr 10:10–19
14. Fuller A, Fan Z, Day C, Barlow C (2020) Digital twin: enabling technologies, challenges and open research. IEEE Access 8:108952–108971
15. Singh M, Fuenmayor E, Hinchy E, Qiao Y, Murray N, Devine D (2021) Digital twin: origin to future. Appl Syst Innov 4(2):36. https://doi.org/10.3390/asi4020036

16. Grieves M, Vickers J (2017) Digital twin: mitigating unpredictable, undesirable emergent behaviour in complex systems, Transdisciplinary Perspect Complex Syst, pp 85–113
17. GE announces name for its Advanced turboprop: The GE Catalyst^{TM} engine, GE Aerospace News, www.geaerospace.com/news/press-releases/regional-business-engines/ge-announces-name-its-advanced-turboprop-ge-catalysttm. Accessed September 2024
18. Digital Twin (2024), Rolls-Royce @ www.rolls-royce.com/innovation/digital/digital-twin.aspx. Accessed September 2024
19. Skywise (2024), Airbus aircraft, www.aircraft.airbus.com/en/services/enhance/skywise. accessed September 2024
20. Sustainable Aerospace Together (2024), Boeing, www.boeing.com/content/dam/boeing/boeingdotcom/principles/sustainability/sustainability-report/2023/assets/2023-Boeing-Sustainability-Report.pdf. Accessed September 2024
21. Pierce G, Heeren J, Hill T (2023) Orion SysML model, digital twin, and lessons learned for artemis I,. 33rd Annu INCOSE Intl Symp 33(1):290–304
22. The Digital Twin (2024), Siemens Global, www.siemens.com/global/en/company/stories/industry/the-digital-twin.html. Accessed September 2024
23. Digital Twins (2024), Bosch connected industry, www.bosch-connected-industry.com/de/en/iiot-insights/digital-twins. Accessed September 2024
24. Olivier R (2021) The future of cardiology: exploring the longer-term horizons and barriers for innovation across cardiac care, Philips, www.philips.co.uk/cardiologysolutions. Accessed September 2024
25. www.siemens-healthineers.com. Accessed September 2024
26. Living Heart (2024), Dassault Systèmes, www.3dexperiencelab.3ds.com/en/projects/life/living-heart/. Accessed September 2024
27. QCar (2024), Quanser, www.quanser.com/products/qcar/. Accessed September 2024
28. QLabs Virtual QCar (2024), Quanser, https://www.quanser.com/products/qlabs-virtual-qcar/. Accessed September 2024
29. Self-Driving Car Studio (2024), Quanser, www.quanser.com/products/self-driving-car-studio/. Accessed September 2024
30. Piaget J (1955) The construction of reality in the child. Routledge & Kegan Paul
31. Shabani K (2016) Applications of Vygotsky's sociocultural approach for teachers professional development. Cogent Educ 3(1):1252177
32. Yasnitsky A, Van der Veer R (2016) Revisionist revolution in Vygotsky studies (eds), Routledge
33. Learning Theories (2024), Social development theory, www.learning-theories.org/doku.php?id=learning_theories:social_development_theory. Accessed September 2024
34. Teach-nology.com (2024), Piaget's theory on constructivism, https://www.teach-nology.com/currenttrends/constructivism/piaget/. Accessed September 2024
35. Rayner R, Kerwin K, Valentine N (2022) Robot-assisted teaching—the future of education? Springer International Publishing, EcoMechatronics, pp 329–357

Correction to: Mechatronic Futures

Peter Hehenberger and David Bradley

Correction to: P. Hehenberger and D. Bradley (eds.), *Mechatronic Futures*, https://doi.org/10.1007/978-3-031-83571-1

The original version of the book was updated with updated figures in Chapters 1, 4, 7, 14 and 15. The book and the chapters have been updated.

The updated version of this book can be found at
https://doi.org/10.1007/978-3-031-83571-1

P. Hehenberger and D. Bradley (eds.), *Mechatronic Futures*,
https://doi.org/10.1007/978-3-031-83571-1_22

The manufacturer's authorised representative in the EU is Springer Nature Customer Service Centre GmbH, Europaplatz 3, 69115 Heidelberg, Germany. If you have any concerns regarding our products, please contact ProductSafety@springernature.com

Printed and bound by CPI Group (UK) Ltd, Croydon, CR0 4YY
23/09/2026
02221532-0007